AF307901

ASTRONOMY AND
ASTROPHYSICS LIBRARY

 ASTRONOMY AND
ASTROPHYSICS LIBRARY

Series Editors: M. Harwit, R. Kippenhahn, V. Trimble, J.-P. Zahn

Tools of Radio Astronomy
By K. Rohlfs

Physics of the Galaxy and Interstellar Matter
By H. Scheffler and H. Elsässer

Galactic and Extragalactic Radio Astronomy 2nd Edition
Editors: G. L. Verschuur and K. I. Kellermann

Observational Astrophysics
By P. Léna

Astrophysical Concepts 2nd Edition
By M. Harwit

The Sun An Introduction
By M. Stix

Stellar Structure and Evolution
By R. Kippenhahn and A. Weigert

**Relativity in Astrometry, Celestial Mechanics
and Geodesy**
By M. H. Soffel

The Solar System
By T. Encrenaz and J.-P. Bibring

Physics and Chemistry of Comets
Editor: W. F. Huebner

Supernovae
Editor: A. Petschek

Astrophysics of Neutron Stars
By V. M. Lipunov

Gravitational Lenses
By P. Schneider, J. Ehlers and E. E. Falco

Vladimir M. Lipunov

Astrophysics of Neutron Stars

Translated by R. S. Wadhwa

With 108 Figures

Springer-Verlag

Berlin Heidelberg New York
London Paris Tokyo
Hong Kong Barcelona
Budapest

Dr. Vladimir M. Lipunov

Sternberg State Astronomical Institute, Universitetski Prospect 13, 119899 Moscow, Russia

Translator

Professor Ram S. Wadhwa

28-3-224 Menzhinsky Street
129281 Moscow, Russia

Volume Editor

Gerhard Börner

Max-Planck-Institut für Physik und Astrophysik
Institut für Astrophysik
Karl-Schwarzschild-Strasse 1
W-8046 Garching, Fed. Rep. of Germany

Series Editors

Martin Harwit

The National Air and Space Museum
Smithsonian Institution
7th St. and Independence Ave. S.W.
Washington, DC 20560, USA

Virginia Trimble

Astronomy Program
University of Maryland
College Park, MD 20742, USA
and Department of Physics
University of California
Irvine, CA 92717, USA

Rudolf Kippenhahn

Rautenbreite 2
W-3400 Göttingen
Fed. Rep. of Germany

Jean-Paul Zahn

Université Paul Sabatier
Observatoires du Pic-du-Midi
et de Toulouse
14, Avenue Edouard-Belin
F-31400 Toulouse, France

Cover picture: Computer simulation of the gravitational deflection of light at a neutron star. The various positions of two diametrically opposed circular emission regions during the star's rotation are indicated by coloured dots. The left-hand picture shows the result without taking gravitational light deflection into consideration; in the right-hand picture the deflection is included, so the star appears larger and a part of the rear side is visible. (Courtesy Dr. H.-P. Nollert, Department of Theoretical Astrophysics, University of Tübingen.)

Title of the original Russian edition: *Astrofizika neĭtronnykh zvezd*
© Nauka, Moscow 1987

ISBN-13: 978-3-642-76352-6   e-ISBN-13: 978-3-642-76350-2
DOI: 10.1007/978-3-642-76350-2

Library of Congress Cataloging-in-Publication Data. Lipunov, V. M. (Vladimir Mikhaĭlovich). [Astrofizika neĭtronnykh zvezd. English] Astrophysics of neutron stars / Vladimir M. Lipunov ; translated by R. S. Wadhwa. p. cm. – (Astronomy and astrophysics library) Includes bibliographical references and index. ISBN 0-387-53568-3 1. Neutron stars. 2. Astrophysics. I. Title. II. Series. QB843.N4L5713 1992 523.8'874–dc20 91-33138

Typesetting: K + V Fotosatz, W-6124 Beerfelden
Production Editor: I. Kaiser

55/3140-5 4 3 2 1 0 – Printed on acid-free paper

Preface to the English Edition

The existence of neutron stars was not only a brilliant theoretical prediction, but also one of the most unexpected and astonishing discoveries of all heavenly bodies. Twenty-five years after the remarkable event of their discovery, neutron stars, which are the densest, the most strongly magnetized, and the most rapidly rotating bodies in the Galaxy, remain objects of intense interest.

This book is a revised and enlarged version of the original Russian edition. The last five years were marked by the discovery of a supernova in the closest galaxy and dozens of X-ray sources and millisecond pulsars, which apparently confirm the validity of the basic ideas underlying these discoveries. The author has concentrated on the astrophysical manifestations of neutron stars, which are believed mainly to be associated with the nature of their interaction with their surroundings. Naturally, this approach does not leave much room for a detailed description of the internal structure of these stars. Fortunately, there exists an excellent monograph by S. L. Shapiro and S. A. Teukolsky (*Black Holes, White Dwarfs, and Neutron Stars*, Wiley, New York 1985) which deals mainly with the purely physical problems. Moreover, the publication of such a book in the West partly makes amends for the lack of information about the work being done by Soviet scientists in this field.

It is gratifying to note that the present book is being published in Germany, a country whose scientists have made a significant contribution to the theoretical and experimental investigations of neutron stars. The discovery of X-ray spectral gyrolines in the pulsar spectra observed by the group led by Prof. J. Trümper, the pioneering work of Dr. G. Börner and his colleagues on the structure of the magnetosphere of accreting neutron stars, and the ever-fresh and original ideas of Prof. W. Kundt about the evolution and astrophysical manifestations of relativistic objects are just a few examples. This list can be extended to cover many other highly regarded studies.

The author would like to take this opportunity to thank all his colleagues for many valuable comments. Special thanks are due to Prof. N. I. Shakura, Drs. K. A. Postnov, M. E. Prokhorov, and E. Osminkin who took upon themselves the arduous task of carefully checking the entire manuscript. The author would also like to place on record his sincerest gratitude to Dr. R. S. Wadhwa for a meticulous translation by virtue of his excellent command of many languages, including the language of physics, and to Dr. H.-U. Daniel whose initiative is mainly responsible for the publication of the monograph in its present form. Finally, it gives the author great pleasure to acknowledge the invaluable help rendered by his wife Dr. N. A. Lipunova but for whose untiring efforts this book would have never seen the light of the day.

Moscow, February 1992   *V. M. Lipunov*

Preface to the Russian Edition

The discovery of radiopulsars in 1967 and of X-ray pulsars a few years later opened a new and rapidly developing field in modern astrophysics — the astrophysics of neutron stars. The huge body of observational data collected during the last couple of decades has considerably broadened our understanding of the processes occurring in the vicinity of neutron stars. Above all, this concerns neutron star manifestations such as X-ray pulsars, transient X-ray sources, X-ray bursters and gamma bursters.

This monograph is devoted to an analysis of these problems. The observational results and the most persistent theoretical views are presented. Since the description is largely theoretical, the structure of the monograph is based on the idea that astrophysical properties of neutron stars are mainly determined by the nature of their interaction with their surroundings. The author supports this idea. Even though this approach is neither widespread nor generally accepted, it is quite effective. For example, the existence of X-ray pulsars was predicted just on the basis of this approach. As a matter of fact, the theory of the interaction of neutron stars with their surroundings predicts the existence of new types of neutron stars that have not yet been discovered. Hence such objects have also been considered in the book.

Naturally, it has not been possible to give equal coverage to all the questions. This is partially compensated for by other monographs and extensive review articles that have been written on the topics not covered in this book.

Contents

1. Theoretical and Observational Principles of the Astrophysics of Neutron Stars 1
  1.1 Prediction 1
  1.2 Accretion 4
  1.3 Rotation and Magnetic Field 6
  1.4 Radiopulsars 7
  1.5 New Ideas 10
  1.6 X-Ray Pulsars 12
  1.7 X-Ray Bursters 17
  1.8 Bursts and Other Sources of Gamma Rays 20
  1.9 General View 22

2. Structure of Neutron Stars 24
  2.1 Equilibrium of Stars 25
  2.2 Exact Equilibrium Equations for Cold Stars 28
  2.3 Physical Conditions Inside Neutron Stars 32
  2.4 Parameters of Neutron Stars 33
  2.5 Mass of Neutron Stars 34
  2.6 Rotational Effects 37

3. Fluid Dynamics of Accretion 39
  3.1 Spherically Symmetric Accretion 42
  3.2 The Role of Radiation and Ejection 45
  3.3 Spherical Accretion to a Neutron Star Without a Magnetic Field 49
  3.4 Capture of Matter by a Moving Star 52
  3.5 Fluid Dynamics of Cylindrical Accretion 54
  3.6 Disk Accretion 57
  3.7 Luminosity and Spectrum of Accretion Disks 62
  3.8 Supercritical Disk Accretion 65
  3.9 Accretion in Binary Systems 66
      3.9.1 Overflow Through the Inner Lagrangian Point 69
      3.9.2 Accretion from Stellar Wind 70
  3.10 Two-Stream Accretion 78
  3.11 Accretion of Magnetic Fields 79

4. Classification of Neutron Stars 82
  4.1 Magnetic Dipole 83
  4.2 Stopping Radius 86
  4.3 Stopping Radius in the Supercritical Case 90
  4.4 The Effect of a Magnetic Field 92
  4.5 Gravimagnetic Parameter 92
  4.6 Corotation Radius 94
  4.7 Nomenclature 97
  4.8 Critical Periods. The $p-y$ and $p-L$ Diagrams 101

5. Boundaries. Magnetospheres of Slowly Rotating Neutron Stars ... 105
  5.1 Physical Conditions in the Alfvén Zone 106
  5.2 Formulation of the Problem 110
  5.3 Simple Configurations 111
  5.4 Magnetosphere in Spherically Symmetric Accretion 115
  5.5 Pascal's Pressure Law 119
    5.5.1 Two-Dimensional Solutions 120
    5.5.2 Three-Dimensional Solutions 123
  5.6 A Dipole Confined by an Ideally Conducting Disk 124
    5.6.1 Two-Dimensional Model 125
    5.6.2 Three-Dimensional Problem 128
    5.6.3 Dipole Rotation 130
  5.7 Magnetosphere in a Plane-parallel Plasma Flow 133
    5.7.1 Two-Dimensional Solution 133
    5.7.2 Three-Dimensional Solution 135
  5.8 Two-Stream Accretion 135

6. Accreting Neutron Stars 138
  6.1 Boundary Stability 139
    6.1.1 Spherically Symmetric Accretion 139
    6.1.2 Disk Accretion onto a Magnetized Neutron Star 145
    6.1.3 Torsion of an Accretion Disk by Magnetic Forces 149
    6.1.4 Magnetosphere Boundary Stability
        for Two-Stream Accretion 152
  6.2 The Polar Column 153
  6.3 Spin-up, Spin-down and Induced Precession
     of Accreting Stars 158
    6.3.1 Spin-up Torque 158
    6.3.2 Spin-down Torque 159
    6.3.3 Analytical Model of Torques Applied
        to a Magnetized Accreting Star 160
    6.3.4 Equilibrium Period 162
  6.4 Observed Properties of X-Ray Pulsars 163
  6.5 Energy Parameters of Pulsars and Transport
     of Matter in Binary Systems 164
  6.6 Spectrum and Magnetic Fields 167

6.7   Periods of X-Ray Pulsars and Their Variation 169
        6.7.1   Equilibrium of X-Ray Pulsars 171
        6.7.2   Magnetic Fields of X-Ray Pulsars 174
        6.7.3   Reasons Behind the Average Spin-up of X-Ray Pulsars 176
        6.7.4   Rapid Fluctuation of Periods and Internal Structure
                of Neutron Stars 180
6.8   Variability of X-Ray Sources. Transients 180
6.9   Generation of Relativistic Particles 183
6.10   X-Ray Bursters 184
        6.10.1 Localization and Spatial Distribution 186
        6.10.2 Periodic Variations of X-Ray Flux. X-Ray Eclipses ... 187
        6.10.3 Luminosity and Spectra of Bursters 187
6.11   Nuclear Burning at the Surface of Neutron Stars.
        Spherically Symmetric Model 189
6.12   Accretion to X-Ray Bursters 194
        6.12.1 Accretion for $\mu < \mu_{\min}$ 195
        6.12.2 Accretion in a Weak Magnetic Field ($\mu \gtrsim \mu_{\min}$) 196
        6.12.3 Nonstationary Spherically Symmetric Accretion 198
6.13   Spinning-up of Weakly Magnetized Neutron Stars 200
6.14   Low-Mass X-Ray Sources. "Noisars" 204
6.15 $p - y$ Diagram for Accreting Neutron Stars 206

7. The "Propeller" Regime 208
7.1   Quasistatic Shells 210
        7.1.1   Supersonic Propeller 212
        7.1.2   Subsonic Propeller 214
        7.1.3   Very Rapid Propeller 215
        7.1.4   Nongravitating Propeller 216
7.2   Spinning-down in the Boundary Layer 217
7.3   Two-Stream Flow Formation due to the Propeller Effect 218
        7.3.1   Stationary Flow from Disks 218
        7.3.2   Time-Dependent Solution 219
7.4   Dead Disks and Accumulator Disks 220
7.5   Nonstationary Disk Accretion.
        Model of Transient X-Ray Sources 222
7.6   Relativistic Propeller 224
7.7   Objects That Can Become Propellers 224
        7.7.1   Binary Systems 224
        7.7.2   Single Neutron Stars 225

8. Ejecting Stars 227
8.1   Observed Characteristics of Radiopulsars 227
        8.1.1   Periods and Their Variation 228
        8.1.2   Pulse Structure 229
        8.1.3   Spectrum and Luminosity 230
        8.1.4   Distribution of Pulsars in Space 232

8.1.5 Spatial Velocity of Radiopulsars 233
8.1.6 Pulsars and Binary Systems 234
8.2 Radiopulsars as Ejecting Neutron Stars 234
8.3 Pulsar Electrodynamics and Generation
    of Relativistic Particles 236
    8.3.1 Vacuum Approximation 237
    8.3.2 Magnetosphere in the Presence of Plasma 239
8.4 Mechanisms of Radiation 243
8.5 Caverns Around Neutron Stars 245
    8.5.1 Caverns in Binary Systems 245
    8.5.2 Caverns Around a Single Neutron Star 248
    8.5.3 Effect of Relativistic Wind
          on Accretion Flow Parameters 250
8.6 Change in Radiopulsar Period 251
    8.6.1 Spin-down of Pulsars and Their Magnetic Fields 251
    8.6.2 Spin-up Episodes and Internal Structure
          of Neutron Stars 254
8.7 Evolution of Radiopulsars 255
    8.7.1 Origin and Age of Pulsars 256
    8.7.2 Evolution of the Radiopulsar Period 258
8.8 Spatial Velocities of Radiopulsars 259
8.9 Ejecting Stars in Binary Systems 262
    8.9.1 Radiopulsars Forming Pairs with Degenerate Stars .. 262
    8.9.2 "Reflection" Effect 265
    8.9.3 Observational Evidence of the Existence
          of Ejecting Stars in Binary Systems 266

9. Supercritical Regimes 267
9.1 Superaccretor 268
    9.1.1 Accretion Pattern 268
    9.1.2 Neutrino Pulsar 270
    9.1.3 Spin-up and Spin-down 271
9.2 Superejectors and Superpropellers 272
9.3 Is SS 433 a Superaccretor? 272
9.4 Other Candidates 275

10. Stars with an Anomalously Low Value
    of Gravimagnetic Parameter 277
10.1 Georotators 277
10.2 Binary Magnetic Systems (Magnetors) 279

11. Evolution of Stars 280
11.1 Normal Stars 280
    11.1.1 Single Stars 280
    11.1.2 Binary Stars 282
11.2 Evolution of Neutron Stars 287

11.2.1 Evolution Equation 287
11.2.2 Statistical Description of the Ensemble
of Neutron Stars 290
11.3 Neutron Star Tracks 292
11.4 Numerical Simulation of the Joint Evolution
of Normal and Neutron Stars 293
11.4.1 Computational Method 294
11.4.2 Evolutionary Tracks 295
11.4.3 Simulation of X-Ray Pulsars (Stage IIA)
and the Choice of Optimal Parameters 297
11.4.4 Abundance of Different Types of Systems
in the Galaxy 298
11.4.5 Physical Characteristics of Neutron Stars
at Various Stages of Evolution 299
11.4.6 Two Types of Radiopulsars 300
11.5 Possible Candidates 301
11.5.1 "Runaway" Stars 301
11.5.2 The SS 433 Object 301
11.5.3 "Single" Wolf-Rayet Stars 302
11.5.4 Collapse Anisotropy 302
11.5.5 Other Numerical Models 303

Appendix 304
Magnetohydrodynamic Instabilities 304
Rayleigh-Taylor (RT) Instability 304
Commutation Instability 307

References 311

Subject Index 321

1. Theoretical and Observational Principles of the Astrophysics of Neutron Stars

1.1 Prediction

Neutron stars were predicted theoretically by L. D. Landau in the beginning of 1930s.

Equilibrium in ordinary stars is established by the pressure forces emerging as a result of the thermal motion of ions and electrons. In order to sustain such an equilibrium over a time longer than the cooling time of a star (thermal time), additional sources of energy are required. (For example, the cooling time of the Sun is of the order of 10^7 years.) Hence it can be concluded that stars are not eternal. This became quite clear by the beginning of the 1940s when the main part of the theory of the internal structure of stars was formulated.

However, there are objects whose equilibrium does not depend on the heat stored in them (our planet is one such object). One can naturally wonder as to what the ultimate fate of stars will be. Can a star not remain stable after all the inner sources of heat in it have been exhausted? Some remarkable ideas on this subject were put forth by Tsiolkovskii in 1893. While studying the processes associated with the compression of stars, he suggested that for a very high density of stellar matter, the elasticity will increase sharply and no longer obey the laws of an ideal gas. The compression ceases, and the stars begin to cool and harden. According to Tsiolkovskii, the elasticity jump must be accompanied by a further decrease in the temperature which may turn out to be sufficient for the formation of mist and crust on the Sun, which will then die away, preserving the energy reserves for unknown purposes. This, however, opens the floodgates to the realm of hypotheses, some of which are so daring that it is better to keep silent.

Only after a lapse of 35 years, it was found that something of this sort must actually occur. The English physicist R. Fowler (1926) showed that white dwarfs (stars having an insufficient luminosity and hence a small size) are objects whose equilibrium is ensured by the equality of gravitational forces and the forces of pressure of the degenerate electron gas.

The pressure of a degenerate electron gas, according to the laws of quantum mechanics (the Pauli exclusion principle) does not vanish even at absolute zero. This leads to the inevitable conclusion that after the exhaustion of the reserves of internal (thermonuclear) energy, all stars are compressed to a size of a few thousand kilometers and are transformed into white dwarfs.

In reality, however, this is not true. It was shown independently by Chandrasekhar (1931), and Landau (1932) that the mass of a white dwarf cannot be indefinitely large. Only stars having a very small mass can be transformed into white dwarfs. The maximum value of the mass of a white dwarf is called the Chandrasekhar limit. The existence of this limit can be explained as follows.

The pressure of a cold degenerate electron gas obeys the law $P \propto \varrho^{\gamma}$, where $\gamma = 5/3$ for a nonrelativistic gas and $4/3$ in the relativistic case. The pressure gradient of the electron gas has the same order of magnitude as the gravitational force:

$$\frac{\varrho^{\gamma}}{R} \sim \frac{M^{\gamma}}{R^{3\gamma+1}} \sim \frac{GM^2}{R^5} ,$$

where R and M are the radius and mass of the star.

It can be seen that $R \propto M^{(\gamma-2)/(3\gamma-4)}$. For a nonrelativistic gas, $R \propto M^{-1/3}$. With increasing mass, the radius of a white dwarf decreases. (This is a characteristic feature of degenerate stars.) As the mass of a white dwarf increases, the electrons are packed more densely and their energy increases in accordance with the Pauli exclusion principle. The electron gas becomes relativistic and γ assumes the value $4/3$. In this case, both forces in the equilibrium equation change identically with radius so that they become equal for one, and only one, value of the mass which is known as the Chandrasekhar limit. Exact calculations (see below) of the maximum mass show that

$$M_{\mathrm{Ch}} \approx \frac{3.1}{\mu_{\mathrm{e}}^2} \left[\frac{hc}{G} \right]^{3/2} \approx 5.83 \, \mu_{\mathrm{e}}^{-2} M_{\odot} \tag{1.1}$$

where μ_{e} is the average number of nucleons per electron. For heavy nuclei, $\mu_{\mathrm{e}} \to 2$, and hence the Chandrasekhar limit tends to $1.46 \, M_{\odot}$ [by the way, (1.1) shows that the mass of stars can be expressed through only the fundamental constants h, c, and G].

What will happen to a star that does not have an internal source of energy and whose mass exceeds the Chandrasekhar limit? Apparently, Landau was the first scientist to raise this question and to find an answer to it.

The star must collapse until the atomic nuclei come in contact and a giant atomic nucleus of density $10^{14} - 10^{15}$ g/cm^3 and size ~ 10 km is formed.

A year later, the American astronomers Baade and Zwicky (1934) suggested that the supernova explosions are just a catastrophic process of formation of superdense stars consisting of neutrons. Neutrons were discovered in 1932 by Chadwick.

First calculations of the structure of neutron stars were carried out by the American physicists Oppenheimer and Volkoff (1939). It should be noted that these investigations were inspired by the pioneering work of Landau (1932). On the basis of the simple physical arguments used by Landau to derive the upper limit of the mass of a white dwarf, which was called the Landau limit by Oppenheimer and Volkoff, the latter were able to draw an important conclusion about the existence of the upper limit for the mass of neutron stars M_{OV}

(Oppenheimer-Volkoff limit) without any preliminary computations. Four years earlier, Eddington (1935) reached practically the same conclusion, but did not attach any importance to his results. He believed that the existence of black holes was an absurdity, and assumed that the stars somehow avoid a total gravitational collapse. (We can now state that Eddington was right to a certain extent: stars indeed resist a collapse and most of them are transformed into white dwarfs or neutron stars.)

Calculations made by Oppenheimer and Volkoff showed that the structure of neutron stars depends significantly on the equation of state of matter at nuclear densities (which have not been determined so far). The nature of elasticity of the neutron matter has an especially strong effect on the value of the critical mass. For example, in the earlier model of an ideal degenerate neutron gas employed by Oppenheimer and Volkoff, it was found that $M_{OV} \approx 0.7\,M_\odot$, i.e., less than the Chandrasekhar limit (!). Results of latest calculations based on different models give $M_{OV} \approx (1.5-3)\,M_\odot$. For a larger mass, the star irreversibly collapses into a state which was termed as a black hole by Wheeler (this term first appeared in the 1960s).

Thus, it was established by the end of the 1930s that after the exhaustion of nuclear sources of energy, stars are converted into white dwarfs, neutron stars, or black holes depending on their mass.

It is expedient to recall another important work published by Landau before the Second World War (1938). In this paper, Landau put forth the hypothesis that dense neutron cores may exist in stars. Accretion of stellar matter onto a core could lead to the liberation of a large amount of energy. At present, we know that most stars shine on account of thermonuclear energy. However, the idea of neutron cores and of accretion onto them both acquired significance in the 1970s (Bisnovaty-Kogan, Lamzin 1984).

Thus, in spite of the fact that the existence of neutron stars was predicted more than fifty years ago, they were discovered only in 1967. Why?

Let us consider the arguments put forth by astronomers before the war. Visible light was the only range of electromagnetic waves accessible to investigations. How bright can a neutron star be in this range? Let us assume that a neutron star emits radiation as a black body. At a temperature equal to that of the Sun, for example, its luminosity will be lower by a factor of 10^{10} (according to the ratio of surface areas). The absolute magnitude of the Sun is $\sim 5^m$, which means that at a distance of 10 parsec such a neutron star would appear as a tiny star of magnitude 30^m. Such a star cannot be seen even through modern terrestrial telescopes. Obviously, nobody wanted to study objects that cannot be observed.

A lull of about twenty years reigned in the investigations of neutron stars, and interest in these objects was revided only at the end of the 1950s. The development of the physics of elementary particles and low temperature physics led to an understanding of many new phenomena. Efforts were directed at investigations in the field of high energies and densities. At this stage, the banal (though largely theoretical) concept of a neutron star as a physical laboratory with unprecedented potentialities was put forth.

In 1959, the Soviet physicist Migdal put forth the idea that the matter of a neutron star must be superfluid. Although an exact microscopic quantum theory of superfluidity of the neutron liquid has not been developed so far, it can be assumed that such a superfluid consists of paired neutrons and is reminiscent of ^{3}He in its properties. A ^{3}He atom contains an odd number of particles and, like a neutron, is a Fermi particle. The superfluidity of ^{3}He sets in at a temperature 0.00265 K (at a pressure of 30 atm). At a density $\sim 10^{14}$ g/cm^3, the critical temperature is found to be of the order of 10^{11} K. If a neutron star can be "cooled" to a temperature below 10^{11} K, its matter becomes superfluid!

At about the same time, the Soviet astrophysicists Ambartsumyan and Saakyan (1960) calculated the internal structure of neutron stars by taking into account new data on elementary particles. However, the astrophysics of neutron stars was actively developed only after the discovery of first X-ray sources and quasars.

1.2 Accretion

The advent of the space age made possible investigations of the Universe using X-rays probes to which the Earth's atmosphere is completely opaque. The density of air is 10^{-3} g/cm^3, and the height of the atmosphere is $\sim 10^6$ cm. Taking the value $\kappa_T \approx 0.2$ g/cm^2 for the Thomson scattering cross section of air, we find that the optical depth of the atmosphere is $\tau_T = \varrho H \kappa_T \approx 200$. Obviously, X-ray counters must be placed above the Earth's atmosphere.

On June 18, 1962, a group of American astrophysicists (Giacconi et al. 1962) launched an Aerobee rocket carrying three Geiger counters. They discovered that the brightest source in the sky in the energy band $1-10$ keV was Scorpius X-1 (abbreviated as Sco X-1). The rocket rose to a height of 225 km and remained in flight for 350 s. One of the Geiger counters went out of order, while the remaining two counters continued to scan the sky, rotating about their axes. Both counters revealed a bright source in the direction of the constellation Scorpius. The X-ray flux diagram is shown in Fig. 1.1.

A year later, Schmidt (Mount Palomar, USA) identified the spectral lines of quasars. Quasars were found to be the most powerful objects in the Universe (with a luminosity of $10^{46}-10^{47}$ erg/s).

These discoveries spurred the theorists to investigate new energy sources. In 1964, the Soviet physicist Zel'dovich and the American astrophysicist Salpeter independently predicted the existence of such a source in the form of accretion of matter onto relativistic stars.

The high efficiency of the accretion of a gas can be explained by the following simple argument. Suppose that a relativistic object has a radius R_x at a surface at which the entire kinetic energy of the matter falling onto it is converted into radiation. Further, let us suppose that the matter falls freely so that a unit mass has a kinetic energy GM_x/R_x. If the amount of substance falling

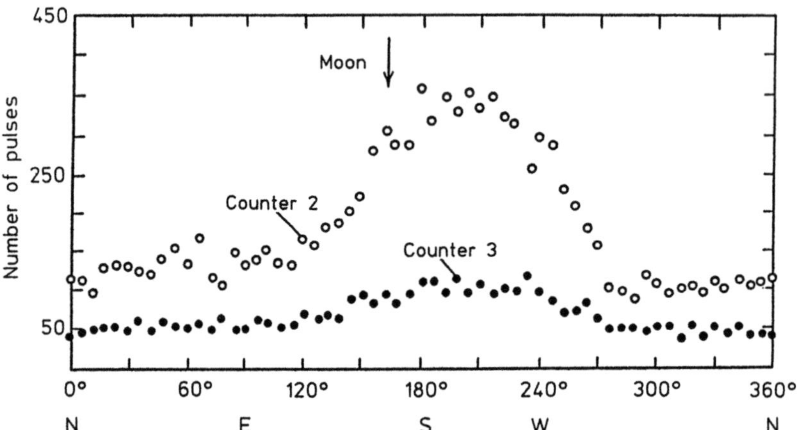

Fig. 1.1. X-ray flux from the source Sco X-1 (Giacconi et al. 1962)

per unit time on the surface is $\dot{M}g$, the total luminosity of the accreting star is obviously given by

$$L = \dot{M}\frac{GM_x}{R_x} \equiv \eta \dot{M}c^2 \tag{1.2}$$

where M is the mass of the relativistic star and η is the accretion efficiency. Introducing the gravitational radius

$$R_g = \frac{2GM_x}{c^2} \tag{1.3}$$

we can write

$$\eta = \frac{1}{2}\frac{R_g}{R_x} . \tag{1.4}$$

For a neutron star, $R_x = 10\,\text{km}$ and $R_g = 3\,\text{km}$, which means that $\eta \approx 10\%$. This is about 100 times the efficiency of nuclear fusion reactions.

In his work, Zel'dovich (1964) emphasized for the first time that conditions for observing a relativistic star are especially favorable when this star is a part of a binary system in which the companion star can supply the accreting matter. In the same year, the Japanese astrophysicists Hayakawa and Matsuoka (1964) discussed the emergence of X-rays in binary systems formed by ordinary stars.

The fact that the accretion of matter in a binary system actually leads to a strong observable effect can be easily verified on the basis of the following simple estimates. Normal stars are capable of losing matter at the rate of $10^{-5}\,M_\odot$ per year. Suppose that only 0.1% of this matter is captured by a relativistic star. An accretion rate of $10^{-8} M_\odot$ per year at an efficiency of 10% results in the emergence of a source with a luminosity $\sim 10^{38}$ erg/ s $\approx 4 \times 10^5 L_\odot$, as in (1.2). Obviously, such a huge amount of energy can be carried away from the surface of a tiny relativistic star only by high-energy quanta, i.e., by X-rays.

It was suggested by Novikov and Zel'dovich (1966) and independently by Schklowskii (1967) that the X-ray source Sco X-1 is an accreting neutron star in a close binary system. Ironically, the brightest source also turned out to be one of the most difficult objects to observe directly. A considerable amount of time had to pass before it was proved that it is a close binary system. The nature of the degenerate star has not been established unambiguously so far.

An important ideological and psychological barrier was surmounted by the Soviet scientists Amnuel' and Guseinov (1968) who pointed out the need to take into account the magnetic field of neutron stars. It follows from general considerations (see below) that neutron stars must have strong magnetic fields, and the accreting matter under real conditions is a good conducting plasma. The way in which matter falls onto a star may change dramatically in a magnetic field. Thus, even a spherical incidence in the vicinity of a neutron star may lead to an anisotropy in the distribution of matter, and hence of radiation as well. As a result of the rotation of neutron stars, the radiation appears to an observer as periodically pulsating.

Everything pointed towards a "planned" discovery of neutron stars being accreting objects.

But fate willed otherwise.

1.3 Rotation and Magnetic Field

Baade and Zwicky (1934) associated the formation of neutron stars with supernova explosions. Hence the interest of the astronomers towards the remnants of supernovae is quite understandable.

The remnant of a supernova explosion observed by Chinese astronomers in 1054 proved to yield the most information. The nebula emission mechanism was explained in 1953 by Schklovskii on the basis of a synchrotron mechanism. Ginzburg and Gordon independently predicted the polarization of the radiation in the 1950s. Indeed, the polarization was soon observed by Dombrovskii (1954) and Vashakidze (1954). Calculation of the concentration and energy of relativistic electrons in the Crab nebula, made by Pikel'ner in 1956, directly indicated a constant pumping of energy $\sim 10^{38}$ erg/s into the nebula.

The source of this energy was discovered in 1964 by Kardashev, who put forth several new ideas about the process of formation of neutron stars and their properties. The most important among these ideas was the spinning up of a collapsing object and the enhancement of the magnetic field strength.

It is well known that as a body is compressed while conserving its rotational angular momentum, its rotational energy increases:

$$E_{\text{rot}} = \frac{I\omega^2}{2} \sim R^{-2} \tag{1.5}$$

where $I\omega$ is the angular momentum, ω is the angular velocity of rotation, and R is the radius of the star. The energy increases on account of the work done by the gravitational force.

The increase in the magnetic field strength during collapse under conditions of a complete freezing-in follows the law of conservation of the magnetic flux:

$$BR^2 = \text{const} \tag{1.6}$$

where B is the magnetic field strength at the surface of the star. The increase in the magnetic field during collapse was first considered by Ginzburg (1964). As a result of compression by a factor of 10^5, the magnetic field increases by a factor of 10^{10}. For a magnetic field of $1-100$ Oe at the surface of a normal star, we obtain as a result of compression a field of $10^{10}-10^{12}$ Oe at the surface of a neutron star. This effect is augmented by an additional field enhancement due to a twisting of the field lines upon rotation of the star.

Strictly speaking, the magnetic field does not increase, but rather disappears when a star collapses. An observer located at a fixed distance from a collapsing star may find (with the help of a magnetometer) that the magnetic field decreases at the point of observation. This is quite obvious, since the dipole moment μ is proportional to BR^3, and hence a decrease in $\mu \propto R$ is observed during a collapse. Higher multipoles disappear even more rapidly, and the magnetic field is "cleansed". During a collapse into black holes, the principles of the general theory of relativity also become significant and hence the magnetic field vanished not as $R \to 0$ but as $R \to R_g$ (Ginzburg 1964).

Thus, it was proposed by Kardashev that the rotation energy of a neutron star, immediately after its formation, is the source of the Crab nebula emission, the energy being transferred by the magnetic field. It is worth noting that the rapid rotation of superdense stars was considered by Hoyle et al. (1964), and Tsuruta and Cameron (1966), while the strong magnetic fields of neutron stars were investigated by Hoyle et al. (1964) and Woltjer (1964). It was suggested by Pacini (1967) that the rapidly rotating magnetic field of a neutron star can accelerate particles "stuck" to the magnetic field lines up to ultrarelativistic velocities. While Pacini's paper was still in press, English radioastronomers discovered the neutron star sheerly by accident.

1.4 Radiopulsars

According to Ter Haar, the discovery of neutron stars is an example of serendipity. While observing the flickering of radio sources, J. Bell, a graduate student of A. Hewish, observed a radiosource with a period of 1.377 s at a fre-

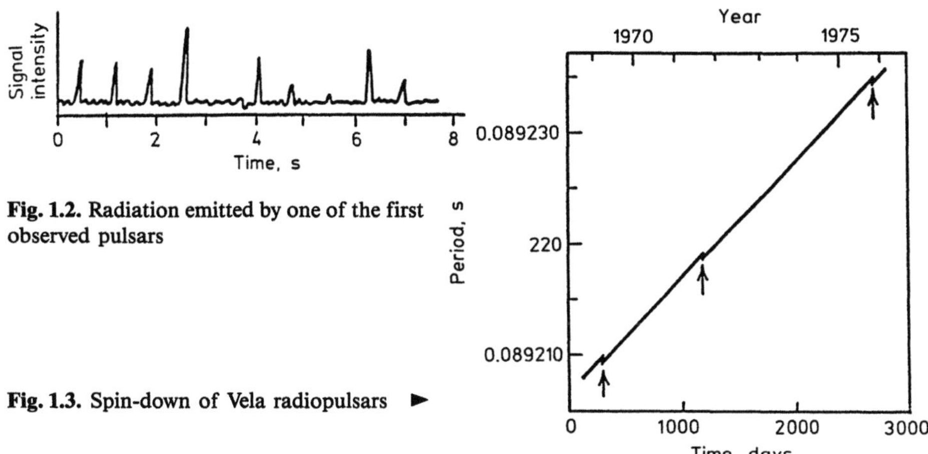

Fig. 1.2. Radiation emitted by one of the first observed pulsars

Fig. 1.3. Spin-down of Vela radiopulsars ▶

quency of 81.5 MHz (Fig. 1.2). After a brief but lively discussion [see, for example, Zel'dovich and Novikov (1971)], the interpretation of this effect proposed by Gold (1968) was universally accepted. Gold's idea was a continuation of the ideas put forth by Kardashev and Pacini. It should be recalled that the pulsating nature of the neutron star radiation was also indicated by Amnuel' and Guseinov.

Two observational facts suggested that a radiopulsar is a magnetized neutron star losing its rotational energy. These facts are a high stability of the period, and a slow increase in the pulsar period (Fig. 1.3). The latter result is quite natural, since it is rotational energy which is emitted by the source.

However, the decisive argument in favor of the rotating neutron star model was the discovery of a radiopulsar in the Crab nebula. Among all known pulsars, this pulsar has the lowest period of 0.033 s. Only a neutron star could rotate at such a high speed.

The fact that rotation causes the emission of energy by pulsars can be verified from the following simple estimate made by Gold (1968). It should be recalled that according to Pikel'ner's analysis (1956), a constant energy injection into the Crab nebula at the rate of 10^{38} erg/s must take place.

The loss of rotational energy of a body is accompanied by a change in its frequency:

$$L_{\text{rot}} = I\omega\dot\omega = 4\pi^2 I p^{-3}\dot p \tag{1.7}$$

where $p = 2\pi/\omega$ is the period of rotation. The observed variation in the pulsar period in the Crab nebula is $\dot p \approx 4\times10^{-13}$. Substituting this value into (1.7), we find that the flux necessary for feeding the nebula is obtained if we put $I \approx 10^{45}$ g cm^2. This value is in excellent agreement with that expected for the neutron stars, viz. $2\times10^{33}\times10^{12} \approx 10^{45}$ g cm^2 (!).

In the model proposed by Pacini (1967) and Gold (1968, 1969), the magnetic field of a neutron star was assumed to be a dipole field, and the loss of the rotational energy of neutron stars was assumed to be exactly the same as the loss of energy by a rotating magnet in the form of magnetic dipole radiation (Landau, Lifshitz 1973):

$$L_{\mathrm{m}} = \frac{2}{3}\frac{\mu^2 \sin^2 \beta}{c^3}\omega^4 \tag{1.8}$$

where μ is the magnetic dipole moment at the magnetic pole and β is the angle between the rotational axis and the magnetic dipole axis. For a magnetized sphere, the magnetic dipole moment is given by

$$\mu = \frac{B_0 R_x^3}{2} \tag{1.9}$$

where B_0 is the magnetic field strength at the magnetic pole. For $L_{\mathrm{m}} \approx 10^{38}$ erg/s and a neutron star radius of $R_x = 10^6$ cm, we find that the magnetic field at the surface of a pulsar is $B_0 \approx 10^{12}$ Oe, which is the same as that expected for a neutron star.

Thus, it was established beyond doubt that Hewish and coworkers had discovered the neutron stars. For this discovery, Hewish was awarded the Nobel Prize in physics for the year 1974.

Soon after the discovery of radiopulsars, glitches (marked by arrows in Fig. 1.3) were observed. The monotonic increase in the period of the Vela pulsar (PSR 0833-45) was suddenly interrupted by a comparatively small but sharp decrease in the period $\Delta p/p \approx 2\times 10^{-6}$ (Downs 1981). Later, the same effect was also observed in the pulsar of the Crab nebula.

This phenomenon was interpreted by Ruderman (1969), who pointed out that in the upper layers of a neutron star, the crystalline state should be energetically more accessible than the liquid state. Neutron stars are then covered by a solid crust (one cannot help recalling the ideas of Tsiolkovskii). In this case, the glitches in the spin-ups could be attributed to a sharp change in the moment of inertia of the crust, say, due to starquakes. A characteristic feature of this effect is that the glitch (which lasted for several days) was followed by a peculiar type of relaxation lasting several weeks during which the derivative of the period gradually reverted to its original value. This proved the validity of Migdal's idea (1959) about superfluidity inside the neutron stars!

In 1969, Goldreich and Julian observed that a rotating magnetized neutron star is reminiscent of a unipolar inductor. In the reference frame attached to the rotating neutron star having a magnetic field B, an electric field E is produced:

$$E \approx \frac{\omega R_x}{c} B_0 \ .$$

For a pulsar with period $p = 1$ s and a magnetic field $B_0 = 10^{12}$ Oe, $E \approx 2\times 10^8$ V cm^{-1}. The force exerted by this electric field on an ion or an electron is billions of times larger than the gravitational force. Such a field is capable of accelerating particles to ultrarelativistic velocities.

The nonthermal radio emission of radiopulsars is of special interest (Ginzburg 1971). The brightness temperature of most radiopulsar radiation has a fantastically high value of 10^{27} K. Hence there is no doubt that this radiation is coherent (Ginzburg, Zheleznyakov 1970a,b).

1.5 New Ideas

The discovery and study of radiopulsars showed that the collapse of normal stars results not only in supernova explosions and the formation of neutron stars as predicted by Baade and Zwicky, but also to a spin-up and generation of strong magnetic fields (Fig. 1.4).

The last two parameters led to the treatment of neutron stars as a qualitatively new class of objects in the Universe. Neutron stars, which are entirely devoid of internal energy sources (see, however, Bisnovaty-Kogan et al. 1975), nevertheless possess strong gravitational and magnetic fields and are therefore capable of manifesting themselves actively and of developing by evolution, albeit in a completely different sense.

This circumstance, of exceptional importance, was explained by the Soviet astrophysicist Shvartsman (1970, 1971). The concept of evolution and the approach developed by him were found to be very effective. A completely new class of astrophysical objects, viz., X-ray pulsars, was predicted without any precise calculations.

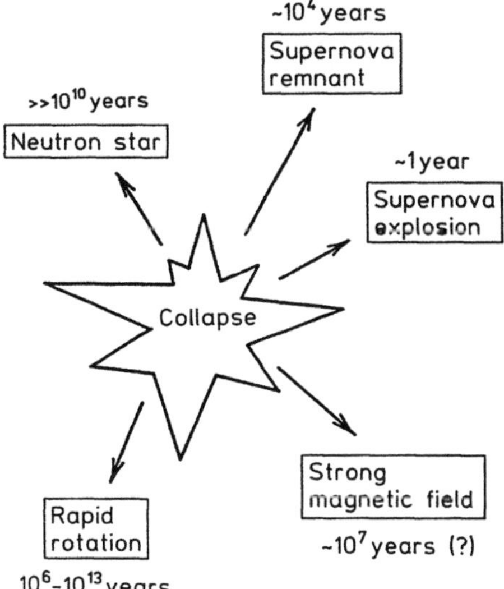

Fig. 1.4. Consequences of collapse

Let us imagine that a newly born neutron star exists in a binary system together with a normal star. In the beginning, the neutron star which has a high rotational frequency serves as a powerful source of electromagnetic radiation and relativistic particles just like a radiopulsar. The pressure of particles ejected by the neutron star speeds up the surrounding plasma and a cavern is formed around the pulsar. However, the emission power decreases as ω^4 with time and the cavern "collapses". The plasma falls within the light cylinder and the pulsar is extinguished. However, accretion is still not possible as before, and is inhibited by the rapidly rotating magnetic field [this effect was later called the propeller effect by Illarionov and Syunyaev (1975)]. The neutron star continues to spin-down. Finally, the accreting matter approaches the surface of the neutron star and a new stage, known as the accreting neutron star stage, sets in. It was shown by Zel'dovich and Shakura (1969) that an accreting neutron star emits radiation in the X-ray range. According to the idea put forth by Amnuel' and Guseinov (1968), the radiation must pulsate. Consequently, X-ray pulsars must be observed in binary systems.

In order to distinguish an accreting pulsar from an ejecting pulsar, an "experimentum crucis" was proposed by Shvartsman (Fig. 1.5). The accretion matter in a binary system always possesses an angular momentum relative to the neutron star. This angular momentum is produced by the orbital motion. (This can be verified easily by going over to a reference frame attached to the neutron star.) The neutron star must experience a spin-up as a result of the accretion matter falling on its surface. This means that unlike an ejecting pulsar, an accreting pulsar must (or at least may) spin-up, and its energy liberation is not associated in any way with rotation.

The X-ray radiation emitted by an accreting neutron star must be subjected to several periodic modulations (Guseinov 1970): (1) short pulsations due to the rotation of the neutron star, and (2) variations with an orbital period associated with the eclipses of the X-ray source and with the variation of the accretion rate due to a change in the separation between the components of the binary system.

Such sources were indeed discovered soon after on the American satellite "Uhuru". However, Soviet scientists made a decisive contribution towards the prediction of the properties of X-ray pulsars and the interpretation of their nature. All publications concerning the accretion theory of neutron stars

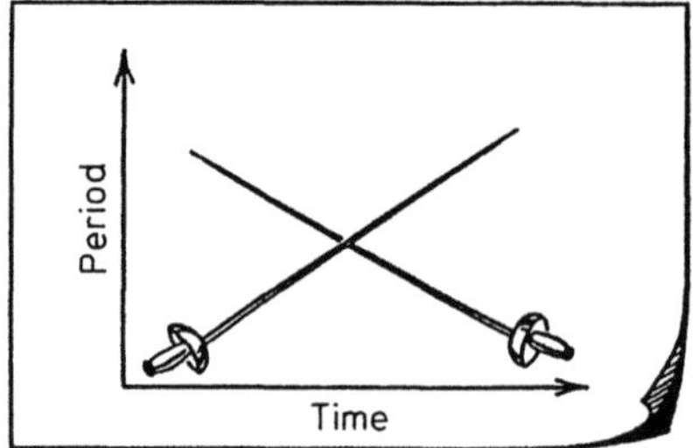

Fig. 1.5. Experimentum crucis

before the launching of Uhuru were made be Soviet scientists (Novikov, Zel'dovich 1966; Shklovskii 1967; Amnuel', Guseinov 1968; Zel'dovich, Shakura 1969; Bisnovaty-Kogan, Fridman 1969; Amnuel', Guseinov 1969; Guseinov 1970; Shvartsman 1970, 1971; Shakura 1972; Amnuel', Guseinov 1972).

1.6 X-Ray Pulsars

The first specialized X-ray satellite Uhuru[1] was launched on December 12, 1970. It carried two systems of X-ray detectors with a total area of about $840\,cm^2$ for detecting radiation in the $2-20$ keV range with an angular resolution of a few degrees. A rapid rotation with a period of 12 min and a slow variation of the orientation of the satellite orbit in space allowed scanning the whole sky. The experiments were carried out at a sensitivity of 10^{-10} erg/cm^2 s.

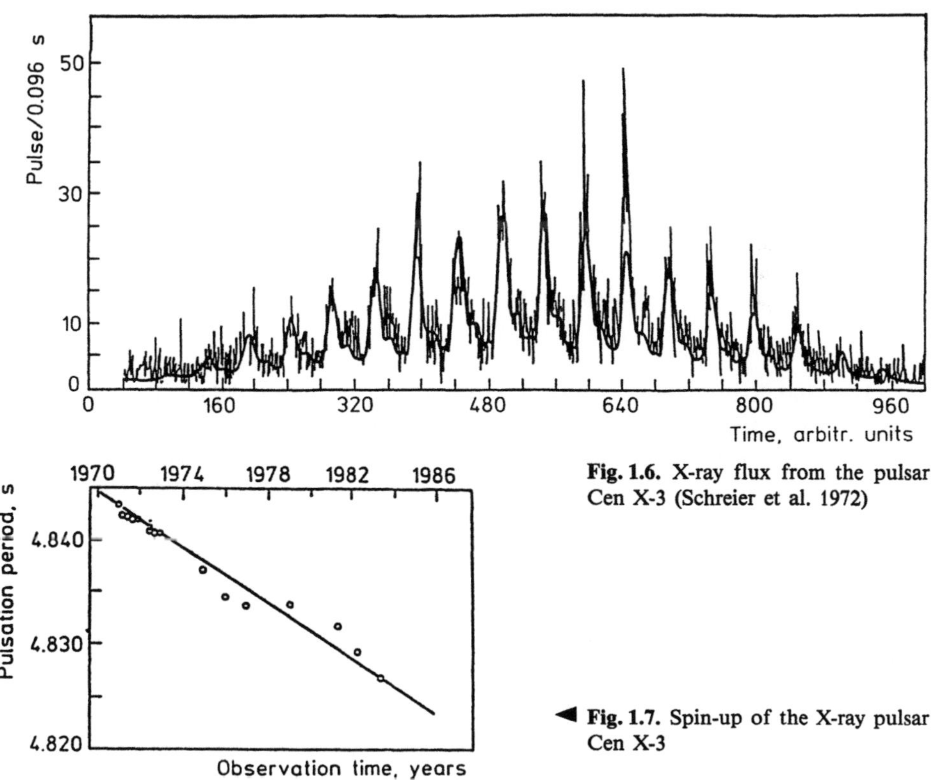

Fig. 1.6. X-ray flux from the pulsar Cen X-3 (Schreier et al. 1972)

◄ Fig. 1.7. Spin-up of the X-ray pulsar Cen X-3

---

[1] The satellite was launched off the coast of Kenya. "Uhuru" in Swahili means "freedom". The satellite was named so in honor of the 10th anniversary of Kenya's independence.

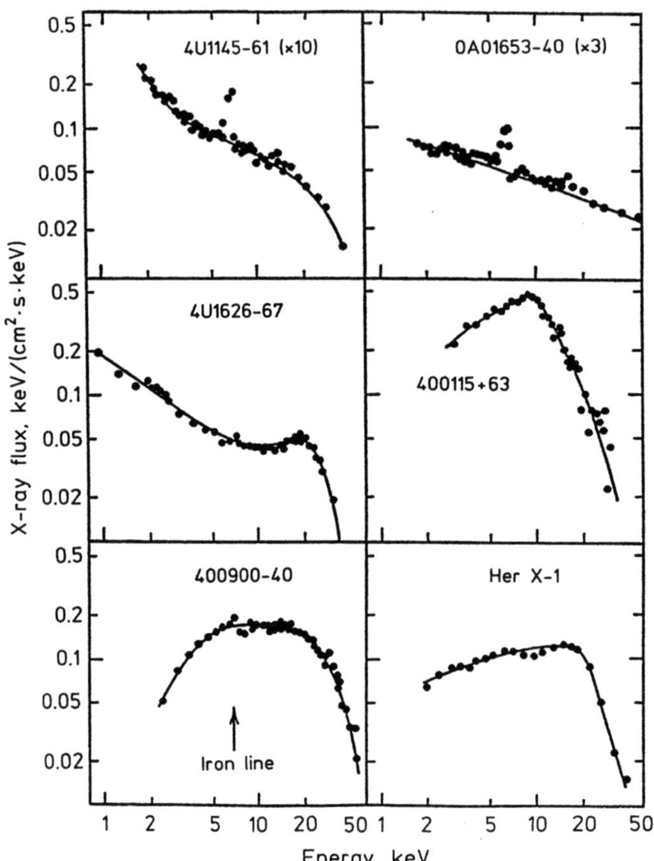

Fig. 1.8. Spectra of X-ray pulsars (Rappaport, Joss 1983)

Over 300 X-ray sources were discovered during 429 days of observation. Most of these sources are concentrated in the Milky Way plane (Forman et al. 1978). The most impressive among these was the discovery of X-ray pulsars in binary systems (Fig. 1.6) (Schreier et al. 1972), whose behavior was found to be exactly the same as predicted by Shvartsman: these pulsars were spun up and not down (Fig. 1.7)! This means that these pulsars are accreting neutron stars. The X-ray pulsars Cen X-3 and Her X-1 were among the first to be observed. The periods of these pulsars are 4.8 and 1.24 s, respectively. Their emission spectrum differs strongly from the nonthermal radiopulsar spectrum and is clearly reminiscent of the thermal radiation of a plasma at a temperature of 10^8 K (Fig. 1.8).

The fact that X-ray pulsars belong to binary systems was established from X-ray eclipses and from the variation of the pulse repetition period due to orbital motion, i.e., from the effects mentioned by Guseinov (1970).

Most of the companion stars of X-ray pulsars were found to be blue supergiants (for pulsars of the type Cen X-3) or light stars which fill their

Roche lobes (pulsars of the type Her-X-1). Both types of binary systems are characterized by intensive gas flow in the form of stellar wind or jets.

The discoveries made on board Uhuru created an impetus for a large number of theoretical works devoted to the accretion of matter onto magnetized neutron stars and to the accretion of matter in binary systems in general. X-ray pulsars were interpreted as accreting neutron stars by Pringle and Rees (1972), Davidson and Ostriker (1973), Lamb et al. (1973), Shakura and Syunyaev (1973), and Davidson (1973). Gnedin and Syunyaev observed in 1974 that the emission spectrum of accretion matter in a strong magnetic field must contain spectral lines corresponding to transitions between Landau levels. It is well known (Landau, Lifshitz 1974) that in the nonrelativistic approximation, the energy levels of an electron in a magnetic field differ from one another by $\hbar\omega_g$, where

$$\omega_g = \frac{eB}{mc}$$

is the gyrofrequency. For a field $B \approx 10^{12}$ Oe, the corresponding transition energy is ~ 10 keV, i.e., it lies in the X-ray range.

In 1976, Trümper and colleagues observed a spectral detail in the spectrum of the X-ray pulsar Her X-1 (Fig. 1.9) with the help of an X-ray detector hoisted on a balloon (Trümper et al. 1978). The position of this line corresponds to a magnetic field of $(3-5)\times 10^{12}$ Oe.

The theory of the evolution of a neutron star in a binary system, essentially along the line presented in the pioneering works of Shvartsman, was subsequently carried on by Bisnovaty-Kogan and Komberg (1974), Illarionov and Syunyaev (1975), Fabian (1975), Lipunov and Shakura (1976), Kundt (1976), and Savonije and van den Heuvel (1977).

It was emphasized by Davidson and Ostriker (1973) that during its evolution, an X-ray pulsar comes to a quasi-equilibrium state in which the angular momentum of the neutron star does not change on the average. The critical test for the verification of this assumption was the prediction of spin-down episodes in X-ray pulsars (Lipunov, Shakura 1976), which were discovered soon after in five X-ray pulsars.

In the 1970s, many rockets, balloons and satellites carrying detectors were launched. To date, about 20 X-rays pulsars have been discovered with the help of these detectors (Table 6.2).

What makes us believe that the X-ray pulsars are indeed neutron stars? Of course, there are no doubts concerning the pulsars with very short periods. For example, the period of the X-ray pulsar A 0538-66 is 0.069 s (which is just about twice the period of the pulsar in the Crab nebula). Only a neutron star can rotate with such a period. Indeed from the equality of centrifugal and gravitational forces

$$\frac{GM}{R^2} \approx \omega^2 R$$

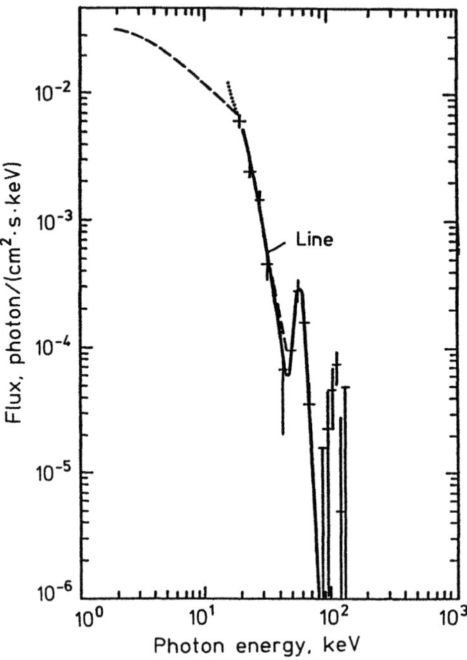

Fig. 1.9. Spectral detail discovered by Trümper et al. (1978) in the spectrum of the X-ray pulsar Her X-1 and interpreted as a gyroline in a magnetic field of strength $(3-5) \times 10^{12}$ Oe

we obtain the following estimate for the minimum period:

$$p_{\min} \approx 2\pi \sqrt{\frac{R^3}{GM}} \approx \frac{2\pi}{\sqrt{G\varrho}} \ . \tag{1.10}$$

For neutron stars, $M \approx 1 M_\odot$, $R = 10^6$ cm, $p_{\min} \approx 10^{-3}$ s while for white dwarfs $R = 10^8$ cm and $p_{\min} \approx 1$ s.

However, most X-ray pulsars have a period much longer than 1 s, and could indeed be white dwarfs. The situation became more perplexing when X-ray radiation from white dwarfs was discovered in the mid 1970s (this radiation was also found to be pulsating!). A decisive test was needed to prove, with a minimum number of assumptions, that an X-ray pulsar is a neutron star and not a white dwarf.

In 1977, Rappaport and Joss suggested that a comparison of the theoretical and observed values of the spin-up of X-ray pulsars could serve as such a test. Within the framework of certain assumptions concerning the accretion regime and the magnetic field strength, the neutron star model provides a better agreement between theory and observation.

A simpler and more reliable test was devised at a later stage (Lipunov 1981). It was shown that there exists a universal upper limit for the value of spin-up, which depends only on a combination of the two observed quantities, viz. the radius and X-ray luminosity, and is independent of the assumptions concerning the nature of accretion or the magnitude of the magnetic field

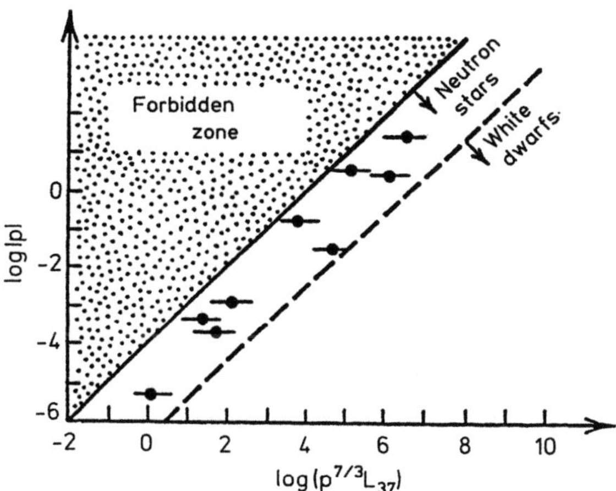

Fig. 1.10. Dependence of the X-ray pulsar spin-up on a combination of observed characteristics: L_{37} is the X-ray luminosity (in units of 10^{37} erg/s) and p is the period (in s). Observational data are shown by *points*. *Straight lines* show the maximum possible spin-up for a neutron star and a white dwarf (Lipunov 1981 a)

strength (Chap. 6). The only parameter in this test is a factor depending on the structure of the star (to be precise, on its mass, moment of inertia and radius). A comparison of the upper limit of the spin-up with the observed values proved beyond doubt that X-ray pulsars whose spin-up was measured are accreting neutron stars (Fig. 1.10). Not a single observed pulsar showed a spin-up exceeding the limiting value for neutron stars. This is an independent confirmation of the fact that we are dealing with accreting spin-ups.

Apart from X-ray pulsars, the so-called transient (nova-like) X-ray sources were also observed. The first transient source was discovered in 1967, when a

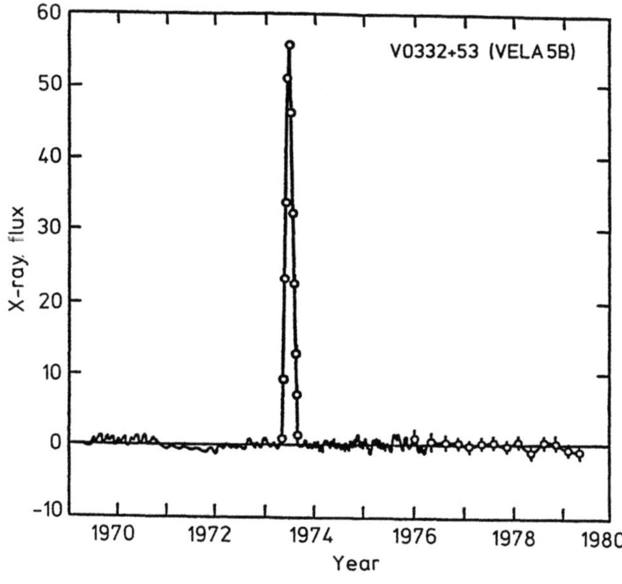

Fig. 1.11. Data accumulated over many years of observation from the specialized satellite "Vela 5 B" of a transient X-ray source V 0332+53 (Terrell, Priedhorsky 1984)

hitherto unknown X-ray source brighter than Sco X-1 was detected from a rocket. The source died away a few months later. The X-ray flux recorded from one of the brightest transient sources ever observed is presented in Fig. 1.11. This source, called V 0332+53, was discovered by a satellite from the "Vela" series. One such satellite (Vela 5 B) observed this source regularly over the period 1969–1979, and thus an X-ray flux curve (Fig. 1.11) could be obtained over an unprecedented length of time and superb signal-to-noise (Terrell, Priedhorsky 1984). A specialized X-ray satellite "Ariel" was launched in 1974 for observing such transient sources and was quite successful. It should be emphasized that the class of transient sources does not form a homogeneous group and apparently contains some types operating on entirely different mechanisms. However, some of these sources turned out to be X-ray pulsars and are therefore undoubtedly accreting neutron stars, at least for this reason.

1.7 X-Ray Bursters

In 1975, an entirely new class of X-ray sources, called X-ray bursters, was discovered (Grindlay et al. 1976). Instruments mounted on the Dutch satellite ANS recorded a burst of X-ray radiation lasting just 20 s coming from the direction of the globular cluster NGC 6624 (Fig. 1.12, 1.13). This source belonged to the class of X-ray sources called galactic bulge sources. A characteristic feature of such sources is their spatial distribution: they are concentrated towards the center of the Galaxy like the spherical component stars and globular clusters. X-ray bursters (over 20 such sources have been discovered so far, Table 6.5) are in the softer X-ray spectral region than X-ray pulsars are [see also the review by Lewin and Clark (1980)]. Their approximation to thermal spectra gives temperatures of the order of several keV.

The bursts from the sources are quasiperiodic and are set against a more or less constant background of X-ray flux (Fig. 1.12). Two types of burst can be distinguished. For bursts of the first type, the typical recurrence time varies between a few hours and several days, and the emission spectrum becomes softer with the evolution of the burst. Most bursters reveal only this type of burst. Bursts of the second type are characterized by a very short recurrence period ranging from tens of seconds to tens of minutes. Bursts of this type (as well as of the first type) are observed in the rapid burster MXB 1730-335[2] (Fig. 1.14). X-ray bursts of the first type were soon explained and associated with thermonuclear bursts on the surface of weakly magnetized neutron stars (Maraschi, Cavaliere 1977). The decisive argument in favor of this model was that for most bursters, the ratio of energy released in a burst to the energy liberated between bursts was approximately the same and equal to 1/100. This situation is analogous to the one in which the radiation between bursts is the result of accretion onto the surface of a neutron star with $\eta \approx 10\%$ [see (1.2)],

---

[2] MXB is an abbreviation for Massachusetts Institute of Technology, X-ray Bursters.

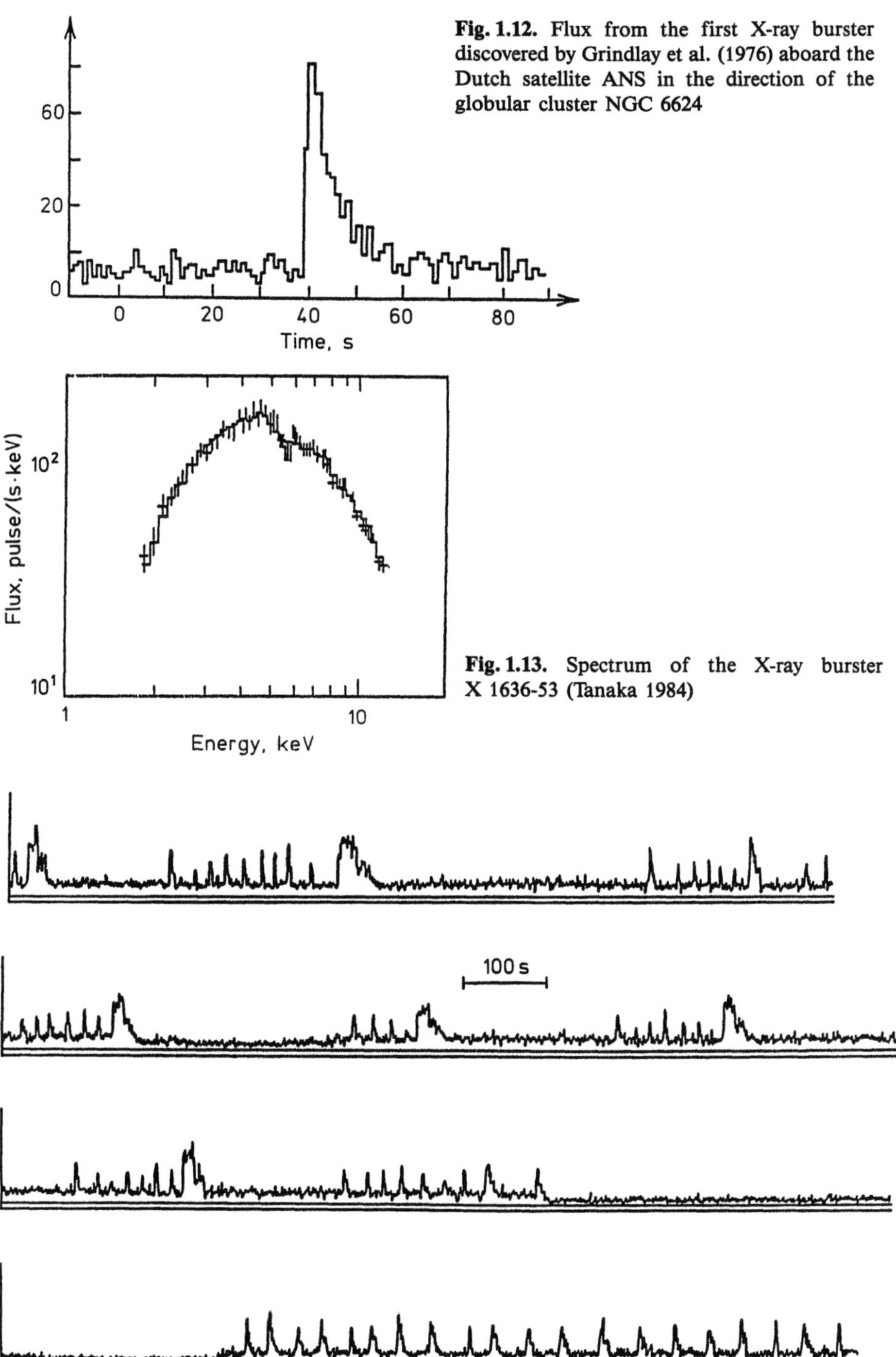

Fig. 1.12. Flux from the first X-ray burster discovered by Grindlay et al. (1976) aboard the Dutch satellite ANS in the direction of the globular cluster NGC 6624

Fig. 1.13. Spectrum of the X-ray burster X 1636-53 (Tanaka 1984)

Fig. 1.14. Radiation from fast X-ray burster MXB 1730-335

while the energy released during a burst is the energy of thermonuclear fusion of the entire accumulated matter with an efficiency $\eta \approx 0.1\%$.

The neutron star model is also in good agreement with the observed fluxes and with the burster spectra. The luminosity of most bursters is estimated at about 10^{37} erg/s, while the emission temperature is $\sim 10^7$ K ≈ 1 keV. Applying the Stefan-Boltzmann law, we obtain

$$R_x = \sqrt{\frac{L_x}{4\pi\sigma T^4}} \ .$$

Hence $R_x \approx 10^6$ cm, which is in excellent agreement with the predictions of the theory.

Numerical and analytical calculations showed that the thermonuclear burst of helium in accumulated matter can indeed explain the observed properties of bursters [for details, see the review by Ergma (1982)].

In addition to the neutron star model, Bahcall and Ostriker (1975) proposed a model of a massive black hole with a mass $\sim 100-1000\, M_\odot$. However, this idea finally faded away after the launching of the X-ray observatory "Einstein" in 1979. The apparatus mounted on this space observatory had not only a record-high sensitivity, but also the best angular resolution ($\sim 1''-2''$) ever achieved for the X-ray range. Observations revealed that the position of the X-ray sources does not coincide with the centers of globular clusters (which was to be expected in the case of bursters having a large mass). The magnitude of the average "spread" led to a value of $\sim 2\, M_\odot$ for the mass (Lightman et al. 1980).

After the launching of this specialized observatory, the potentialities of observational X-ray astronomy rose sharply. It is sufficient to mention that the overall number of discovered X-ray sources rose by more than an order of magnitude. A special role in the understanding of the processes of formation and the internal structure of neutron stars was played by the quest for stellar X-ray sources in the remnants of supernova explosions. The radiation must be of pulsar type (as in the Crab nebula emission), and it must include the thermal radiation from neutron stars that have not yet cooled down after their formation [see the Einstein satellite review by Giacconi (1981)].

First calculations of neutron star cooling were carried out by Bahcall and Wolf (1965), and by Tsuruta and Cameron (1965). These works were based on the ideas of Chiu and Salpeter (1964) on the possibility of thermal X-ray emission by hot neutron stars. The cooling rate of the neutron stars depends considerably on the state of the matter in the star. Interpretation of new data from the space observatory Einstein led to additional restrictions on the temperature of neutron stars (Helfand 1981).

1.8 Bursts and Other Sources of Gamma Rays

The discovery of gamma-ray bursts is another example of an unexpected event.

In 1967, short (of the order of a few seconds) bursts of 1 MeV gamma rays were recorded on several "Vela" satellites launched by the United States to monitor nuclear tests. These recordings were analyzed only after a few years and the discovery became known only in the early 1970s (Klebesadel et al. 1973).

For a long time, nothing was known about the nature of these sources. The observation of gamma-ray bursts is quite a peculiar process. A narrow-field detector is unsuitable for this purpose since gamma-ray bursts occur very rarely (perhaps even less frequently than a few times per year). A wide-field detector fails to provide information about the position of the source. The only redeeming feature is the explosive nature of the effect. If a burst is registered simultaneously by several satellites, the direction of the gamma-ray source can be determined from the delay time.

The first experiment which threw light on the nature of gamma-ray bursts (experiment "Konus") was carried out by a group of Soviet scientists led by E. P. Mazets. The experiment was performed in 1978 on three satellites "Venera-11", "Venera-12" and "Prognoz-7". Owing to the very high sensitivity of the detector, about 150 gamma-ray bursts were observed during $1\frac{1}{2}$ years.

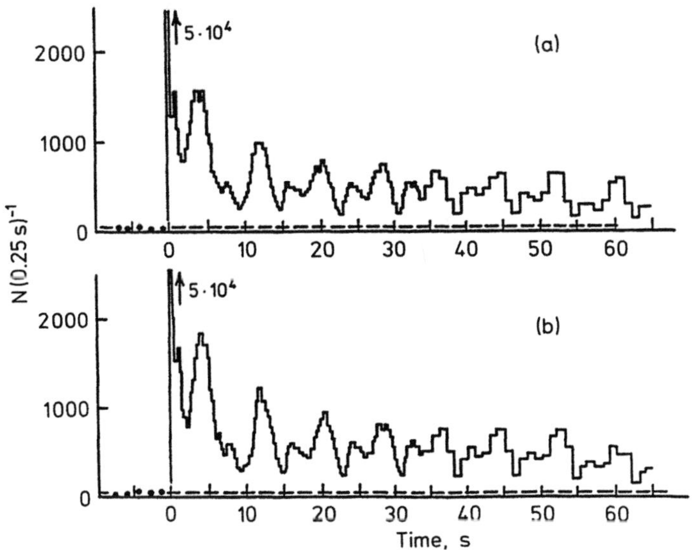

Fig. 1.15a,b. Gamma-ray burst of March 5, 1979. The recording was made on board the Soviet satellites "Venera-12" (a) and "Venera-11" (b) (Mazets et al. 1980)

This was more than the total number of such bursts observed since their discovery.

The most remarkable event took place on March 5, 1979 (Fig. 1.15). In less than 1 ms, the flux rose to 10^{-3} erg/cm^2 s. If such a burst had occurred in the optical range, it could have been observed by the naked eye even in day time. But what is more important is that periodic variations in the radiation with a period of about 8 s were observed. This at once made the neutron star model more favorable than other models (e.g., the black hole model). The pulse shape and the emission spectrum, especially in the pulsating component, were reminiscent of the radiation emitted by X-ray pulsars. The position of this burst was determined with an extremely high degree of accuracy. The burst was superimposed on the remnant of the supernova explosion in the Small Magellanic Cloud. However, this was apparently a coincidence. At a distance of ~ 50 kiloparsec from the Magellanic Cloud, the luminosity of the source is found to be $\sim 10^{44}$ erg/s, which is comparable with the luminosity of the whole galaxy. This result is absolutely incompatible with the fact that somewhat weaker bursts were also registered subsequently from this source. However, if the superposition on the remnant is just accidental, it means that the burst occurred from nowhere!

Although to date several models have been put forth to explain this phenomenon, it remains as enigmatic as ever. Detailed investigations of constant gamma-ray sources in the range of about 100 MeV were started after the launching of the specialized American satellite "COS-B". "Point" sources were also discovered in addition to the diffuse sources. (The quotation marks are used to emphasize the fact that the angular resolution of the apparatus is 0.5°.) The brightest of these sources was observed in the constellation Gemini and was named GEMINGA[3]. This object drew considerable attention in the following years (see Nature 310, August 1984). Some of the numerous models that appeared subsequently associate GEMINGA with a neutron star.

Another range in which neutron stars are observed is the gamma range of ultrahigh energies $\gtrsim 10^{15}$ eV. Quanta in this range can be detected on the Earth's surface via the secondary Cherenkov radiation (Stepanyan 1984). Recent communications indicate that gamma quanta with energies $\gtrsim 10^{15}$ eV are emitted not only be active radiopulsars like the Crab nebula pulsar (e.g. Cyg X-3), but also by the classical binary X-ray systems of the type Vela X-1.

It should be noted that the gamma-astronomy of ultrahigh energies is still in its initial stage and most of the results should be viewed with caution. After

---

[3] The term GEMINGA was coined by Italian astronomers and can be interpreted in two ways, viz. a combination of GEMINI and GAMMA, or "non-sense" in the Milanese dialect.

all, the total number of quanta registered in this range during the last 10 years is just a few hundred.

1.9 General View

The discovery of neutron stars was one of the most remarkable events in astrophysics during the 1960s and 1970s. Together with events like the discovery of relict radiation and quasars, it comprises what is known as the modern revolution in astronomy.

By the end of the 1980s, over 500 radio pulsars, about 20 X-ray pulsars, and about the same number of X-ray bursters had been registered. Several hundreds of gamma-ray bursts, which are also apparently neutron stars, have been discovered. The overall view of our knowledge of neutron stars is presented in Table 1.1.

Table 1.1. Observed characteristics of neutron stars

Source	Observed number	Total number in the Galaxy	Emitted energy [erg/s]	Reliability of identification as a neutron star
Radio pulsars	~ 500	$\sim 10^5$	$10^{31} - 10^{38}$	+
X-ray pulsars	~ 20	~ 100	$10^{33} - 10^{39}$	+
X-ray bursters	~ 30	100	$10^{37} - 10^{38}$	+ −
Gamma-ray bursters	~ 300	?	?	− +
SS 433-type sources	1	?	?	− +
GEMINGA	1	?	?	− −
Bulge sources like Sco X-1, Cyg X-2, and QPO (noisars)	~ 30	~ 100	$10^{38} - 10^{37}$	+ −

Neutron star investigations cover two aspects. Firstly, they involve the study of physical processes occurring in the vicinity of specific sources or types of sources. Secondly, the investigations are aimed at establishing an evolutionary connection between different types of neutron stars and between neutron stars and normal stars.

The colossal diversity in the behavior of different types of objects discovered in the last few decades suggests that it is hopeless to make any attempts to describe all the observed phenomena from some common viewpoint. It is certainly not an easy task and it would be frivolous to assume that a single "formula" can provide an interpretation of all observed effects.

Nevertheless, we shall endeavor to reduce the problem to the smallest number of parameters. The subsequent description (with the exception of Chap. 2) will be based on just this principle. The material will reflect present-day concepts about the nature of the interaction of neutron stars with the surrounding matter. By the "evolution" of a neutron star, we shall mean a slow variation in the nature of this interaction.

2. Structure of Neutron Stars

The equilibrium of cold self-gravitating matter can only be sustained by essentially quantum effects.[1] For example, if the density of matter is less than $\sim 10^{11}$ g/cm^3, equilibrium can be ensured by the repulsive forces of degenerate electrons (Fowler 1926). However, it was first shown by Chandrasekhar (1931) and Landau (1932) that if we consider a very large amount of a substance, the configuration becomes unstable. The physical reason behind this instability is that the electrons become relativistic and the equation of state of the degenerate gas becomes close to the relativistic gas equation $P \propto \varrho^{4/3}$. This is a polytropic equation with exponent $n = 3$. According to the Lane-Emden theory of polytropic stars (see, for example, Schwarzschild 1958; Zel'dovich, Novikov 1971), this equation is critical as the configuration becomes unstable for $n = 3$. Thus, the maximum mass of a degenerate dwarf (Chandrasekhar limit) is determined from the condition that the electrons become relativistic (see below).

The maximum mass was found to be close to the mass of the Sun, which is an ordinary star. Stars with a much larger mass are also known to exist (after all, a star does not know anything about the Chandrasekhar limit when it is born!). What will happen to this star when the energy sources in it are exhausted?

It was proposed by Landau (1932) that superdense stars, whose equilibrium is sustained by nuclear forces, may exist in nature. Baade and Zwicky (1934) termed these stars neutron stars and attributed their origin to the supernova bursts.

The size of a neutron star is $R_x \approx 10$ km, and the gravitational potential $GM/R_x \approx 0.1 \, c^2$. This means that we must use the general theory of relativity to study the structure of a neutron star (Eddington 1935). First steps in this direction were taken by Oppenheimer and Volkoff (1939).

In this chapter, we shall consider some intrinsic properties of neutron stars, attaching due importance to the observed properties (i.e., those which are important in their manifestation). We shall confine ourselves to a brief account of this problem. For a more detailed theory of the structure of neutron stars, we refer the reader to Zel'dovich and Novikov (1971), and to a recent monograph by Shapiro and Teukolsky (1985).

---

[1] For the present, we neglect rotational effects.

2.1 Equilibrium of Stars

Equilibrium of stars is maintained by the equality of the gravitational force and the pressure gradient. Any equilibrium state of a star corresponds to an extremum of its total energy. The total energy of a star is the sum of its gravitational energy E_g and the kinetic energy E_k of its particles

$$E = E_g + E_k \; . \tag{2.1}$$

If a star consists of an ideal gas, its kinetic energy E_k is the thermal energy of the star. For a given mass M and an average density ϱ, the total energy of a star is given by

$$E = \varepsilon_k M - C_1 M^{5/3} \varrho^{1/3} \; , \tag{2.2}$$

where C_1 is a constant depending on the distribution of matter in the star, and ε_k is the mean kinetic energy of one gram of this matter. For an ordinary gas, the polytropic equation of state has the form

$$P = A \varrho^{\gamma} \; , \tag{2.3}$$

while the energy ε_k is given by

$$\varepsilon_k = C_2 \varrho^{\gamma-1} + C_3 \; , \tag{2.4}$$

where C_2 is a constant depending on the entropy, and C_3 is a constant depending on the chemical composition. Substituting ε_k into (2.2), we obtain

$$E = -C_1 M^{5/3} \varrho^{1/3} + C_2 M \varrho^{\gamma-1} + C_3 M \; . \tag{2.5}$$

The minimum total energy E corresponds to the stable equilibrium of the star while the maximum energy corresponds to unstable equilibrium. In the equilibrium state, the kinetic and gravitational energies of the star are of the same order.

It follows from the last expression that for $\gamma < 4/3$, the $E(\varrho)$ curve does not have any minimum, and hence the stable equilibrium of a star consisting of such a gas is not possible (the elasticity of the substance is low). Figure 2.1 shows the qualitative dependence of E on ϱ.

In degenerate stars, ε_k is replaced by the kinetic energy of the degenerate electrons. The transition to the equation of state with $\gamma = 4/3$ occurs at the instant when the electrons become relativistic. This allows us to determine the maximum mass of a degenerate configuration.

Suppose that the thermal energy is always equal to zero (cold star). Let us estimate the total kinetic energy of a degenerate Fermi gas. A gas becomes degenerate when a cell in the (six-dimensional) phase space of coordinates and momenta formally contains more than one particle. Fermions are particles with a half-integral spin and obey Pauli's exclusion principle: two fermions

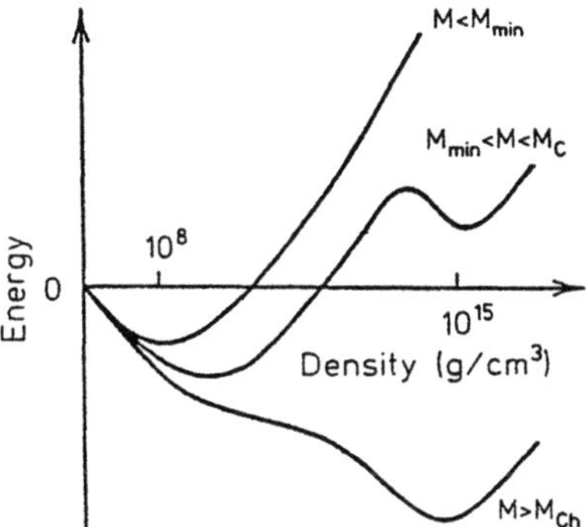

Fig. 2.1. Qualitative dependence of the total energy of a star E on its mean density ϱ for different masses

cannot exist in the same state, for example in a cell having a volume $\sim h^3$. Obviously, we cannot pack the particles in such a way that the product of their momentum Δp and the distance Δx between them is less than the value corresponding to Heisenberg's uncertainty relation

$$\Delta p \Delta x \sim \hbar \ . \tag{2.6}$$

Taking the cube of this relation, we obtain the volume occupied by one particle of the degenerate electron gas.

Let us consider a star of radius R and mass M, whose equilibrium is maintained by the pressure of the relativistic gas. The separation between neighboring particles is estimated by

$$\Delta x \approx \mu^{1/3} R / (M/m_b)^{1/3} \ , \tag{2.7}$$

where $\mu = N_b/N_f$ is the ratio of the number of baryons to the number of degenerate fermions and m_b is the mass of a baryon. From the uncertainty relation (2.6), we obtain the momentum of the particle:

$$\Delta p \approx \hbar (M/m_b)^{1/3} \mu^{-1/3} R^{-1} \ . \tag{2.8}$$

Since the energy of a relativistic particle is of the order of $\Delta p c$, we find that the total kinetic energy of fermions in a star is approximately equal to

$$E_k \approx E_F \approx \frac{\hbar c}{R} \left(\frac{M}{m_b}\right)^{4/3} \mu^{-4/3} \ , \tag{2.9}$$

where E_F is the Fermi energy. By the way, it follows from this relation that $\varepsilon_k = E_k/M \propto M^{-1/3} \propto \varrho^{1/3}$. In other words, a comparison with (2.4) shows that $\gamma = 4/3$ in the equation of state for a relativistic Fermi gas. However, the kinetic and gravitational energies in the equilibrium state are of the same order.

This means that

$$\mu^{-4/3} \frac{\hbar c}{R} \left(\frac{M}{m_b}\right)^{4/3} \approx \frac{GM^2}{R} \ . \tag{2.10}$$

This leads to the approximate critical mass for which a degenerate gas becomes relativistic and the star collapses:

$$M_{Ch} \approx \mu^{-2} \left(\frac{\hbar c}{Gm_b^2}\right)^{3/2} m_b \ . \tag{2.11}$$

For a degenerate gas consisting of heavy nuclei, $\mu = 2$ and, following Landau (1932), we find that $M_{Ch} \simeq 1.5\, M_\odot$.

It should be observed that while deriving (2.11), we have not specified anywhere the nature of the Fermi particles. This shows that the critical mass of a degenerate electron gas and a degenerate neutron gas must be of the same order. However, the radii of the degenerate configurations are found to be quite different. The minimum radius is found from the condition of relativistic degeneracy, when the Fermi energy for a particle becomes of the same order as the rest energy $m_f c^2$.

Substituting (2.9) into the relation

$$\frac{E_k}{N_f} \approx m_f c^2 \ , \tag{2.12}$$

we obtain the radius of a degenerate star at the stability limit (Shapiro, Teukolsky 1985):

$$R_{min} \approx \frac{\hbar}{m_f c} \left(\frac{\hbar c}{Gm_b}\right)^{1/2} \approx \begin{cases} 5 \times 10^8 \text{ cm}, \ m_f = m_e, \\ 3 \times 10^5 \text{ cm}, \ m_f = m_n \ . \end{cases}$$

The above qualitative analysis shows that a star having a mass M_{Ch} inevitably contracts to a size at which the atoms disintegrate and the nuclei come in contact. As a result of the effects associated with the general theory of relativity, the $E(\varrho)$ curve shown in Fig. 2.1 acquires a secondary minimum near $\varrho \gtrsim 10^6$ g/cm^3. In addition to the GTR effects, the neutronization of matter is also quite an important process in which the nuclei capture electrons and a proton is transformed into a neutron:

$$(z, A) + e^- \rightarrow (z-1, A) + \nu \ ,$$

where z is the nuclear charge and A the atomic mass. Neutronization at the center of the star sets in at densities of $\sim 10^9 - 10^{10}$ g/cm^3.

Following Landau (1938), let us estimate the minimum mass of a neutron star. For a neutronization reaction to take place, a certain amount of energy must be spent. For example, the transformation

$$^{16}O + 8e^- \rightarrow 16_1 n^0$$

requires an expenditure of 7×10^{18} erg per gram of the substance. Hence, in order to transform the entire amount of matter of a star of mass M into a neutron mixture, we must spend an amount of energy equal to $7 \times 10^{18} M$ erg. However, compression to the size of a neutron star ($R \simeq 10$ km) leads to the liberation of gravitational energy equal to $\sim 10^{53} M^{5/3}$ erg. Obviously, the transformation of the entire substance into neutron matter requires a "sample" of quite large mass, and

$$M_{\min} \approx 6 \times 10^{31} \text{ g} \approx 0.03 \, M_\odot \ .$$

This estimate shows that, in principle, neutron stars may be very "light". It must be remembered, however, that as long as the mass of a star is less than the Chandrasekhar limit, the degenerate dwarf state is more advantageous from the energy point of view. Hence the formation of "light" neutron stars requires the expenditure of an additional amount of energy.

2.2 Exact Equilibrium Equations for Cold Stars

To begin, let us consider the nonrelativistic theory of Lane-Emden polytropic spheres. The system of equations describing the structure of a star has the form

$$\frac{dP}{dr} = -\frac{GM_r}{r^2} \varrho \ , \tag{2.13}$$

$$\frac{dM_r}{dr} = 4\pi r^2 \varrho \ , \tag{2.14}$$

$$P = A\varrho^\gamma \ , \quad \gamma \equiv 1 + 1/n \ . \tag{2.15}$$

Here, M_r is the mass of the star enclosed within a sphere of radius r. The first of these equations is the hydrostatic equilibrium equation, while the second establishes a connection between M_r and the density ϱ. The third equation is the equation of state. Dividing the first equation by ϱ/r^2 and differentiating after substituting (2.15), we obtain

$$\frac{1}{r^2} \frac{d}{dr} \left(\frac{r^2}{\varrho} \frac{dP}{dr} \right) = 4\pi G\varrho \ . \tag{2.16}$$

We introduce dimensionless variables

$$\varrho = \varrho_c \theta^n \; ,$$

$$r = a \xi \; ,$$

$$a = \left[\frac{(n+1) A \varrho_c^{(1/n-1)}}{4 \pi G} \right]^{1/2} \; , \tag{2.17}$$

where $\varrho = \varrho_c$ for $r = 0$, i.e., the central density of the star. In terms of the new variables, we can present the Lane-Emden equation in the form

$$\frac{1}{\xi^2} \frac{d}{d\xi} \xi^2 \frac{d\theta}{d\xi} = -\theta^n \; . \tag{2.18}$$

The boundary condition can be written as

$$\theta(0) = 1 \; . \tag{2.19}$$

It is not possible to obtain an analytical solution for $n = 3$. However, it was shown by Emden that a regular solution exists for $n = 3$ only if

$$\alpha \equiv \xi_1^2 |\theta'(\xi_1)| \approx 2.01824 \; . \tag{2.20}$$

This allows us to obtain the exact value of the Chandrasekhar limit. Indeed, the total mass of a star is given by

$$M = \int_0^R 4 \pi r^2 \varrho \, dr = 4 \pi a^3 \varrho_c \int_0^{\xi_1} \xi^2 \theta^n \, d\xi \; ; \tag{2.21}$$

where ξ_1 is the dimensionless radius of the star. Substituting (2.18) into this relation, we obtain

$$M = 4 \pi \left[\frac{(n+1) K}{4 \pi G} \right]^{3/2} \varrho_c^{(3-n)/2n} \alpha \; . \tag{2.22}$$

For an ideal and completely relativistic electron gas (Zel'dovich, Novikov 1971) we obtain

$$A = 1.2435 \times 10^{15} \text{ c.g.s units} \; . \tag{2.23}$$

In this case, we obtain from (2.22)

$$M_{\text{Ch}} \approx 1.457 \left(\frac{2}{\mu_e} \right)^2 M_\odot \tag{2.24}$$

or, in a general form,

$$M_{\text{Ch}} \approx \frac{3.1}{\mu_e^2} \left(\frac{\hbar c}{2 \pi G} \right)^{3/2} . \tag{2.25}$$

A consideration of relativistic effects and rigid rotation changes this quantity by less than a few percent (Tassoul 1978). However, differential rotation may considerably alter the situation. Approximate calculations show that the Chandrasekhar limit may rise to $\sim 3 M_\odot$ (Ostriker, Bodenheimer 1968).

The above theory can be used only for very light neutron stars in which the neutron gas can be treated as ideal. Qualitatively, it is clear that the limiting mass of a neutron star in such a model may turn out to be even smaller than in the case of a degenerate electron configuration: after all, the relativistic effects decrease the stability of a star. This is exactly what was observed by Oppenheimer and Volkoff (1939) in their calculations.

Taking into consideration the general theory of relativity, we can present the hydrostatic equilibrium equation in the form (Tallman-Oppenheimer-Volkoff equation)

$$\frac{dP}{dr} = -\frac{GM_r \varrho}{r^2} \left(1 + \frac{P}{\varrho c^2} \right) \left(1 + \frac{4 \pi r^2 P}{M_r c^2} \right) \left(1 - \frac{2 G M_r}{r c^2} \right)^{-1} , \tag{2.26}$$

where M_r is the mass as before:

$$M_r = \int_0^r 4 \pi r^2 \varrho(r) dr . \tag{2.27}$$

Together with the equation of state

$$P = P(\varrho) \tag{2.28}$$

and the boundary condition

$$P(R_x) = 0 \tag{2.29}$$

equations (2.25, 26) can be used to completely calculate the structure of a non-rotating neutron star.

Two types of GTR effects can be considered in (2.25). Firstly, the metric curvature exerts an influence. Secondly, the pressure makes a contribution to the right-hand side (the pressure plays the same role as a weight).

The main difficulty in constructing a model of a neutron star is that we do not know the exact relation between P and ϱ, especially if the density is higher than that of the nuclear matter. Naturally, the form of the equation of state considerably affects the upper limit of the mass of a neutron star (Oppenheimer-Volkoff limit). Hence the determination of the masses and radii of neutron stars from observations is of fundamental importance in the construction of the theory of nuclear matter.

Before describing the newest concepts of the properties of the matter constituting neutron stars, let us consider a few analytical solutions of the Tallman equation for an idealized equation of state.

For the case of an incompressible liquid, $\varrho = $ const and as long as $P \ll \varrho c^2$, the Oppenheimer-Volkoff limit is given by (Brecher, Caporaso 1977)

$$M_{\mathrm{OV}} = \frac{c^3}{G^{3/2}} \left(\frac{3}{32\,\pi} \right)^{1/2} \varrho_{\mathrm{m}}^{-1/2} \approx 11.4 \left(\frac{\varrho}{10^{14}\,\mathrm{g/cm^3}} \right)^{-1/2} M_{\odot} \ . \tag{2.30}$$

For the nuclear density $\varrho_{\mathrm{nucl}} = 3 \times 10^{14}\,\mathrm{g/cm^3}$, the maximum mass of a neutron star is found to be $M_{\mathrm{OV}} \approx 6.6\,M_{\odot}$. Formula (2.30) can be obtained in the general form from elementary considerations. The gravitational energy of a neutron star is several tens percent of the total energy $GM^2/R \approx 0.1\,Mc^2$. Substituting the radius from the relation $M = (4/3)\pi R^3 \varrho_{\mathrm{nucl}}$, we obtain $M \sim c^3/(G^{3/2}\varrho_{\mathrm{nucl}}^{-1/2})$. For

$$P = \alpha \varrho c^2 \tag{2.31}$$

an exact analytical solution exists (Oppenheimer, Volkoff 1939; Misner, Zapolsky 1964; Brecher, Caporaso 1976):

$$\varrho(r) = \frac{c^2}{G} \left[\frac{\alpha}{2\pi(\alpha^2 + 6\alpha + 1)} \right] r^{-2} \ . \tag{2.32}$$

Although this solution has a singularity at the center $(r \to 0)$, the mass of the star is finite:

$$M = \frac{c^3}{G^{3/2}} \sqrt{\frac{2}{\pi}} \left[\frac{\alpha}{\alpha^2 + 6\alpha + 1} \right]^{3/2} \varrho_{\mathrm{m}}^{-1/2} \ , \tag{2.33}$$

where ϱ_{m} is called the "fitting" density up to which the equation of state is still known. The maximum is attained at $\alpha = 1$ and is determined by the maximum density ϱ_{m} for which the assumptions concerning the equation of state of the substance are still valid. For a softer equation of state

$$P = (\varrho - \varrho_{\mathrm{m}})c^2 \tag{2.34}$$

and a finite central density, the solution is a superposition of the two solutions presented above (Brecher, Caporaso 1977). In this case, the mass of the star is found to be

$$M = (6\sqrt{2\pi})^{-1/2}(c^3/G^{3/2})\varrho_{\mathrm{m}}^{-1/2} \approx 3.0 \left(\frac{3 \times 10^{14}\,\mathrm{g/cm^3}}{\varrho_{\mathrm{m}}} \right)^{1/2} M_{\odot} \ . \tag{2.35}$$

It should be recalled that in the approximation of an ideal degenerate neutron gas, the Oppenheimer-Volkoff limit is equal to $0.8\,M_{\odot}$.

All this is illustrated quite elegantly by the dependence of the maximum mass of a neutron star on the properties of the matter constituting it.

2.3 Physical Conditions Inside Neutron Stars

The following pattern emerged from a detailed analysis of the structure of neutron stars. The radius of a neutron star of solar mass is about $10-16$ km. The surface of a neutron star is a hard crust of thickness $1-7$ km whose density increases inwards from 10^6 to 10^{10} g/cm^3. The properties of the substance in the crystalline crust have been studied quite extensively (Baym, Pethick 1979). Beyond the crust, the crystalline structure is destroyed and matter (comprised mainly of free neutrons) passes into liquid phase. The pattern is not quite clear at places where the density increases to values close to the density of nuclear matter ($\sim 2.8 \times 10^{14}$ g/cm^3). The possibility of a hard core cannot be ruled out in this case. It was first pointed out by Migdal (1959) that the neutron liquid within a neutron star must be superfluid. Subsequently, it turned out that this property plays a significant role in a number of processes occurring in a neutron star and observed on the Earth. Hence we shall consider the properties of the superfluid state of matter in somewhat greater detail. Superfluidity (superconductivity) is a macroscopic quantum-mechanical effect. One of the manifestations of superfluidity is the complete disappearance of viscosity (or resistance in the case of superconductivity). This means that a superfluid liquid set into motion relative to a vessel will practically never come to rest. The phenomenon of superfluidity, however, is not confined to the hydrodynamics of a liquid with zero viscosity. For example, the phenomenological theory of superfluid helium devised by Landau is a two-liquid fluid mechanics. Superfluidity was first observed in ^{4}He. Helium cooled to 4.22 K under normal pressure passes from a gaseous to a liquid state. If cooling is continued, a second-order phase transition occurs at 2.19 K and is accompanied by a sharp decrease in the heat capacity. This new state was called ^{3}He II. In 1938, Kapitsa observed that the motion of liquid ^{3}He II in a narrow capillary, or its passage through an orifice, is characterized by a complete absence of viscosity. This phenomenon was called superfluidity.

The neutron matter in a star is divided into bound pairs, similar to the Cooper pairs in a superconductor. The superfluidity of this substance is similar to the superfluidity of ^{3}He (since ^{3}He is also a fermion). It was first observed by Ginzburg and Kirzhnits (1964) that vortices must exist in the superfluid component of a rotating neutron star.

The concept of superfluid cores in neutron stars has been confirmed not only in observations (Chap. 8), but also on a laboratory scale in the creation of an artificial "neutron" star (Tsakadze, Tsakadze 1975). Cylindrical or spherical vessels filled with superfluid helium were magnetically suspended and allowed to rotate freely. Among other things, the irregularities observed in the period of a pulsar were also simulated in these experiments.

Verification of the theory of cooling of neutron stars (Nomoto, Tsuruta 1981) is an important step towards the understanding of the physical conditions prevailing in them. Novikov and Perevodchikova (1984) have considered the possible consequences of proton decay.

2.4 Parameters of Neutron Stars

Among the parameters of neutron stars that can be observed, the most important ones are mass M, radius R_x, moment of inertia I, and the Oppenheimer-Volkoff limit M_{OV}. All these parameters were calculated theoretically under different assumptions concerning the equation of state [see, for example, the review article by Canuto (1977) and the monograph by Shapiro and Teukolsky (1985)]. Figure 2.2 shows the mass, radius, and density for the most realistic equations of state [see Baym, Pethick (1979) for details]. Table 2.1 presents the values of the Oppenheimer-Volkoff limit for these calculations. The difference between the R and TNI models is illustrated in Fig. 2.3.

It can be seen that both models are characterized by the existence of a hard crust of thickness of ~ 0.1 the radius of the star. The substance constituting the crust is a mixture of nuclei and electrons for low densities ($\varrho \lesssim 4 \times 10^{11}$ g/cm^3). For higher densities (inner crust: $4.3 \times 10^{11} < \varrho < 2 \times 10^{14}$ g/cm^3), this is a mixture of a nuclear lattice and the superfluid neutron liquid. Most of the substance (in terms of mass and moment of inertia) is a superfluid liquid consisting of protons and neutrons ($\varrho \gtrsim (2-6) \times 10^{14}$ g/cm^3).

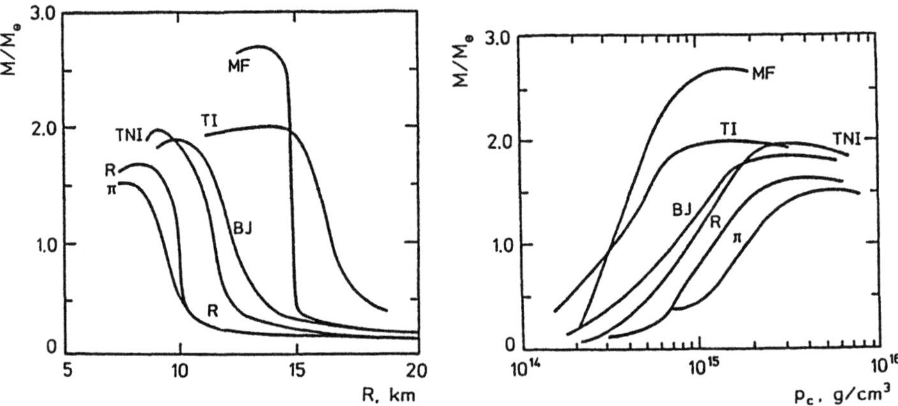

Fig. 2.2. Results of numerical computations of neutron star properties using different models for the superdense matter (Baym, Pethick 1979)

Table 2.1. Oppenheimer-Volkoff limit for different equations of state

Equation of state	Notation	$M_{OV}/M_\odot$ without rotation	$M_{OV}/M_\odot$ with rotation
Pion condensate	π	1.5	?
Reid's equation	R	1.6	?
Bethe-Johnson equation	BJ	1.9	2.16
Three-nucleon interaction approximation	TNI	2.0	?
Tensor interaction approximation	TI	2.0	?
Mean field approximation	MF	2.7	3.18

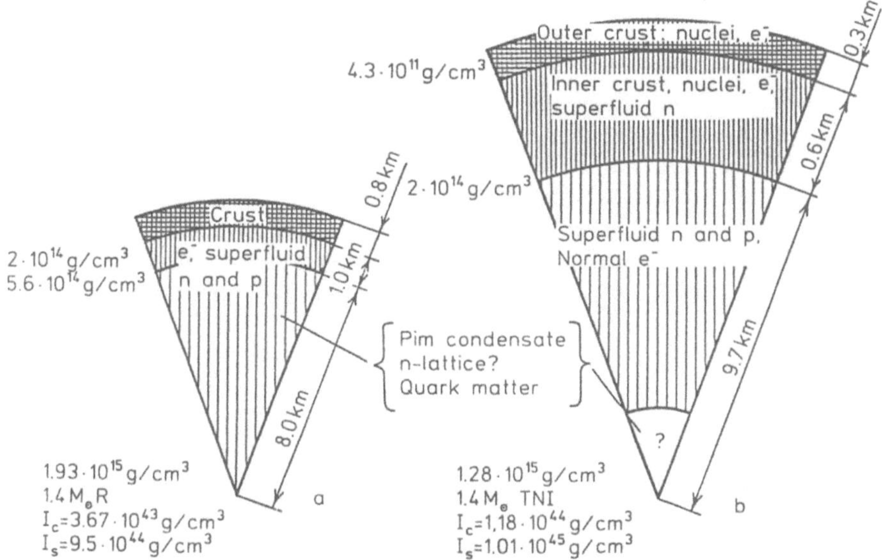

Fig. 2.3. Internal structure of a neutron star for two equations of state: *a* R and *b* TNI

The properties of the central regions have not been clearly established: it is not clear whether this material is a π-condensate or a solid neutron core or a quark core.

All the calculations presented here concern non-rotating neutron stars only.

2.5 Mass of Neutron Stars

Some of the observed neutron stars form a part of a binary system. This happy coincidence allows us to "weigh" such stars. In some cases, this has been done with extremely high accuracy: paradoxical though it may seem, the highest accuracy in astronomical mass measurements has been attained for neutron stars (error $\sim 4\%$)!.

Before proceeding to a description of the observational data, let us consider the scale of the masses of real neutron stars. A naive answer would sound like this: the mass of a neutron star must lie between the Chandrasekhar limit and the Oppenheimer-Volkoff limit ($M_{\mathrm{Ch}} < M_x < M_{\mathrm{OV}}$). This is not true. Even if we assume that a neutron star of minimum mass is a result of the total collapse of a degenerate core having a limiting mass M_{Ch}, the mass of the neutron star measured by a distant observer will be less than M_{Ch} due to the gravitational mass defect.

The mass of a neutron star that can be measured with the help of far removed test objects is called the Tallman mass. A star is a stable object.

Energy is liberated during the formation of a star from infinitely removed particles. Hence the mass of such a star will be smaller than the rest mass of the particles constituting it. The difference between these two masses is known as the mass defect. Obviously, the mass defect in neutron stars is due both to nuclear and gravitational interactions.

The M_r mass appearing in the Tallman-Oppenheimer-Volkoff equation (2.26) and defined by (2.27) includes the local energy connected with the motion and interaction of particles, as well as the gravitational energy. Naturally, this mass is different from the ordinary Newtonian mass since the local volume in the Schwarzschild metric differs from the classical expression:

$$dV = \left(1 - \frac{2GM_r}{rc^2}\right)^{-1/2} 4\pi r^2 dr \ . \tag{2.36}$$

For neutron stars, the mass defect may be $10-20\%$. Consequently, a neutron star formed by a limiting white dwarf may have a smaller mass. Besides, the matter released during the formation of a neutron star also cannot be discarded. Calculations for the collapse of while dwarfs show that a star may shed a sufficiently large mass and even completely disintegrate (Imshennik, Nadezhin 1982). Hence it would be more appropriate to present the expected mass values of neutron stars in the form

$$0.03\, M_\odot \approx M_{\min} < M_x < M_{OV} \approx 2-3\, M_\odot \ . \tag{2.37}$$

However, it should be emphasized that light neutron stars can apparently be formed only as a result of evolution of low-mass binaries (Chap. 11). In massive binary systems, the minimum mass of a star must be close to the Chandrasekhar limit (if we take into account the mass defect). The expected mass range is in excellent agreement with observations. Outstanding accuracy was attained in the measurement of the mass of the radiopulsar PSR 1913+16 (often known as the Taylor or Hulse-Taylor pulsar), which is a member of a binary system. Owing to the high stability of the period, relativistic effects (periastron advance) can be observed in this system. This enables us to determine the mass of both components:

$$M\,(\text{pulsar}) \qquad = (1.41 \pm 0.06) M_\odot \ ,$$

$$M\,(\text{companion}) = (1.41 \pm 0.06) M_\odot \ .$$

It was mentioned in Chap. 1 that two more radiopulsars in binary systems are known to exist at present, but their mass has not been determined accurately so far. At the same time, about 20 X-ray pulsars in binary systems are known. The stability of the pulse shape and periods of the X-ray pulsars is much worse but the possibility of observation of optical companions allows us to obtain quite reliable estimates for the mass of the neutron stars in such systems. Obviously, the mass of the X-ray pulsars can be determined by the spectral observation of their optical companions since the velocity with which the optical component moves around the center of mass is determined by the force of attraction of the neutron star.

Suppose that the semi-major axis of a binary system is given by $a = a_0 + a_x$, where a_0 and a_x denote the distance of the optical and neutron star, respectively, from the center of mass. In the case of circular orbits, the half-amplitude, v_0 of the variations of the optical star velocity, determined from the spectrum, is the projection of the orbital velocity on the line of sight: $v_0 = 2\pi a_0 T^{-1} \sin i$. Supplementing these relations by Kepler's third law, we obtain the system of equations

$$a = a_0 + a_x \, , \tag{2.38}$$

$$M_0 a_0 = M_x a_x \, ,$$

$$\frac{G(M_0 + M_x)}{a^3} = \left(\frac{2\pi}{T}\right)^2 .$$

It follows that there exists an algebraic combination of the values of mass of the components and the angle of inclination i of the orbit of a binary system, which can be presented in terms of the observable quantities only:

$$f_0(M) \equiv \frac{(M_x \sin i)^3}{(M_x + M_0)^2} = \frac{T v_0^3}{2\pi G} . \tag{2.39}$$

The function $f(M)$ is called the mass function. It is quite clear from symmetry considerations that by measuring the half-amplitude of variations of the radial velocity of a neutron star, we can obtain the mass function for an X-ray pulsar in the form

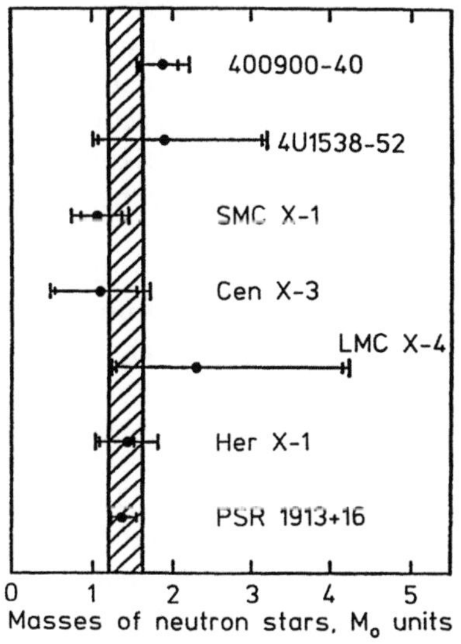

Fig. 2.4. Masses of neutron stars (Rappaport, Joss 1983)

$$f_x(M) = \frac{(M_0 \sin i)^3}{(M_x + M_0)^2} = \frac{T v_x^3}{2 \pi G} \ .$$

(2.40)

By measuring the velocities v_0 and v_x, and having additional information about the angle of inclination i, we can obtain the masses of the components. Figure 2.4 presents the results of such measurements for seven X-ray pulsars (Rappaport, Joss 1983).

2.6 Rotational Effects

Following the discovery of the millisecond pulsar (Backer et al. 1982), it became obvious that the Galaxy contains neutron stars whose equilibrium depends considerably on rotational effects. Detailed numerical calculations of the structure of rotating neutron stars were carried out recently by Friedman et al. (1985). The change in the internal structure of a rotating star as compared to a stationary star is associated not only with the appearance of a centrifugal force, but also with the emergence of the purely relativistic effect of "entrainment of the reference system", which is characteristic, for example, for the Kerr metric.

The most interesting effects caused by the rotation of a star include (a) a change in the Oppenheimer-Volkoff limit; (b) the emergence of a bifurcation-type instability corresponding to the emergence of a "triaxiality" and, consequently, to the emission of gravitational waves; (c) deviation from the "mass-radius" relation.

Figure 2.5 shows the dependence of the rotational frequency on the ratio of the rotational energy to the binding energy of a neutron star. The letters indicate the equations of state of the substance as worked out by Arnett and Bowers (1977). It should be observed that C and D are the versions given by the Bethe-Johnson (BJ) model, A is Reid's equation (R), L corresponds to the mean field (MF) approximation and, finally, M is the tensor interaction (TI) approximation. Table 2.1 shows the values of the Oppenheimer-Volkoff limit for these equations of state. It can be seen that rotation increases the maximum mass of a neutron star by $10-20\%$.

According to Arnett and Bowers, the instability associated with the rotation emerges first at higher harmonics, and critical frequencies do not differ significantly from Roche's critical frequency (Fig. 2.5 and Sect. 6.1).

After the publication of the Russian edition of this monograph, a new millisecond pulsar PSR 1957+20 was discovered. This discovery may throw some light on the problem concerning the choice of the equation of state for the neutron matter (Fruchter et al. 1988). This pulsar has a period of 1.607 ms, and a variation in the period $\dot{P} < 10^{-20}$, which corresponds to a magnetic moment $\mu < 10^{26}\,\mathrm{G\,cm}^3$. Moreover, this pulsar was found to be a component of a binary system with an orbital period of $T \simeq 9.17$ hours.

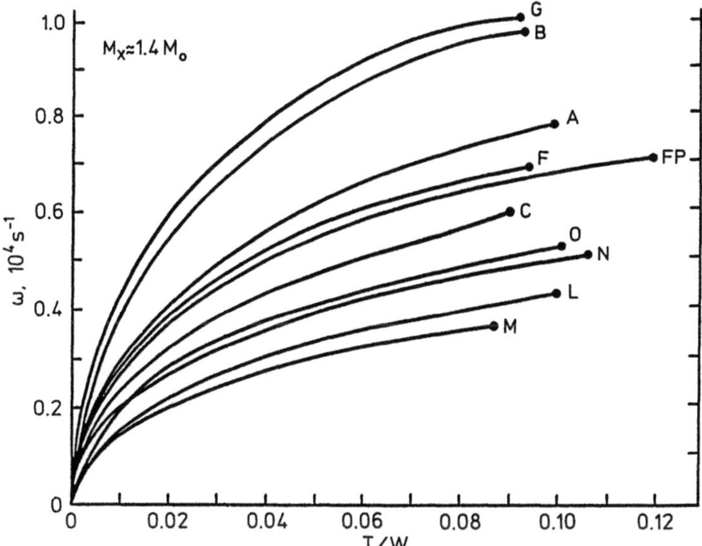

Fig. 2.5. Dependence of the rotational velocity of a neutron star on the ratio of the rotational energy to the binding energy of the star (Friedman et al. 1985)

Above all, the closeness of the period of this newly discovered pulsar to that of the first millisecond pulsar PSR $1937+21$ ($P = 1.56\,\mathrm{ms}$) is astonishing. This circumstance, which was actually predicted by Lipunov and Postnov (1984), can be explained naturally if we assume that

(a) both pulsars attain a bifurcation frequency in the course of their evolution, and

(b) the real equation of state of neutron stars is quite stiff and is correctly described by the mean-field model, i.e., it corresponds to L (Lipunov, Postnov 1988; Friedman et al. 1988).

More detailed argumentation is presented in Sect. 6.13. This conclusion means that the Oppenheimer-Volkoff limit for nonrotating neutron stars is $M_{OV} \simeq 2.7\,M_\odot$, while the maximum rotational frequency does not exceed $700-800\,\mathrm{Hz}$.

Only a few weeks after the publication of the above mentioned works, the discovery of a superfast pulsar in the supernova remnant 1987a with a rotational frequency of 2000 Hz was reported (Middleditch et al. 1989). It would appear that this nullifies the connection with the real equation of state of neutron stars attained as a result of some hard work, and some authors even hastily retreated to their previous positions.

However, subsequent observations failed to confirm the existence of pulsations in the remnant of supernova 1987a. We believe that if the observed pulsations are real, they are not associated directly with the rotation of the neutron star.[2]

---

[2] As a matter of fact, while this manuscript was being prepared, it has become clear that the detected signal was actually due to a faulty video camera.

3. Fluid Dynamics of Accretion

Investigations of the gasdynamic flow of matter in a gravitational force field were started in the 1940s by Hoyle, Bondi, and McCray in connection with the problem of interaction of ordinary stars with interstellar matter. The accretion process was also studied by Gurevich (1953) and Lebedinskii (1953) during the forties and fifties in connection with the formation and evolution of stars. However, as a rule, the influence of accretion of interstellar matter on the evolution and observational properties of normal stars is insignificant.

The interest in the accretion theory rose sharply in the sixties, when it became clear that accretion is the most effective mechanism of liberation of energy in relativistic stars. A large body of theoretical data has been accumulated on this subject. Several monographs and reviews have been devoted to the problem (Zel'dovich, Novikov 1971; Syunyaev 1978; Shapiro, Teukolsky 1985; Gorbatskii 1977).

In this chapter, we shall confine ourselves to a description of the main results of investigations of the accretion process without taking into consideration the effect of the intrinsic magnetic field of the accreting neutron star.

It became clear soon after the appearance of the first publications on this subject (Bondi, Hoyle 1944; Bondi 1952) that the nature of accretion of matter which does not have an angular momentum is mainly determined by the relation between the velocity of sound a_∞ in the matter, and the velocity v_∞ of a star relative to the medium (and vice versa). Accretion of matter having an angular momentum can lead to the formation of accretion discs.

The following four accretion regimes have been investigated in detail and are frequently realized in practice (Fig. 3.1).

1) Spherically symmetric accretion. The accreting star practically does not move relative to the medium: $v_\infty \ll a_\infty$. The matter constituting the medium does not possess any significant angular momentum.

2) Cylindrical accretion. As before, the angular momentum is small, but the velocity of the star is comparable with, or larger than, the velocity of sound in matter: $v_\infty \gtrsim a_\infty$.

3) Accretion disk. The total angular momentum of matter is sufficient to form an accretion disc.

4) In a number of cases (Lipunov 1980a), a two-stream accretion is realized, when a quasi-sperically symmetric flux of matter exists in addition to the accretion disk. The fluid dynamics of a special case of such an accretion was considered by Kolykhalov and Syunyaev (1979).

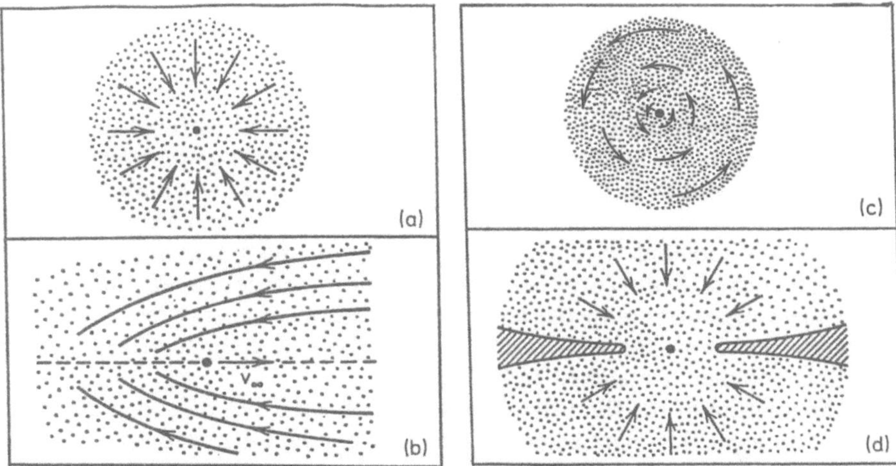

Fig. 3.1a–d. Four accretion regimes: spherically symmetric accretion (**a**), cylindrical accretion (**b**), disk accretion (**c**), and two-stream accretion (**d**)

We shall not take into consideration relativistic effects whose contribution, in general, is not small ($\sim 10\%$). This is justified by the fact that, for example, the influence of magnetic fields is much more significant, but is taken into account with a lower accuracy.

The applicability of the fluid dynamic description is determined by the relation between the mean free path l and the characteristic dimension of the problem under consideration. In the case of accretion, the self-gravitation of the accreting matter is low, and hence the characteristic dimension of the problem is equal to the range of gravitational interaction R_G (we shall frequently refer to it as the gravitational capture radius). This quantity is equal to the characteristic distance at which the kinetic energy of matter becomes comparable to its gravitational energy: $(a_\infty^2 + v_\infty^2)/2 = GM/R_G$. Thus, the range of gravitational interaction can be defined as

$$R_G = \frac{2GM}{v_\infty^2 + a_\infty^2} \ . \tag{3.1}$$

The condition of applicability of the equations for a continuous medium is $l \ll R_G$.

In the absence of radiation, the complete system of equations describing the accretion process can be presented in the form

$$\frac{\partial v}{\partial t} + v \nabla v = -\frac{1}{\varrho} \nabla P - \nabla \varphi \ , \tag{3.2a}$$

$$\frac{\partial \varrho}{\partial t} + \mathrm{div}\,(\varrho v) = 0 \ , \tag{3.2b}$$

$$P = P(\varrho) ,\tag{3.2c}$$

$$\nabla^2 \varphi = -4\pi G(\varrho+\varrho_*) .\tag{3.2d}$$

The first of these equations is the Euler equation in a gravitational field, while the second equation is the continuity equation. We assume that matter does not appear or vanish at $0 < r < \infty$. The third equation is the equation of state for the isoentropic case, while the fourth equation is Poisson's equation for a gravitational potential, φ, ϱ_* being the density of the gravitating body.

In order to take into consideration the radiation processes, we must supplement these equations with the law of conservation of energy (see below). In accretion problems, the self-gravitation of a gas is usually neglected and the gravitating center is assumed to be a point.

The solution of the fluid-dynamic system of equations must contain information on local flow parameters like density, temperature, velocity, as well as on the integral parameters. The most important among these are the accretion rate $\dot{M}$ which does not vary with time in the stationary case and determines the total flow rate of mass to the accreting star. It follows from general considerations that the accretion rate can be presented in the form

$$\dot{M} = \sigma_G \varrho_\infty v_\infty.\tag{3.3}$$

where σ_G denotes the cross section of gravitational interaction (or capture).

The accretion rate $\dot{M}$ depends most significantly on whether or not the medium is gasdynamic (Zel'dovich, Novikov 1971). In order to illustrate this, let us estimate the accretion rate for a collisionless gas to a star of radius R_x moving with a velocity v_∞. In this case, the angular momentum of each particle is preserved and the accretion rate of matter is determined by the maximum impact parameter l_{max} for a particle that can still arrive at the surface of the star:

$$l_{max} \approx R_x \frac{v_p}{v_\infty} .\tag{3.4}$$

This relation becomes exact when v_p (parabolic velocity at the surface of the star) tends to infinity. Obviously, the accretion rate will be given by

$$\dot{M} = \pi l_{max}^2 \varrho_\infty v_\infty \approx \pi R_G^2 \varrho_\infty v_\infty (R_x/R_G) .\tag{3.5}$$

Comparing this expression with (3.3), we obtain the capture cross section for a collisionless accretion

$$\sigma_G \text{ (collisionless)} \approx \pi R_G^2 \frac{R_x}{R_G} .\tag{3.6}$$

In the case of a continuous medium, the law of conservation of angular momentum is not satisfied for an individual particle, and it can be assumed

that all particles whose kinetic energy is lower than the gravitational energy
will be captured. Thus, the ratio of the gravitational capture cross sections for
collisionless and continuous media is given by

$$\frac{\sigma_G \text{ (collisionless)}}{\sigma_G \text{ (continuum)}} \approx \frac{R_x}{R_G} .$$

For a neutron star, $R_x \approx 10^6$ cm, and $R_G \approx 10^{12} v_7^{-2}$ cm (where $v_7 = v_\infty /$
10^7 cm/s is the characteristic velocity relative to the star). Under typical conditions, the accretion rate for a gas is about a million times the accretion rate
of noncolliding particles. Such a strong difference in continuous and collisionless accretion regimes determines the importance of the fluid dynamic investigations for the flow of matter in a gravitational field.

3.1 Spherically Symmetric Accretion

The solution for stationary spherically symmetric accretion was obtained by
Bondi (1952). Under the assumption that the flow is isoentropic, Euler's equation (3.2a) contains the Bernoulli (energy) integral. Assuming that the equation of state has the form of Poisson's adiabat ($P \propto \varrho^\gamma$), we can present the
Bernoulli integral in the form

$$\frac{v^2}{2} + \frac{\gamma}{\gamma-1} \frac{P}{\varrho} - \frac{GM}{R} = \text{const} = \varepsilon_0 . \tag{3.7}$$

The left-hand side is the sum of the kinetic energy, enthalpy, and gravitational
energy of 1 g of accreting substance.

The continuity equation (3.2b) can be presented in the form

$$\dot{M} = 4\pi R^2 \varrho v = \text{const} . \tag{3.8}$$

These two equations (3.7, 8) completely describe any steady-state spherically
symmetric flow. Such a classification of flows is given in the monograph by
Zel'dovich and Novikov (1971). We shall consider only the case of accretion,
i.e., the falling of matter. In this case, the boundary conditions are specified
at infinity:

$$\varepsilon_0 = \frac{\gamma}{\gamma-1} \frac{p_\infty}{\varrho_\infty} = \frac{a_\infty^2}{\gamma-1} . \tag{3.9}$$

Let $a_s = \sqrt{r P/\varrho}$ be the velocity of sound. Substituting the boundary conditions, we obtain the following system of equations:

$$\frac{v^2}{2} + \frac{a_s^2}{\gamma-1} = \frac{GM}{R} + \frac{a_\infty^2}{\gamma-1} \; , \tag{3.10}$$

$$v = \frac{\dot{M}}{4\pi\varrho_\infty R^2} \left(\frac{a_\infty}{a_s}\right)^{2/(\gamma-1)} \; .$$

In the v, a_s plane, these equations describe a family of ellipses and hyperbolas, respectively. For given boundary conditions and a given R, the system (3.10) contains three unknown parameters, v, a_s, and $\dot{M}$, the last of these being generally a constant which must be determined in the course of the solution of the problem. Several methods have been proposed for determining this quantity (Zel'dovich, Novikov 1971; Shapiro, Teukolsky 1985). For the sake of variety, we proceed as follows. We shall consider a flow for which the motion of the plasma is subsonic at large distances ($R \to \infty$) and supersonic at small distances ($R \to 0$). At a certain critical point $R = R_B$, the rate of accretion of matter becomes equal to the velocity of sound: $v = a_s$. It can be easily verified that ellipses and hyperbolas corresponding to the Bernoulli equation and the continuity equation touch at this point. In general, ellipses and hyperbolas touch at the bisector in the system of coordinates (a_s, v). Let us find the value of $\dot{M}$ for which the stationary accretion occurs and is accompanied by a transition from the subsonic to the supersonic regime. At the point of intersection of a hyperbola and an ellipse with the bisector, the velocity of sound is equal to

$$a_s^2 = 2\frac{\gamma-1}{\gamma+1}\left(\varepsilon_0 + \frac{GM}{R}\right) \; , \tag{3.11a}$$

$$a_s^2 = \left[\frac{\dot{M}}{4\pi\varrho_\infty}a_\infty^{2/(\gamma-1)}\right]^{2[\gamma-1/(\gamma+1)]} R^{-4[\gamma-1/(\gamma+1)]} \; . \tag{3.11b}$$

This system of equations is obtained from the system (3.10) as a result of the substitution $v = a_s$. The dependence of a_s on R using (3.11) is shown in Fig. 3.2. It can be seen that for $\gamma < 5/3$, the solution being sought corresponds to the point of contact of curves 1 and 2.

Differentiating both equations in (3.11) with respect to R and equating the results, we find that at the critical point

$$a_s^2 = \frac{1}{2}\frac{GM}{R_B} \; . \tag{3.12}$$

Substituting this value into (3.11) and solving the system of equations in $\dot{M}$, we obtain

$$\dot{M} = \pi\left(\frac{2}{5-3\gamma}\right)^{5-3\gamma/[2(\gamma-1)]} \frac{GM}{a_\infty^3}\varrho_\infty \; . \tag{3.13}$$

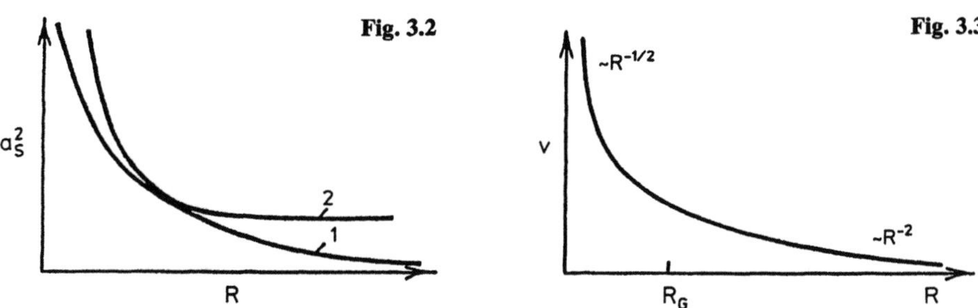

Fig. 3.2. Qualitative dependence of a_s^2 on distance according to (3.11)

Fig. 3.3. Dependence of the free-fall gas velocity on the distance from the star for Bondi's spherically symmetric flow

The dimensionless factor in this formula tends to unity for $\gamma \to 5/3$. The sub-sonic-supersonic transition takes place at a distance

$$R_B = \frac{5-3\gamma}{4} \frac{GM}{a_\infty^2} . \tag{3.14}$$

For $v_\infty = a_\infty$, we obtain from (3.3) the radius of gravitational capture:

$$R_G \approx \frac{2GM}{a_\infty^2} .$$

As expected, accretion of a gas is much more effective than the accretion of a collisonless medium.

Let us consider some properties of the spherically symmetric flow. For $\gamma < 5/3$, there exists a critical point $R = R_B$ at which the velocity of motion passes through the velocity of sound. For $\gamma = 5/3$, the critical point is formally situated at the origin $R = 0$, so that the motion is subsonic everywhere. This can be understood quite easily. Under adiabatic flow, half the gravitational energy passes into the thermal energy of the gas, so that $a_s^2 \approx GM/r$, while the other half passes into the kinetic energy of motion: $v^2 = GM/r$. For $\gamma < 5/3$, the matter is practically in the state of free fall after passing through the critical point (Fig. 3.3):

$$\begin{cases} v \approx \sqrt{\dfrac{2GM}{R}} , \\[2ex] \varrho \approx \dfrac{\dot{M}}{4\pi\sqrt{2GM}} R^{-3/2} \end{cases} \qquad R \ll R_B . \tag{3.15}$$

The asymptotic form for the density of the kinetic energy $\varrho v^2/2$ can be presented as

$$\frac{\varrho v^2}{2} \to \frac{\dot{M}\sqrt{2GM}}{4\pi} R^{-5/2} \; . \tag{3.16}$$

If a rigid wall is placed in the way of the incident flow, the dynamic pressure exerted by the flow on it would be of the order of (3.16).

The physical reason behind the practically free fall of a gas for $R \ll R_{\mathrm{B}}$ is that the flow is supersonic and the underlying layers do not influence the substance entrained by this flow.

3.2 The Role of Radiation and Ejection

Radiation can be taken into account qualitatively by putting $\gamma < 5/3$. However, we can also solve the problem by supplementing the system (3.2) of gasdynamic equations by the second law of thermodynamics. Shvartsman (1971) was the first to propose such an approach. For the time being, we shall disregard the energy liberation at the surface of the star.

The variation in energy of 1 g of substance is

$$d\varepsilon = -PdV + dQ \; , \tag{3.17}$$

where V is the specific volume, and dQ is the heat liberated by 1 g of the substance. For a monatomic gas (or a completely ionized plasma), (3.17) assumes the form

$$\frac{3}{2\mu} R_{\mathrm{u}} \frac{dT}{dt} = R_{\mathrm{u}} \frac{T}{\mu\varrho} \frac{d\varrho}{dt} - \alpha_{\mathrm{ff}} T^{1/2} \varrho + \frac{dQ}{dt} \tag{3.18}$$

where R_{u} is the universal gas constant. The second term on the right-hand side describes the energy loss in free-free radiation (for completely ionized hydrogen plasma, $\alpha_{\mathrm{ff}} \approx 5 \times 10^{20}$ erg/g s), while the third term describes possible losses through radiation in other processes. Substituting $vdt = dR$, and considering that $\varrho \propto R^{-3/2}$, we obtain the equation for temperature distribution in steady-state accretion flow:

$$\frac{dT}{dR} = -\frac{T}{R} + B\frac{\sqrt{T}}{R} + \frac{2\mu}{3R_{\mathrm{u}}} \frac{dQ'}{dR} \; , \tag{3.19}$$

where B is a certain constant. If the radiation does not play a significant role, we automatically obtain $T \propto R^{-1}$, which means that the thermal energy follows the gravitational energy.

Let us write down the solution of (3.19) for the case when there are no additional radiation losses besides the free-free transitions. We shall assume that at a distance $R = R_{\mathrm{G}}$ the temperature of the matter is $T = T_\infty$. As a result, we obtain

$$T = [\text{const} \cdot \ln (R/R_G) + T_\infty^{1/2}]^2 \ . \tag{3.20}$$

It can be seen that the temperature decreases in this case. This situation is called the cooling flow.

In order to estimate the contribution from radiation processes in the temperature variation along the flow, we must compare the radial free-fall time

$$t_R \approx \frac{R}{v_R} \approx R^{3/2} \tag{3.21}$$

with the cooling time

$$t_{\text{cool}} \approx \frac{(3/2) R_u T}{dQ/dt} \approx T^{1/2} \varrho^{-1} \approx R \ , \tag{3.22}$$

where we have taken into consideration cooling due to free-free losses and used the majorizing approximation for the behavior of temperature and density: $T \propto 1/R$ and $\varrho \propto R^{-3/2}$. Comparing t_R and t_{cool}, we observe that the relative role of the cooling processes decreases as the star is approached.

Let us consider the applicability of the gasdynamic approximation. The mean free path in the plasma determined by the Coulomb collisions is given by (Pikel'ner 1966)

$$l = \frac{(kT)^2}{ne^4 \Lambda_c} \approx 10^{12} T_4^2 n^{-1} \text{ cm} \ , \tag{3.23}$$

where $T_4 = T/10^4 \text{ K}$, and n is the number density of the substance. If the radiation does not play a significant role, $T \propto R^{-1}$ and, hence, the mean free path $l \propto R^{-2}$, i.e., it increases more rapidly. However, this does not mean that the gasdynamic approximation is not applicable. A weak magnetic field muddles the trajectories of particles and the medium may be assumed to be continuous (Shvartsman 1971 b).

Unlike a black hole, in which the entire radiation is due to the energy losses in the accretion flow, a neutron star which has a solid surface and a magnetic field mainly radiates as a result of collisions. Suppose that the substance is stopped at a certain distance R_{st}. We assume that the entire energy is transformed into radiation as a result of this process. In this case, the luminosity will be given by (1.2):

$$L = \dot{M} \frac{GM}{R_{st}} \ . \tag{3.24}$$

It can be easily shown that the ratio of the luminosity of the accretion flow due to free-free radiation during free fall to the energy liberated during collision is equal to the ratio of the radial free-fall time to the characteristic cooling time:

$$\frac{L_{ff}}{L} \approx \frac{t_R}{t_{cool}} \ .$$ (3.25)

As a rule, this ratio is much smaller than unity in the case of accretion to neutron stars. Energy liberation at the stopping distance turns out to be the main factor. In this case, the radiation passes through the accretion flow and may influence its dynamics in the case of a sufficiently large luminosity.

Let σ be the cross section of interaction of the emitted radiation with matter. The force acting on the incident particles is $\sigma L/(4\pi R^2 c)$. It depends on distance in the same way as the gravitational force GMm_p/R^2. For a certain critical value of the luminosity $L = L_{Ed}$ called the Eddington limit, these forces become equal:

$$L_{Ed} = \frac{4\pi GMm_p c}{\sigma} \ .$$ (3.26)

In the case of Thompson scattering, $\sigma = \sigma_T$ and $\kappa_T = \sigma_T/m_p \approx 0.4\ \mathrm{cm^2/g}$ (for hydrogen); this leads to the estimate

$$L_{Ed} \approx 1.3 \times 10^{38}\, m\ \mathrm{erg/s}\ , \quad \text{where}\quad m = M/M_\odot\ .$$ (3.27)

Obviously, the accretion luminosity cannot be larger than the Eddington limit because otherwise the accretion would come to a stop. It should be emphasized that all this is valid under the assumption of spherical symmetry. The critical accretion at the Eddington luminosity limit was considered by Shakura (1974).

The Eddington limit plays a fundamental role in the case of accreting stars. A comparison of (3.24) and (3.26) shows that the Eddington luminosity limit corresponds to the critical accretion rate

$$\dot{M}_{cr} = \frac{4\pi R_{st} m_p c}{\sigma} \ .$$ (3.28)

Thus the rate of accretion to a star is limited to a value determined only by the stopping radius and interaction cross section.

For the Thompson cross section, it is convenient to introduce the following estimate:

$$\dot{M}_{cr} \approx 10^{18}\, R_6\,\mathrm{g/s} \approx 1.5 \times 10^{-8}\, R_6 M_\odot/\mathrm{year}\ ,$$ (3.29)

where $R_6 = R_{st} 10^6\ \mathrm{cm}$.

This estimate shows that the value of the critical accretion rate is not distinguished in any way, and hence "sub" and "super" critical accretion stars must be observed in nature.

Let us consider an instructive example for the estimate for the optical thickness of the accretion flux:

$$\tau = \int\limits_{R_{st}}^{\infty} \kappa \varrho \, dR = 2 \frac{c}{v_p} \frac{\dot{M}}{\dot{M}_{cr}} \, . \tag{3.30}$$

Here, κ is the absorption coefficient per gram of the substance as before, and v_p is the parabolic velocity at the stopping radius. In the case of accretion to a neutron star in the absence of a magnetic field, $R_{st} = R_x$ and $v_p \approx c/3$, so that $\tau \approx 6(\dot{M}/\dot{M}_{cr})$. Thus the optical thickness in the subcritical regime is always small.

Another effect which influences the accretion rate is associated with the heating of the substance in the vicinity of the gravitational capture radius R_G (Mestel 1954). It follows from Bondi's formula (3.13) that the accretion rate depends most strongly on the velocity of the star, and hence on the temperature of the matter in the surrounding medium: $\dot{M} \propto a_\infty^{-3} \propto T_\infty^{-3}$. Obviously, heating of matter leads to a peculiar self-regulation (Shvartsman 1970b): as the luminosity increases, so does the temperature; this results in a decrease in the accretion rate and luminosity.

It can be asked whether heating of matter leads to a more stringent constraint on the accretion rate than the Eddington limit (Buff, McCray 1974). Bisnovaty-Kogan and Blinnikov (1980) numerically solved the equations of steady-state spherically symmetric accretion with X-ray heating. It was found that steadystate may set in for any value of luminosity, right up to the Eddington limit. Thus, there is no "thermal" limit for luminosity.

Ejection of stellar matter may act as a much more effective constraint on accretion (Shvartsman 1970c). Suppose that the mean free path of the ejected particles is much smaller than the characteristic length of the interaction, i.e., the particle is completely stuck in the accreting matter and imparts all its momentum to this matter. Denoting by L_{ej} the power carried away by the ejected particles and by v_{ej} the velocity of these particles, we can write the pressure exerted by them on the accreting matter in the form

$$P_{ej} = \frac{L_{ej}}{4\pi R^2 v_{ej}} \, . \tag{3.31}$$

Note that $P_{ej} \propto R^{-2}$, while the dynamic pressure of the accreting matter is $P_a \propto R^{-5/2}$ [see (3.16)]. If a star ejects a particle before the onset of accretion, we must equate the pressures at the gravitational capture radius:

$$\frac{L_{ej}}{4\pi R_G^2 v_{ej}} = \varrho v_\infty^2 = \frac{\dot{M}}{4\pi R_G^2} v_\infty \, .$$

Hence the critical value $L_{ej}(cr)$ of luminosity of the ejected particles obstructing accretion can be presented in the form

$$L_{ej}(cr) = \dot{M} v_\infty v_{ej} = \frac{L}{\eta} \left(\frac{v_\infty v_{ej}}{c^2} \right) \, .$$

Here we have made use of (1.2). It can be seen from the above relation that $L_{ej} \ll L$ for a relativistic star. An ejected wind of infinitely low luminosity is capable of stopping accretion.

3.3 Spherical Accretion to a Neutron Star Without a Magnetic Field

Spherically symmetric accretion to a neutron star without a magnetic field was first considered by Zel'dovich and Shakura (1969). Following this work, we shall describe the main physical phenomena associated with the appearance of a solid surface in the path of the accreting flux.

As the particles of the accreting flux collide with the surface of a neutron star, they give away their kinetic energy and come to rest. The stopping of particles may be due to collision processes, as well as to the excitation of plasma instabilities. The kinetic energy is transformed into radiation. Obviously, the temperature will be determined by the balance between the two processes of heating due to deceleration and cooling due to radiation. In the adiabatic approximation (slow cooling), the characteristic temperature can be estimated from Hugoniot's collision adiabat (Zel'dovich, Raizer 1966):

$$kT \approx \frac{mv^2}{2} \ .$$

For $v \approx c/2$, we obtain $T \approx 10^{12}$ K. This is the upper estimate for the temperature. In contrast, if a neutron star emits radiation like a black body, the equality

$$\dot{M}\frac{GM}{R_x} = 4\pi R^2 \sigma T^4$$

leads to the lower temperature estimate:

$$T \approx 10^7 \, \dot{M}_{-8}^{1/4} R_6^{-3/4} m^{1/4} \text{K} \ .$$

Such a drastic difference shows that a detailed analysis of the deceleration and radiation processes in the energy transformation must be carried out even for obtaining an order-of-magnitude estimate. Before a collision with the surface, practically all the kinetic energy is concentrated in protons (an electron is lighter by a factor of 1800). The protons colliding with the atmosphere of a neutron star slow down by gradually giving away their energy first to the protons of the atmospheric layer and then, through collisions, to the electrons of this layer as well. The electrons receiving this energy give it away in the form of bremsstrahlung or the reverse Compton effect.

It is convenient to introduce the parameter $y = \int \varrho \, dx$, which indicates the amount of matter through which a particle travels during deceleration. Let y_0

be the total amount of matter required for the particle to come to a complete stop. In this case, the energy liberated per gram of the matter constituting the atmosphere will be approximately equal to

$$W^+ \approx Q/y_0 \, , \quad y<y_0 \, , \tag{3.32}$$

$$W^+ = \theta \, , \qquad y>y_0 \, ,$$

where $\theta = L/4\pi R^2$ is the energy flux per unit surface area of the star. This energy is carried away by bremsstrahlung

$$\dot{W}_{\mathrm{br}}^- \approx 5\times 10^{20}\sqrt{T_e}\varrho \, , \tag{3.33}$$

where T_e is the temperature and ϱ the density of the ionized plasma, as well as by comptonization

$$W_c^- = \frac{4\varepsilon_r c\sigma_T}{m_p} \frac{kT_e}{m_e c^2} \tag{3.34}$$

(ε_r is the emission energy density). Formulas (3.33, 34) are obtained by neglecting the inverse processes which can be taken into account by introducing certain effective temperatures T_1 and T_2, so that the heat balance equation assumes the form:

$$\frac{Q}{y_0} = 5\times 10^{20}\, T_e^{1/2}\varrho\left(1-\frac{T_1}{T_e}\right)+6.5\,\varepsilon\, T_e\left(1-\frac{T_2}{T_e}\right) \, . \tag{3.35}$$

Generally speaking, the values of T_1 and T_2 depend on the spectrum, but in the black body approximation $T_1 = T_2 = (\varepsilon/a)^{1/4}$.

The radiant energy density is determined from the diffusion equation for the emission flux q:

$$q = Q\frac{y-y_0}{y_0} = -\frac{c}{3\kappa_T}\frac{d\varepsilon_r}{dy} \, , \quad y<y_0 \, ,$$

where $\kappa_T = 0.38 \mathrm{\ cm^2/g}$. For $y>y_0$, $q = 0$ and $\varepsilon_r = $ const. Taking into account the boundary condition

$$\varepsilon_r = \frac{\sqrt{3}\,Q}{c} \quad \text{for} \quad y = 0$$

we obtain

$$\varepsilon_r = \frac{Q}{c}\left\{\sqrt{3}+3\kappa_T y_0\left[\frac{y}{y_0}-\frac{1}{2}\left(\frac{y}{y_0}\right)^2\right]\right\} \, , \quad 0<y<y_0 \, , \tag{3.36}$$

$$\varepsilon_{\mathrm{r}} = \frac{Q}{c}\left(\sqrt{3}+\frac{3}{2}\kappa_{\mathrm{T}}y_0\right) \quad y>y_0 \; .$$

Complete thermodynamic equilibrium is established at a depth $y\gg y_0$. The temperature is determined from the condition $\varepsilon_{\mathrm{r}} = a\,T^4$. At the surface $\varrho\to 0$ and $W_{\mathrm{br}}^-\to 0$, the electron temperature is determined by comptonization and is independent of luminosity, since $\varepsilon\propto Q\propto L$. The density distribution of matter is obtained from the hydrostatic equilibrium equation

$$P = \frac{2\varrho kT}{m_{\mathrm{p}}} = \left(\frac{GM}{R_x^2}+\frac{\varrho_0 v^2}{y_0}\right) y \; , \quad 0<y<y_0 \; ,$$

$$P = \frac{2\varrho kT}{m_{\mathrm{p}}} = \frac{GM}{R_x^2}y+\varrho_0 v^2 \; , \quad y>y_0 \; .$$

$$(3.37)$$

Here, in addition to the gravitational force, we have also taken into account the force of dynamic pressure of the incident matter.

By specifying the luminosity L or, which is the same, Q (for a given value of the radius of the star), and the braking distance y_0, we can obtain an idea about the distribution of density and temperature in the atmosphere of a neutron star. Obviously, the most important parameter here is y_0 which depends significantly on the way in which the incident particles are slowed down. The kinetic energy of the incident protons is of the order of several hundred MeV. The characteristic mean free path of such protons in a completely ionized plasma is determined by the Coulomb collisions and corresponds to the mass $y = \varrho l \approx 5-30\ \mathrm{g/cm^2}$ of the substance. In this case, for $y_0 = 20\ \mathrm{g/cm^2}$, $T_1 \approx 1.5\times 10^7$ K at a depth $y\gg y_0$. The luminosity is assumed to be equal to $0.1\,L_{\mathrm{Ed}}$, where the subscript stands for "Eddington". The electron temperature at the surface is $T_{\mathrm{e}} \approx 10^8$ K.

Let us estimate the role of comptonization by introducing a coefficient showing the fraction of energy given by the electrons for comptonization:

$$\eta = \frac{1}{Q}\int_0^{y_0} W_{\mathrm{c}}^- \, dy \; .$$

Of $y_0\approx 20\ \mathrm{g/cm^2}$, $\eta\lesssim 0.05$ for $L = 0.1\,L_{\mathrm{Ed}}$ and $\eta\lesssim 0.01$ for $L = 0.01\,L_{\mathrm{Ed}}$. The value of y_0 can decrease strongly due to the emergence of plasma oscillations: the passage of a beam of charged particles through the plasma induces in it oscillations at a plasma frequency $v_{\mathrm{p}} = \sqrt{\pi n e^2/4 m_{\mathrm{e}}}$, which interact with the beam and slow it down. The decrease in the mean free path is accompanied by an increase in the temperature of electrons. Assuming that the electron temperature $T_{\mathrm{e}}\approx m_{\mathrm{e}} v^2 \approx 10^9$ K, we find from (3.36, 37) that $y_0 = 2\ \mathrm{g/cm^2}$. In this case, the role of comptonization increases sharply ($\eta\approx 0.96$ for $L\approx 0$ and $\eta\approx 0.07$ for $L\approx 0.01\,L_{\mathrm{Ed}}$). Moreover, hard X-ray quanta appear with an

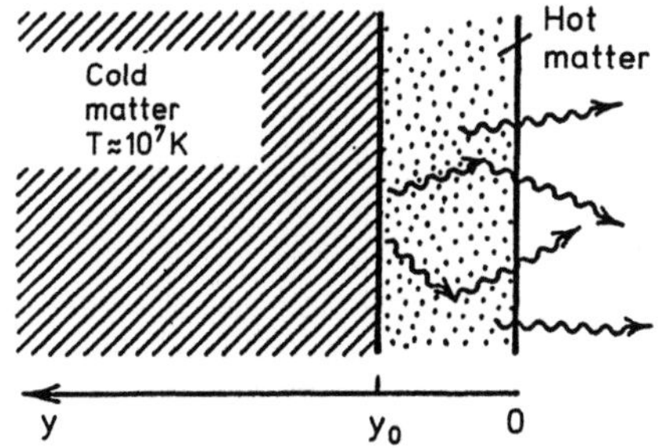

Fig. 3.4. Structure of the braking zone for accreting matter on a neutron star surface

energy of $50-100$ keV. The emerging spectrum will differ strongly from Planck's spectrum.

To understand why the spectrum should differ so strongly from that of blackbody radiation we can use the following argument. The atmosphere of the star is divided into two parts: a cold half-space with a temperature $T \approx 10^7$ K, and a thin hot layer with a temperature $T \approx 10^8 - 10^9$ K depending on the braking distance y_0. The flux emitted by the "cold" half-space is comptonized by the hot electrons of the thin layer (Fig. 3.4). Under the conditions considered above, the coefficient σ_T of scattering by free electrons is much larger than the real absorption coefficient σ_{ff}. The consideration of comptonization turns out to be the most difficult task. Under the conditions when the mean energy of electrons is much higher than the energy of the emitted photons, the change in the spectrum essentially depends on a single parameter called the comptonization parameter: $\kappa = kT_e \Delta t / m_e c^2$, where Δt is the characteristic residence time for quanta in the high-temperature region. Comptonization leads to the pumping of cold quanta to the hard X-ray radiation region with a characteristic exponential cut-off.

Additional radiation may be produced in the gamma region as a result of the decay of π-mesons generated by the collisions of protons in the stellar atmosphere (Zel'dovich, Novikov 1967). Estimates show that the fraction of this radiation is comparatively small ($\sim 10^{-4}$). Nevertheless, it would be quite interesting to observe this type of radiation.

3.4 Capture of Matter by a Moving Star

Chronologically, the first problem to be investigated was the problem of accretion of a gas to a moving gravitating center (Hoyle, Lyttleton 1939; Bondi, Hoyle 1944; McCrea 1953). The gasdynamic problem turned out to be quite complex and, unlike the case of spherical symmetry, it was not possible to obtain exact analytical solutions in this case. In the astrophysics of neutron stars, the most important factor is the accretion rate of the captured substance. It can be stated that although the structure of the accretion flux has not been determined completely so far, a fairly reliable approximate expression exists for the quantity $\dot{M}_c$.

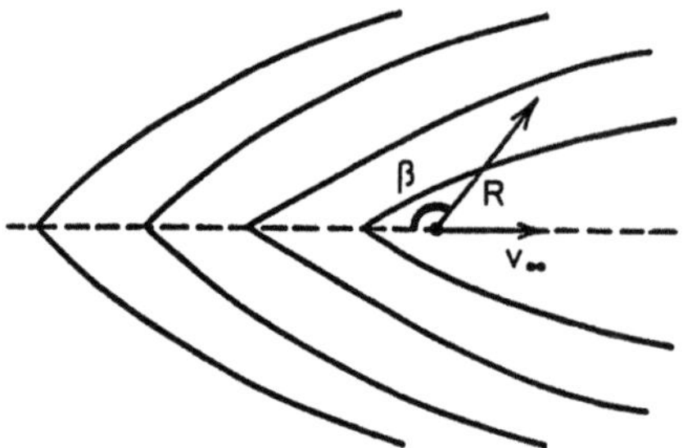

Fig. 3.5. Stream lines for cylindrical accretion

To begin with, let us consider the problem is the powder approximation. We go over to a system of coordinates associated with the gravitating center moving relative to the medium with a velocity v_∞. Suppose that the particles move freely everywhere, but stick to the symmetry axis (accretion axis, Fig. 3.5). In this approximation, the stream lines are hyperbolas. As individual particles move, the angular momentum is conserved relative to the accreting star:

$$|R \times v| = b v_\infty$$

where b is the impact parameter of the particle. Let $v_{||}$ and $v_\perp$ be the parallel and perpendicular components of the velocity at the sticking point. The contribution to the angular momentum on the accretion axis is made only by the component $v_\perp$. Hence,

$$v_\perp R_{\text{col}} = b v_\infty \ .$$

In the case of sticking, the perpendicular component of the velocity vanishes. Hence the kinetic energy of the particles decreases and their total energy may become negative (the particles will be captured). Obviously, only those particles will be captured for which the velocity after the collision is less than the parabolic velocity:

$$v_{||} \leq \sqrt{2GM/R_{\text{col}}} \ .$$

In order to determine the maximum sticking parameter for a captured particle, we take into consideration the law of energy conservation:

$$\frac{1}{2}(v_{||}^2 + v_\perp^2) - \frac{GM}{R_{\text{col}}} = \frac{1}{2} v_\infty^2 \ .$$

It follows from the last two equations that only those particles are captured for which $v_\perp \leq v_\infty$.

Using the equation for the trajectory of a particle, we can easily find the maximum impact parameter $b_{\max}$ of a captured particle and the accretion rate $\dot{M} = \pi b_{\max}^2 \varrho_\infty v_\infty$. Calculations lead to the following quite obvious relation:

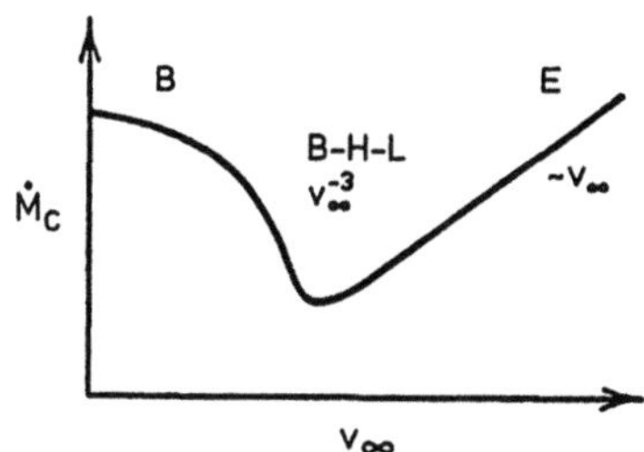

Fig. 3.6. Qualitative dependence of the rate of accretion of substance to a star on the star velocity. B – Bondi's spherical accretion, BHL – Bondi-Hoyle-Lyttleton cylindrical accretion, E – Eddington accretion (capture by geometric flow)

$$\dot{M}_c = \xi_1 \pi \frac{(2GM)^2}{v_\infty^3} \varrho_\infty \tag{3.38}$$

where ξ_1 is a dimensionless coefficient of the order of unity.

It can be seen that for the case under consideration, the gravitational capture cross section is determined by the capture radius R_G: $\sigma_G \approx \pi R_G^2$.

The structure of (3.38) is similar to that of Bondi's formula (3.13). This fact allowed Bondi (1952) to use a convenient analytical expression for the accretion rate even for the case when the velocity of a moving object is comparable with the velocity of sound (Bondi-Hoyle-Lyttleton formula):

$$\dot{M}_c = \xi_1 \frac{(2G\dot{M})^2}{(v_\infty^2 + a_\infty^2)^{3/2}} \varrho_m \ . \tag{3.39}$$

This formula gives the correct asymptotic form for large and small Mach numbers. The only important point to be noted is that for very large velocities v_∞, the gravitational capture radius turns out to be smaller than the star radius R_x. In this case, obviously, the capture cross section will be equal to the geometric cross section (Eddington accretion):

$$\dot{M}_c = \pi R_x^2 \varrho_\infty v_\infty \ , \quad v_\infty \gg 2GM/R_x \ .$$

We can now construct the complete dependence of the accretion rate for an arbitrary relation between three characteristic velocities, viz., the velocity a_∞ of sound in matter, the velocity v_∞ of the moving body, and the parabolic velocity v_p at the surface of the star (Fig. 3.6). It is significant that $\dot{M}_c$ does not become zero under any conditions.

3.5 Fluid Dynamics of Cylindrical Accretion

The powder approximation considered in Sect. 3.4 leads to a formula which describes the accretion rate of a gas towards a moving center quite accurately. However, this formula fails to provide any information about the structure of the gas flow. In principle, using the trajectory equation in the powder approximation, we can obtain an expression for the density of a gas at any point (Spiegel 1970). Such a solution has a singularity on the accretion axis: the density of matter becomes infinite. Obviously, the pressure effects manifested in this case should lead to the formation of a conical shock wave. This can be seen clearly from the example of solution of the linearized gasdynamic equations (Spiegel 1970).

Suppose that the density perturbations are quite small:

$$\varrho = \varrho_\infty + \delta\varrho \ , \quad \delta = \delta\varrho/\varrho_\infty \ll 1 \ .$$

We shall assume the gravitational center to be a point moving with a velocity v_∞:

$$\varrho_* = \dot{M}\delta(R - v_\infty t) \ .$$

We shall also assume that the accretion rate $\dot{M}$ is given by

$$\dot{M} = \xi_1 \pi R_\mathrm{G}^2 v_\infty \varrho_\infty \ .$$

Linearizing the system of equations (3.2), we obtain

$$\frac{\partial v}{\partial t} = -a_\infty^2 \nabla\delta + \nabla\varphi \ ,$$

$$\frac{\partial\delta}{\partial t} + \nabla v = -\xi_1 \pi R_\mathrm{G}^2 v_\infty \delta(R - v_\infty t) \ , \tag{3.40}$$

$$\nabla^2\delta\varphi = -4\pi G(\varrho_* + \delta\varrho_\infty) \ .$$

We introduce the Jeans wavelength through the formula (Zel'dovich, Novikov 1971)

$$k_\mathrm{J}^2 = \frac{4\pi G\varrho_\infty}{a_\infty^2} \ .$$

As a result of simple transformations, the system of Eqs. (3.40) can be reduced to the following equation in perturbations produced by the moving gravitational center:

$$\Box\delta + k_\mathrm{J}^2 a_\infty^2 \delta = -4\pi G\varrho_* + \xi_1 \pi R_\mathrm{G}^2 v_\infty \frac{\partial}{\partial t}\delta(R - v_\infty t) \ , \tag{3.41}$$

where $\Box = \partial^2/\partial t^2 + \nabla^2$ is the d'Alembert operator. Going over to the system of coordinates associated with the moving star (Fig. 3.5) and neglecting the self-gravitation of the gas ($k_\mathrm{J} = 0$), we arrive at the equation

$$(a_\infty^2 - v_\infty^2)\nabla^2\delta = 4\pi GM\delta(r)$$

whose solution is given by

$$\delta = \frac{R_{\mathrm{G}} M_{\mathrm{M}}}{R(I - M_{\mathrm{M}}^2 \sin^2 \beta^{1/2})}$$

where $M_{\mathrm{M}} = v_\infty / a_\infty$ is the Mach number. The obtained solution has a singularity at the surface of the cone:

$$\sin \beta_{\mathrm{sh}} = \frac{1}{M_{\mathrm{M}}} \ .$$

Obviously, the linear approximation is not valid in the vicinity of the cone. The singularity points towards the emergence of a conical shock wave with a cone angle β_{sh}.

It should be noted that even in the linear approximation, a dynamic friction force

$$F_{\mathrm{fr}} = \pi R_{\mathrm{G}}^2 \varrho_\infty v_\infty^2$$

appears and slows down the accreting star. The emergence of such a force was first revealed by Chandrasekhar during a consideration of the motion of a heavy particle in a collisionless medium (Chandrasekhar 1943). The dynamic friction appears due to the fact that the density of the background matter in the wake will be higher than in front of the moving center.

Numerical calculations for the cylindrical accretion were carried out in the seventies by Hunt (1971) and Eadie et al. (1975), who calculated the adiabatic ($\gamma = 5/3$) steady flow for Mach numbers $M_{\mathrm{M}} = 1$, 2, and 4. It was found that a frontal shock wave is formed in front of the gravitating center. For $\gamma = 4/3$, the flow becomes more complicated and contains conical and detached shock waves (Eadie et al. 1975).

These calculations show that the flow pattern depends significantly on the efficiency of the gas cooling mechanism. If the density of the gas near the capture radius is low, the radiation is weak and the case $\gamma = 5/3$ is realized. In this case, a frontal shock wave is formed in front of the star at a distance $\sim R_{\mathrm{G}}$. The temperature in the wake of this shock wave is defined by

$$T_{\mathrm{sh}} = \frac{m_{\mathrm{p}} v_\infty^2}{6\mathrm{k}} \approx 2.5 \times 10^5 v_7^2 \,\mathrm{K} \ .$$

Here,

$$v_7 = v_\infty / 10^7 \,\mathrm{cm/s} \ .$$

For large densities of a substance, the plasma becomes radiative, its compressibility increases, and a conical shock wave is formed in the wake of the star (Illarionov, Syunyaev 1975). The critical accretion rate separating these two regimes is given by the expression (Syunyaev 1978)

$$\dot{M}_{18} \approx \frac{3\sqrt{\pi}}{32}\left(-\frac{m_{\mathrm{p}}}{m_{\mathrm{e}}}\right)^{1/2} v_8 \ .$$

The quest for approximate analytical solutions has attracted considerable interest.

Bisnovaty-Kogan et al. (1979) have obtained self-similar solutions describing the flow in the vicinity of a gravitating center for the polytrope $1.31 < \gamma < 5/3$. An analysis of the obtained results led the authors to two important conclusions: (1) in a real flow with a large Mach number, the shock wave in the leading flow will be an attached wave and not a detached wave; (2) the generator of the cone in the general case will not be a straight line. It can be seen that the investigations of the cyclindrical accretion can by no means be considered as complete, and the details of the pattern still have to be worked out.

3.6 Disk Accretion

In the preceding sections, we have considered the case when matter captured by the gravitational field of a star has zero total angular momentum. However, such a situation is never realized in nature as the interstellar medium is turbulized (Kaplan, Pikel'ner 1979). In binary systems, the matter supplied to one of the stars by the companion has an angular momentum due to the orbital motion (Gorbatskii 1965). The galactic matter possesses an angular momentum due to the differential rotation of the Galaxy (Shvartsman 1971 b), and so on. Estimates show that in many cases the angular momentum is so large that we must take into account not only the gravitational force, but also the centrifugal forces.

Such a situation ultimately results in the formation of an accretion disk. First investigations of the gas dynamics of accretion disks were carried out by Weisszäcker (1948) in connection with the formation of the galaxies. Later, Gorbatskii (1965) (see Gorbatskii 1974) studied the transport of matter in close binary systems. Prendergast (1960) considered gas flows in binary systems under the assumption of noninteracting particles.

Considerable progress in the understanding of disk accretion on relativistic stars was made by Lynden-Bell (1969), Shakura (1972), Pringle and Rees (1972), Shakura and Syunyaev (1973), and Novikov and Thorne (1972). The main difficulty in the construction of the disk accretion theory lies in that we do not know the nature of turbulence in disks and, consequently, the coefficient of dynamic viscosity.

Progress accelerated following the publication of Shakura's work in 1972, in which the only unknown factor in turbulence was the dimensionless parameter α. In the final form, the α-model (or the standard model) of steady-state

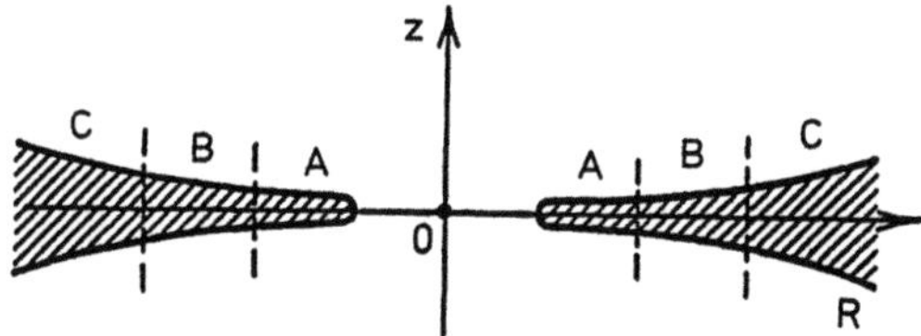

Fig. 3.7. Disk accretion. A, B, and C are three disk regions with different roles of radiation pressure and different opacity mechanisms

disk accretion was constructed by Shakura and Syunyaev (1973). We shall make use of the results obtained in these fruitful investigations.

We shall not formally derive the disk accretion equations from the Navier-Stokes equations. Instead, we shall directly write the simplified equations on the basis of physical concepts. It should be emphasized that the theory of steady-state disk accretion is conceptually different from the cases considered above. While the accretion rate $\dot{M}$ was determined earlier by solving the problem, it is now introduced as an external parameter (specified beforehand).

We assume that the accretion disk is thin, i.e., its characteristic scale along the z-coordinate is $H \ll R$ (Fig. 3.7). We shall also assume that the matter of the disk is in hydrostatic equilibrium along the z-axis, i.e., the pressure gradient is balanced by the vertical component of the gravitational force of the central star (we neglect the self-gravitation of the disk):

$$\frac{1}{\varrho}\frac{dP}{dz} = -\frac{GM}{R^3}z \ .$$

Another obvious circumstance, which is important for understanding the essence of this problem, should be noted. The star exerts a radial force on each particle of the disk, and if we neglect collisions, each particle will move in a circle in a plane inclined to the symmetry plane of the disk. In a gaseous disk, the particles outside the symmetry plane move in a circle, but not in a Keplerian orbit (the center does not coincide with the star). Obviously, the collisions with neighboring particles ensure that the motion will be in a plane coplanar with the symmetry plane of the disk.

We shall not concern ourselves with the vertical structure of the disk. Hence we can write the hydrostatic equilibrium equation, putting $\Delta P = \varrho a^2$ (a is the velocity of sound), and $\Delta z = H$ (half-thickness of the disk). In this case,

$$a = \omega_{\mathrm{K}} H \ , \tag{3.42}$$

where $\omega_{\mathrm{K}} = \sqrt{GM/R^3}$ is the Keplerian angular velocity.

We shall assume that the circular motion in the disk occurs along Keplerian orbits. This statement must be valid to within $(H/R)^2$:

$$v_{\varphi} = \sqrt{\frac{GM}{R}} = \omega_{\mathrm{K}} R \ . \tag{3.43}$$

Equations (3.42) and (3.43) automatically lead to the following important relation:

$$\frac{a}{v_\varphi} \approx \frac{H}{R} \ . \tag{3.44}$$

Hence the thermal energy of the gas in thin disks is much lower than the gravitational energy (the entire energy is "concentrated" in the kinetic energy of rotation). Note that this justifies our assumption that equilibrium along the φ-coordinate does not involve the pressure gradient (the gravitational force is balanced by the centrifugal force).

The radial motion in the disk is due to the friction between adjacent layers and the exchange of the angular momentum between them. The transport of the angular momentum in a disk along the R-coordinate is associated with the moment of viscous forces:

$$\dot{M}\frac{d\omega_K R^2}{dR} = 2\pi \frac{d}{dR} W_{r\varphi}R^2 \ , \tag{3.45}$$

where $W_{r\varphi}$ is the component of the viscous stresses in the disk:

$$W_{r\varphi} = -2\eta H R \frac{\partial \omega_K}{\partial R} \ . \tag{3.46}$$

Here, η is the coefficient of dynamic viscosity averaged over the z-coordinate. The viscous stresses are proportional to the angular velocity gradient and vanish for the motion of rigid bodies.

In the case of isotropic turbulence, the viscosity is given by (Landau, Lifshitz 1953)

$$\eta = \tfrac{1}{3}\varrho\, v_t l_t$$

where v_t and l_t are the characteristic velocity and scale of turbulence pulsations. The quantities v_t and l_t are not known beforehand in an accretion disk. This uncertainty can be avoided by assuming (Shakura 1972)

$$v_t l_t = \alpha a_s H \ , \tag{3.47}$$

where α is a dimensionless parameter called the turbulence parameter. Obviously, in the case of isotropic turbulence, $l_t \lesssim H$. Moreover, supersonic turbulence attenuates rapidly and hence $v_t \lesssim a_s$. Consequently, it is assumed that $\alpha \lesssim 1$.

Equation (3.45) describing the change in the angular momentum can be easily integrated:

$$W_{r\varphi} = -\frac{\dot{M}}{2\pi}\omega_K \left[1 - \left(\frac{R_d}{R}\right)^{1/2} \right] + W_{r\varphi}(\text{in}) \tag{3.48}$$

where R_d is the radius of the inner edge of the disk, and $W_{r\varphi}(\text{in})$ is the component of the tensor of viscous stresses on this edge ($R = R_d$).

While considering accretion to a nonrotating black hole, it is assumed that the inner edge corresponds to the last stable orbit $R_d = 3R_g$ (Kaplan 1949). Subsequent radial motion of matter to the black hole occurs due to the GRT effects, and hence

$$W_{r\varphi}(\text{in}) = 0 \ .$$

The radius R_d of the inner edge for disk accretion to a neutron star is determined by (a) collisions with the rigid surface of the star, (b) the action of the magnetic forces, and (c) the interaction of matter with the ejected relativistic wind. Completely different conditions may be imposed on the viscous stress tensor $W_{r\varphi}$ (Chaps. 6 and 7).

Thus, we have considered three equations which are corollaries of the equation of motion (Stokes equations). Obviously, the continuity equation can be presented in the form

$$\dot{M} = 2\pi\varrho(2H)Rv_r \ , \tag{3.49}$$

where v_r is the radial velocity of motion of matter in a disk. Introducing the surface density

$$\Sigma = 2H\varrho$$

we can write (3.49) in the form

$$\dot{M} = 2\pi\Sigma Rv_r \ . \tag{3.50}$$

Let us assume that $W_{r\varphi}(\text{in}) = 0$ at the inner edge. In this case, we obtain from (3.48) at large distances from the inner edge:

$$W_{r\varphi} \approx -\frac{\dot{M}}{2\pi}\omega_K \ .$$

From the definition of $W_{r\varphi}$ [Eq. (3.46)], we obtain

$$W_{r\varphi} \approx 3\eta H\omega_K - \alpha P II \ . \tag{3.51}$$

From (3.49) and (3.51), we find that

$$\frac{v_r}{v_\varphi} \approx \alpha\left(\frac{H}{R}\right)^2 \ . \tag{3.52}$$

Thus, as expected, the radial motion is an effect having second order of smallness in (H/R).

The transport of mechanical energy in a disk (the angular momentum is transported over the disk) results in the liberation of heat

Table 3.1. System of equations for standard disk accretion [after Shakura and Syunyaev (1976)]

Number	Meaning	Equation
I	Kepler's law	$\omega = \omega_K = (GM/R^3)^{1/2}$
II	Continuity equation	$\dot{M} = -2\pi\Sigma v_r R$
III	Law of variation of angular momentum	$W_{r\varphi} = \dfrac{\dot{M}}{2\pi}\omega\left[1 - \left(\dfrac{R_d}{R}\right)^{1/2}\right] + W_{r\varphi}(\text{in})$
IV	Hydrostatic equilibrium equation	$P = \dfrac{\sum \omega^2 H}{6}$
V	Viscosity tensor	$W_{r\varphi} = \alpha P H$
VI	Liberation of mechanical energy	$Q^+ = -\dfrac{1}{2} W_{r\varphi} R \dfrac{d\omega}{dR}$
VII	Energy losses by radiation	$Q^- = \dfrac{2}{3}\dfrac{\varepsilon_r}{\kappa \sum} c$
VIII	Equation of state	$P = \dfrac{3}{2}\varrho R_u(T_e + T_i) + \varepsilon_r/3$
IX	Absorption cross section	$\sigma\,[\text{cm}^2] = \sigma_T + \sigma_{ff} \approx 6.65 \times 10^{-25}\,\text{n}$ $+ \dfrac{1.8 \times 10^{-25}}{T^{7/2}}$

$$Q^+ = -\frac{1}{2} W_{r\varphi} R \frac{d\omega}{dR} = \frac{3}{4}\omega W_{r\varphi} \; ; \qquad (3.53)$$

where Q^+ is the amount of energy supplied to a unit surface area of the disk
per unit time on each side. This energy is mainly carried away by radiation
(Lyubarskii 1984). Obviously, the energy flux in the diffusion approximation
(Zel'dovich, Raizer 1966) is given by

$$Q^- = \frac{c}{3\kappa\varrho}\frac{d\varepsilon_r}{dz} \approx \frac{2\varepsilon_r c}{3\kappa\Sigma} \; . \qquad (3.54)$$

In the steady state, $Q^- = Q^+$. The final system of equations for disk accretion is presented in Table 3.1.

The system of equations presented in Table 3.1 can be solved algebraically
if one of the terms in VIII and IX can be neglected. Hence an accretion disk
is divided into three zones, each of which is dominated by one term or another.
In the innermost zone (zone A) (Fig. 3.7), the radiation pressure P_r is much
higher than the gas pressure P_g, and the free-free absorption can be neglected.
In the middle zone (zone B) the radiation pressure is already low ($P_r \ll P_g$),

but the Thompson scattering still dominates over free-free absorption: $\sigma_T \gg \sigma_{ff}$. Finally, in the outer zone (zone C), $P_r \ll P_g$ and $\sigma_{ff} \gg \sigma_T$.

The dimensions of the transition layers between zones can be estimated as follows:

$$r_{AB} \approx 50 (\alpha m)^{2/21} \dot{m}^{16/21} \ ,$$

$$r_{BC} \approx 2{,}7 \times 10^3 \dot{m}^{2/3} \ . \tag{3.55}$$

Here $r \equiv R/3R_g$, $\dot{m} = \dot{M}/\dot{M}_{cr}$, $m = M/M_\odot$. For neutron stars, $R_g \approx 5$ km. In this case, $R_{AB} \approx 750$ km for the critical accretion rate. It will be shown below that in many cases the accretion disk is destroyed earlier by the magnetic field. Hence the accretion disks around strongly magnetized neutron stars do not contain zone A.

Before concluding this section, let us briefly consider the important difference between the hydrostatics of accretion disks and the hydrostatics of stars and stellar atmospheres.

From the hydrostatic equilibrium equation IV, we obtain for an isothermal atmosphere

$$P = P_0 e^{-(z/H)^2} \ ,$$

$$\varrho = \varrho_0 e^{-(z/H)^2} \ ,$$

$$H = \left(\frac{kTR^3}{Mm_p\sigma} \right)^{1/2} \ . \tag{3.56}$$

The density and pressure in the disk atmosphere decrease more rapidly than in the stellar atmospheres ($\propto e^{-z/H}$). This is because as we rise above the disk plane, the vertical component of the gravitational force exerted by the central star increases.

The problem on the emergence of thermal instability of disk accretion (Lightman 1974; Lightman, Eardly 1974) was analyzed in detail by Syunyaev and Shakura (1975) (see also Shakura, Syunyaev 1976). It was shown that in the inner zone (zone A), where radiation plays the main role, the process of accretion is unstable. The variation of the radiation emitted by the disk is quite strong (Syunyaev 1972).

3.7 Luminosity and Spectrum of Accretion Disks

The energy released from a unit surface area of the disk on both sides is determined by integrating the expression (3.53). Let us find the amount of energy liberated in an elementary ring of thickness dR in the disk:

$$dL(R) = 2Q^+ 2\pi R \, dR = \frac{3}{2} \dot{M} \frac{GM}{R^2} \left(1 - \sqrt{\frac{R_d}{R}} \right) dR \ . \tag{3.57}$$

It would seem that the quantity $dL(R)$ corresponds to the work done by the gravitational field (to be more precise, to half the work). Indeed, for a slow displacement along R, half the energy is transformed into the kinetic energy of matter moving along the φ-coordinate, while the other half is transformed into heat:

$$dL_{\mathrm{gr}}(R) = \dot{M}\frac{d}{dr}\left(-\frac{GM}{2R}\right)dR = \frac{1}{2}\dot{M}\frac{GM}{R^2}dr\ . \tag{3.58}$$

A comparison of (3.57) and (3.58) shows that at large distances $(R \gg R_{\mathrm{d}})$ in the disk, three times larger amount of energy is released.

Where does this additional energy come from? The paradox is explained quite easily if we consider that a continuous transport of angular momentum and mechanical energy from the inner parts takes place in the disk. Naturally, the total luminosity of the disk is ultimately due to the liberation of half the total gravitational energy of the matter incident of the disk from infinity. Indeed, integrating (3.57), we obtain the total luminosity of the disk in complete accord with the law of energy conservation:

$$L_{\mathrm{d}} = \int\limits_{R_{\mathrm{d}}}^{\infty} \frac{dL(R)}{dR}\,dR = \frac{\dot{M}GM}{2R_{\mathrm{d}}}\ . \tag{3.59}$$

It should be emphasized once again that this relation is valid only for the case when there is no transport of angular momentum at the inner edge of the disk: $W_{r\varphi}(\mathrm{in}) = 0$.

Substituting into (3.59) $R_{\mathrm{d}} = 3R_{\mathrm{g}}$ for a non-rotating black hole and dividing the expression thus obtained by $\dot{M}c^2$, we find that the efficiency of energy liberation during disk accretion to a black hole is $\eta \simeq 1/12$ [see (1.2)], i.e., $\sim 8\%$. The consideration of rotation of the black hole sharply increases the accretion efficiency to $\sim 42\%$ (Bardeen et al. 1972).

For an accreting neutron star, the accretion luminosity of the disk makes a significant contribution to the observed flux only in weak magnetic fields, since the energy liberated at the surface is larger by a factor of (R_{d}/R_x).

Estimates show that the optical thickness of the accretion disk is $\tau \approx \kappa H\varrho \gg 1$ (the only exception to this may be zone A in which the radiation pressure plays a dominant role).

It follows from Fig. 3.8 that the maximum energy is liberated near the inner edge of the disk $R = (25/16)R_{\mathrm{d}}$. In the case of accretion to a black hole or to a neutron star without a magnetic field, the temperature in the disk attains values $T \approx 10^7 - 10^8$ K, and the photons emitted in the central parts of the disk are comptonized on hot electrons, leading to a spectrum of a peculiar shape with an exponential drop in the Wien region (Shakura, Syunyaev 1973).

The universal spectrum (defined below) is formed in cold disks or in the outer parts of accretion disks. Suppose that a disk emits radiation like a black body. In this case, we can write

$$Q^- = \sigma_{\mathrm{SB}} T^4 \tag{3.60}$$

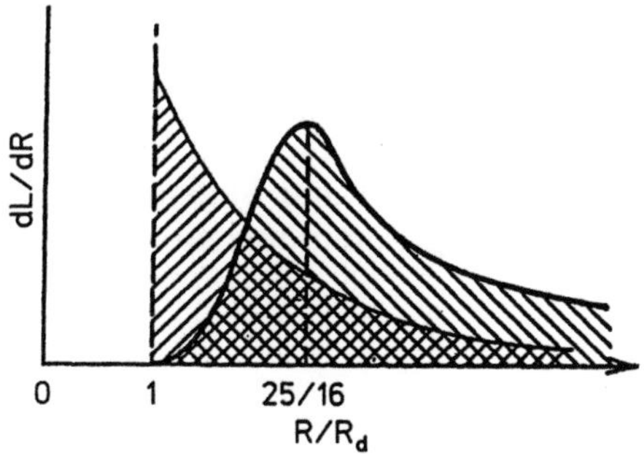

Fig. 3.8. Energy liberation in an accretion disk ring of unit width (*thick line*). The *thin line* shows the variation of the gravitation energy of the accretion matter in the same ring as a function of distance from the gravitational center

where σ_{SB} is the Stefan-Boltzmann constant, $\sigma_{\text{SB}} = 5.67 \times 10^{-5}$ erg/cm^2 s deg^4. Using (3.57), we obtain $Q^+ = [3/(8\pi)]\dot{M}GMR^{-3}(R \gg R_d)$. The energy balance equation $Q^+ = Q^-$ directly gives the radial distribution of temperature in the disk:

$$T(R) = \left(\frac{3}{8\pi\sigma_{\text{SB}}}\dot{M}\frac{GM}{R^3}\right)^{1/4} \sim R^{-3/4} \ . \tag{3.61}$$

The complete spectrum of the disk is a superposition of the blackbody spectra of the rings:

$$I_\nu = 2\pi \int\limits_{R_{\text{in}}}^{\infty} B_\nu[T(R)]R\,dR \ ; \tag{3.62}$$

where R_{in} is the distance starting from which the disk radiates like a black body, and B_ν is Planck's function:

$$B_\nu = \frac{2\pi h}{c^2}\left(\frac{kT}{h}\right)^3 \frac{x^3}{e^x - 1} \ \left(x \equiv \frac{h\nu}{kT}\right) \ . \tag{3.63}$$

Substituting (3.61) and (3.63) into (3.62), we obtain

$$I_\nu = \frac{16\pi^2 R_{\text{in}}^2}{c^2}\left(\frac{kT_{\text{in}}}{h}\right)^{8/3} h\nu^{1/3} \ . \tag{3.64}$$

The disk radiation increases with the cube root of the frequency. This result was first obtained by Lynden-Bell (1969).

The characteristic temperature of radiation can be estimated as

$$\frac{1}{2}\dot{M}\frac{GM}{R_d} = 2\pi R_d^2 \sigma_{\text{SB}} T^4 \ ,$$

whence

$$T = \left(\frac{\dot{M}GM}{4\pi R_d^3 \sigma_{\text{SB}}}\right)^{1/4} \approx 2 \times 10^7 \dot{M}_{18} R_6^{-3/4} m^{1/4} \text{K} \ ,$$

where $\dot{M}_{18} = \dot{M}/10^{18}$ g/s, $R_6 = R_{\mathrm{d}}/10^6$ cm. It can be seen that such a simple estimate is in good agreement with the spectrum of X-ray bursters in which the accretion disks practically reach the surface of the neutron stars (Chap. 6).

The most difficult task is to calculate the spectrum of the inner regions of the accretion disk (Pozdnyakov et al. 1982). Zel'dovich and Shakura (1969) were the first to point towards the necessity of taking Comptonization into account.

3.8 Supercritical Disk Accretion

The accretion equation for a thin disk considered in Sect. 3.6 describes an essentially subcritical accretion regime: $\dot{M} \ll \dot{M}_{\mathrm{cr}}$ [see (3.29)]. For the Thompson scattering cross section, the critical accretion rate is not too high and can be fully realized under real conditions.

In spite of the fact that several publications describe the results of investigation of the supercritical disk accretion, a unified approach to this problem has not been worked out so far. The solution of the problem is complicated by the fact that for $\dot{M} \gtrsim \dot{M}_{\mathrm{cr}}$ the radiation pressure force becomes comparable with the z-component of the gravitational force and the disk can no longer be considered as thin. The problem thus becomes two-dimensional.

A fairly realistic model for supercritical disk accretion was proposed by Shakura and Syunyaev (1973) who also were the first to investigate this problem. They proposed the existence of a self-consistent accretion regime which can be called a dynamic regime with a strongly manifested turbulence.

Other attempts were made to construct "quasi-static" thick accretion disks whose equilibrium along z- and R-coordinates depends strongly on the pressure exerted by matter and radiation (Paczynski, Wiita 1980; Jaroszynski et al. 1980; Abramowicz et al. 1980). However, there are indications that such configurations are essentially unstable (Nityananda, Narayan 1984). Here we shall consider the dynamic model which, in our opinion, has been substantiated most strongly. This model is also supported by observational arguments (Sect. 9.3).

As we approach a gravitational center, the energy release and the force of light pressure increase monotonically. At a certain radius R_{s}, called the spherization radius, the disk luminosity attains its critical value. The spherization radius is determined from the approximate expression $GM\dot{M}/R_{\mathrm{s}} = L_{\mathrm{Ed}}$:

$$R_{\mathrm{s}} = \frac{\dot{M}\sigma_{\mathrm{T}}}{4\pi m_{\mathrm{p}}c} \approx 10^6 \dot{M}_{-8}\,\mathrm{cm}\ . \tag{3.65}$$

For $R > R_{\mathrm{s}}$, the structure of an accretion disk does not differ from the subcritical case. At the spherization radius, the force of light pressure becomes comparable with the vertical component of the gravitational force and the disk becomes thicker, its thickness becoming of the same order as its radius. Under

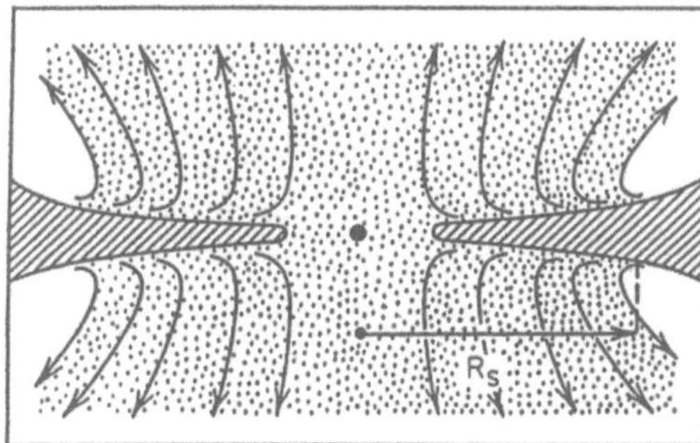

Fig. 3.9. Flow pattern of matter in the case of supercritical disk accretion (Shakura, Syunyaev 1973)

the action of the radial pressure, a part of the matter begins to flow (Fig. 3.9). It was observed by Shakura and Syunyaev (1973) that if the accretion rate decreases according to the law

$$\dot{M}(R) = \frac{R}{R_s} \dot{M}_{cr} \, , \tag{3.66}$$

the total luminosity of the accretion disk will never exceed the critical value considerably. Speaking more precisely, the total luminosity exceeds the Eddington limit by a factor of just about $\ln (R_s/R_d)$.

The main part of accreting matter will flow in the form of a quasi-spherical shell with a velocity of the order of parabolic velocity at the spherization radius. For $R_s \gg R_d$, the optical thickness of the outflowing shell is much larger than unity. Thus, the entire hard radiation generated at the inner edge of the disk is transformed into softer radiation.

Two flows of matter with a subrelativistic velocity may be formed along the disk axis. By the way, if an observer looks in the direction of the disk axis, he may see the harder radiation moving away from the central regions.

The radius of the photosphere in the optical region is determined from the condition $\tau_T \tau_{ff} \simeq 1$, where τ_T and τ_{ff} are the optical thickness according to Thompson scattering and free-free absorption, respectively. For an accretion rate $\sim 10^{-4} M_\odot$ per year, the radius of the photosphere becomes comparable with the radius of a supergiant star ($\sim 10^{12}$ cm). Thus, an accreting disk in the supercritical regime looks like a supergiant star with an anomalously powerful stellar wind.

3.9 Accretion in Binary Systems

It was first mentioned by Zel'dovich (1964) that the most favorable conditions for accretion to a relativistic star are encountered in the case when this star forms a pair with a normal star. This assumption is in excellent agreement with

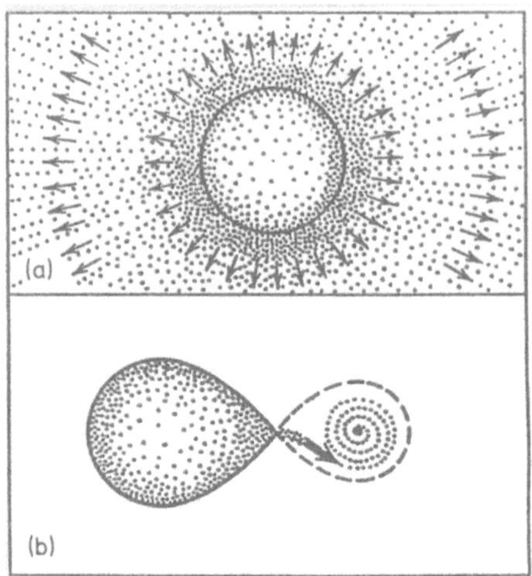
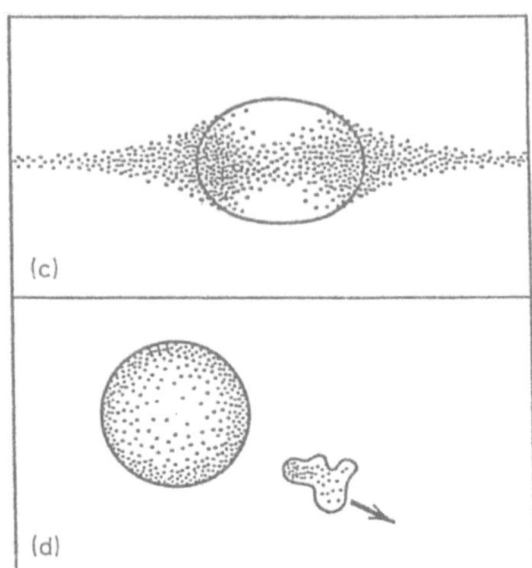

Fig. 3.10a–d. Four regimes of loss of matter by stars

observations: the brightest X-ray sources in the Galaxy are companions of normal stars in binary systems.

The matter accreted by a relativistic star in a binary system is supplied by the neighboring normal star. Hence the accretion regime depends considerably on the nature of efflux by the neighboring star.

It is well known that normal stars can lose matter mainly in two ways (naturally, we are dealing with slow processes only and not with the catalysmic processes like supernova bursts etc.): (1) In the form of a quasi-spherical stellar wind (Fig. 3.10 a); such phenomena are observed practically in all stars, starting from the Sun and ending with the massive supergiants; (2) in the form of gas jets upon the filling of a Roche lobe by a normal star (Fig. 3.10b). Apparently, there exists another flow regime which characterizes rapidly rotating stars (e.g., Be-stars), viz., overflow in the form of disk-shaped shells (Fig. 3.10c). Details for such a regime have not been worked out clearly (both from theoretical and observational points of view) since Be-stars are generally difficult objects both for spectral and photometric investigations. Possibly, there also exists a strongly transient regime in which matter is ejected in the form of individual gas clusters (Fig. 3.10d).

Formally, accretion in a binary system is the flow of gas in the gravitational field of two attracting (not necessarily concentrated) masses. The exact solution of such a problem is not possible in view of unsurmountable technical difficulties. Hence it is inevitable that one has to make certain simplifications. For example, it is quite natural to assume that the masses of both stars are concentrated at points. This is justified in view of the high concentration of matter at the centers of the stars.

In the case of circular orbits in the reference system rigidly connected with a component of a binary system, there exists an effective scalar potential Φ describing the gravitational and centrifugal force. In the plane of the orbit, this potential can be prestented in the form

$$\Phi = -\frac{GM_0}{R_1} - \frac{GM_x}{R_2} + \frac{\Omega^2(x^2+y^2)}{2} \tag{3.67}$$

where $\Omega = 2\pi/T$ is the angular velocity of rotation in the binary system associated with the semi-major axis through Kepler's third law

$$a = \left(\frac{G(M_x+M_0)}{\Omega^2}\right)^{1/3} . \tag{3.68}$$

During the motion of a free particle in a field with potential Φ, the total energy of the particle is conserved:

$$\Phi + \frac{v^2}{2} = \text{const} = E_0 . \tag{3.69}$$

The constant in this equation is determined by the total energy of the particle at a certain instant of time. The exact trajectory of the particle cannot be described analytically, but the region of the possible movement of the particle can always be predicted if we know its energy.

Suppose that a particle with a certain energy E_0 is "emitted" by one of the components. Obviously, as the particle moves away from the gravitational center, its velocity will decrease. If the particle energy E_0 is low, the particle will "turn back" somewhere. At the turning point, $v = 0$. The region of possible motion of the particle is determined by the equality $v^2/2 \geqslant 0$. In this case, we obtain from (3.69) the equivalent relation

$$\Phi \leqslant E_0 .$$

Consequently, the equipotential surface $\Phi = E_0$ (Hill's surface) limits the region of possible trajectories of a moving particle with energy E_0 (Fig. 3.11).

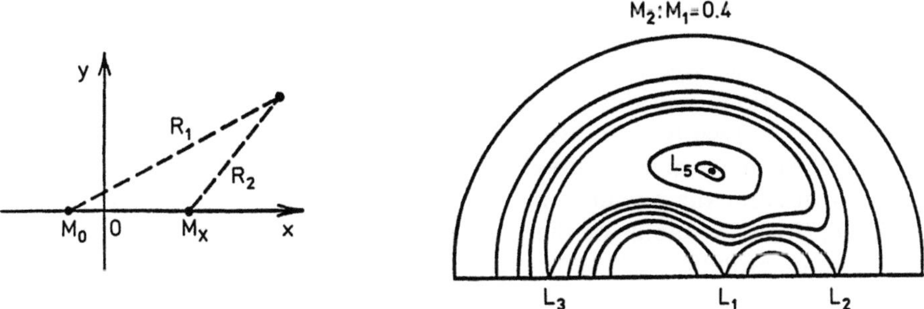

Fig. 3.11. Equipotential surfaces in a binary system

For a certain energy $E_R = \Phi_R$, Hill's surfaces around adjacent stars come in contact and form the Roche lobe. The point of contact (inner Lagrangian point) can be determined from the condition $d\Phi/dx = 0$ (the resultant of all the forces is equal to zero). The shape of the Hill's surfaces is independent of the absolute values of the component masses and depends only on their ratio $(q = M_x/M_0)$.

Let us briefly consider different types of accretion in binary systems.

3.9.1 Overflow Through the Inner Lagrangian Point

At a certain stage of its evolution, a normal star fills its Roche lobe and begins to intensely flow over to the neighboring component through the inner Lagrangian point (Yungel'son, Masevich 1983). At the Lagrangian point, a particle can go over from one Roche lobe to another without losing energy (of course, in reality the flow occurs in the neighborhood of the Lagrangian point). We shall not consider in detail the gas dynamics of a jet and refer the reader to the monographs by Gorbatskii (1974, 1977). The problem is complicated mainly due to the fact that the accretion scale (capture radius R_G) is of the order of the semi-major axis: $R_G \approx a$. However, the situation is eased in view of the following circumstance. The matter flowing from a normal star has an enormous angular momentum due to the orbital motion:

$$k_{orb} \simeq \Omega a^2 \, , \tag{3.70}$$

where k_{orb} is the specific angular momentum of 1 g of the substance. This makes it possible to guess the steady-state solution without considering the nonstationary problem which is a lot more complicated.

Since the angular momentum is conserved as the accreting star is approached, the centrifugal acceleration increases in accordance with the law $v^2/R \propto R^{-3}$, i.e., more rapidly than the acceleration due to gravity. At a certain distance, the matter enters some orbit. The following batches of the outflowing matter have nearly the same initial conditions and hence fall on the same orbit. A ring of increasing density is formed by the gas. Because of the mutual collisions of individual particles or turbulence cells, the angular momentum is redistributed. The ring spreads out into a disk in which the matter rotates differentially. Gradually, the motion turns into steady-state disk accretion on a time scale determined by the viscosity. If viscous stresses disappear at the inner edge of the disk, the following global transport takes place: the matter flows inwards from outside, and the angular momentum is transported outwards from inside.

Thus, we arrive at the following steady-state pattern (Fig. 3.10b): the matter is torn from the neighboring star in the form of a gas jet of thickness $\sim 0.1\, R_0$ (R_0 is the radius of the normal star). The flow rate is determined by the state of evolution of the normal star and the ratio of the masses of the components. For $q = M_x/M_0 \ll 1$, the flow of a star which has left the main sequence occurs on the thermal time scale t_{KH} (Yungel'son, Masevich 1983):

$$t_{\mathrm{KH}} = GM_0^2/R_0L_0 \; . \tag{3.71}$$

In this case, the flow rate of matter is written in the form

$$\dot{M} \approx M_0/t_{\mathrm{KH}} \; . \tag{3.72}$$

Such a turbulent flow is explained by the fact that as the matter is transported from the larger to the smaller star, the components come closer and the Roche lobe becomes smaller, which intensifies the mass transfer. In low-mass binary systems, the components may come closer by emitting gravitational waves (Paczinsky 1967) or magnetic wind.

The gas jet collides with the outer edge of the accretion disk, which is determined from the condition that the entire angular momentum coming from inside is carried away. The size R_{out} of the outer edge is comparable with the size of the Roche lobe of the accreting star (Paczinsky 1977). All or most of the matter transported by the jet falls into the disk where it moves along a sharply twisted spiral towards the compact star. Consequently, in the flow regime through the inner Lagrangian point, the accretion rate of the trapped matter is given by

$$\dot{M}_c \simeq \dot{M}_0 \; . \tag{3.73}$$

Let us estimate the characteristic time t_r in which the matter crosses the disk. Obviously,

$$t_r \simeq \frac{R_{\mathrm{out}}}{v_r} \simeq \frac{a}{v_r} \approx \frac{a}{\alpha v_\varphi (H/R)^2} \approx \frac{T}{2\pi\alpha} \left(\frac{R}{H}\right)^2 \; . \tag{3.74}$$

Here, we have used the approximate relation between v_r and v_φ in the standard disk accretion model [see (3.52)]. Since the disk is thin, i.e., $H \ll R$ and $\alpha < 1$, it must be assumed that within the framework of the α-model, the time of radial motion of matter in an accretion disk is much longer than the period of orbital rotation:

$$\frac{t_r}{T} \gg 1 \; . \tag{3.75}$$

An important consequence of this circumstance is the delay effect which is frequently overlooked. If certain changes occur in the normal component, the accreting star feels their effects only after a time which is much longer than the period of revolution of the binary system!

3.9.2 Accretion from Stellar Wind

The capture of matter by a relativistic star from stellar wind was first studied independently by Shakura and Syunyaev (1973), Tutukov and Yungel'son (1973), and Davidson and Ostriker (1973).

It has been shown by numerous observations that practically all stars emit matter in the form of a quasi-spherical stellar wind (Michalas 1978), although the reasons behind this emission may be different. For cold low-mass stars like the Sun, the emergence of the stellar wind is apparently associated with the dissipation of the energy of convective motion in surface layers. The flow in the case of hot OB-stars is due to their luminosity and the acceleration mechanism is largely radiative (selective absorption in lines). In order to estimate the rate and velocity of flow, we can make use of different semi-empirical dependences. For example, the flow rate of hot stars can be correctly approximated through the relation

$$\dot{M}_0 = \alpha_1 \frac{L_0}{v_\infty c} \, , \tag{3.76}$$

where α_1 is a dimensionless constant of the order of unity [the value $\alpha_1 \simeq 0.8 - 0.4$ is sometimes used (Barlow, Cohen 1977)], v_∞ is the velocity of the stellar wind at infinity (the value $v_\infty = 3 v_p$ is often taken, v_p being the parabolic velocity at the surface of the effusing star).

The acceleration regime of the stellar wind is of considerable significance in the physics of accretion processes. Usually, it is assumed that

$$v_w(r) = v_\infty \left[1 - \left(\frac{R_0}{r} \right)^{\alpha_2} \right]^{\alpha_3} \tag{3.77}$$

where R_0 is the radius of the effusing star. Several attempts to determine the constants α_2 and α_3 from observations have led to conflicting results. As a first approximation, we can use the values $\alpha_2 = 1$, $\alpha_3 = 1/2$:

$$v_w(r) = v_\infty \sqrt{1 - R_0/r} \, . \tag{3.78}$$

The dependence on the radius in this formula is quite logical. The force of the radiation pressure is proportional to $1/r^2$, i.e., it varies in the same way as the gravitational force. Hence the equation of motion of a particle subjected to the action of only these two forces (the radiative force is stronger) has the form

$$\frac{d^2 r}{dt^2} = \frac{\text{const}}{R^2} \, .$$

Integrating, we obtain

$$\frac{v^2}{2} + \frac{\text{const}}{R} = \text{const} \, .$$

Putting $v(R \to \infty) = v_\infty$ and $v(R = R_0) = 0$, we arrive at (3.78). These arguments show that (3.78) is quite rough and does not take into account the gas dynamics of flow, or the fact that the condition of excitation and ionization of atoms varies continuously in the outflowing wind so that the force of pressure cannot vary as $1/r^2$. However, these arguments describe the qualitative aspect of the problem.

For massive OB-stars, the flow rate attains a value of $\dot{M}_0 \simeq 10^{-6} - 10^{-5} M_\odot$ per year, while the flow velocity v_∞ is $1000-3000$ km/s. At the same time, the temperature of the stellar wind is $T \simeq 10^4$ K, so that the velocity of sound is $a_s \simeq 10^5$ km/s. Consequently, the accretion to a rapidly moving gravitational center is realized in such binary systems (Sect. 3.3).

If the flow velocity is much larger than the orbital velocity, i.e., $v_w \gg v_{orb}$, the gravitational capture radius R_G in massive binary systems in much smaller than the semi-major axis:

$$R_G \approx 10^{10} m v_8^{-2} \text{ cm} , \tag{3.79}$$

where $v_8 = v_w/10^8$ cm/s. This allows us to find the accretion rate of the matter captured by a compact star in a simple way. It follows from the continuity equations that

$$\dot{M}_0 = 4\pi a^2 \varrho v_w ,$$
$$\dot{M}_c = \pi R_G^2 \varrho v_w , \tag{3.80}$$

where ϱ and v_w are the density and velocity of the stellar wind at a distance equal to the separation between the stars. From (3.80), we obtain

$$\dot{M}_c = \frac{1}{4} \left(\frac{R_G}{a} \right)^2 \dot{M}_0 . \tag{3.81}$$

In close binary systems (Fig. 3.12) having a period of 1 to 10 days, $a_s \approx 10^{12}$ cm, and hence the compact star captures a $\sim 1/10000$ part of the

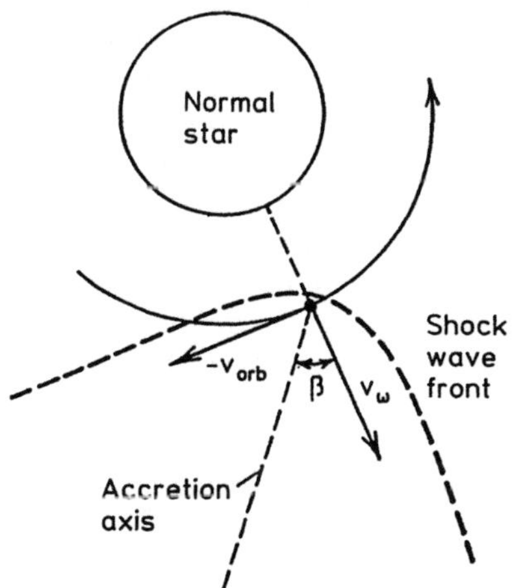

Fig. 3.12. Stellar wind accretion

stellar wind, i.e., $\dot{M}_c \simeq 10^{-10} M_\odot$ per year (for $\dot{M}_0 \simeq 10^{-6} M_\odot$ per year) and the accretion to a neutron star will be accompanied by the release of $\sim 10^{37}$ erg/s (standard luminosity of X-ray pulsars, Table 6.2). If the velocity of stellar wind becomes comparable with the orbital velocity, we can use the generalized Bondi-Hoyle-Lyttleton formula (3.39). In this case, the accretion axis is deflected from the radial direction (Fig. 3.12) by an angle β:

$$\tan \beta = v_{\mathrm{orb}}/v_{\mathrm{w}}(a) \ . \tag{3.82}$$

Using formula (3.39), we obtain

$$\dot{M}_c = \xi_1 \frac{(2GM_x)^2}{4\pi a^2 v_{\mathrm{w}}^4 (1+\tan^2 \beta)^{3/2}} \dot{M}_0 \ . \tag{3.83}$$

With the help fo Kepler's third law (3.67), we can express the capture coefficient $(= \dot{M}_c/\dot{M}_0)$ in terms of the observable quantities:

$$\frac{\dot{M}_c}{\dot{M}_0} = \frac{\xi_1 \tan^4 \beta}{\pi(1+\tan^2 \beta)^{3/2}} q^2 (1+q)^2 \ , \tag{3.84}$$

For $v_{\mathrm{orb}} \ll v_{\mathrm{w}}$ and $q \ll 1$, we obtain the asymptotic form $\dot{M}_c/\dot{M}_0 \approx \beta^4 q^2$.

It is useful to compare the luminosity $L_x = \eta \dot{M} c^2$ of an accreting relativistic star with the luminosity L_0 of an optical star. For this purpose, we use the semi-empirical relation (3.76)

$$\frac{L_x}{L_0} \approx \frac{\xi_1}{\pi} \eta \alpha_1 \left(\frac{c}{v_\infty}\right) \beta^4 q^2 \quad \text{for} \quad \beta \ll 1 \ , \quad q \ll 1 \ . \tag{3.85}$$

Observations by Bradt et al. (1979) give $L_x/L_0 \approx 10^{-1} - 10^{-3}$ for massive X-ray systems, which is in excellent agreement with the obtained estimate.

It should be emphasized that the accretion rate depends strongly on the velocity of the stellar wind at distances equal to the semi-major axis of the binary system: $\dot{M}_c \propto v_{\mathrm{w}}^{-4}$. Formula (3.85) can be used to estimate the velocity of stellar wind at a distance equal to the semi-major axis from the effusing star. Obviously, it is important to know the real law of variation of the velocity of the stellar wind [of the type (3.77)]. For strongly isolated systems, $v_{\mathrm{w}} \rightarrow v_\infty$ and $v_{\mathrm{orb}} \propto a^{-1/2}$, whence $\beta \propto a^{-1/2}$ and, consequently, $L_x/L_0 \propto a^{-2}$.

Let us consider the accretion regime in the vicinity of a compact star, which depends considerably on the average angular momentum of matter falling in the capture region. A rough estimate of the average angular momentum (per gram of the substance) can be obtained as follows (Davidson, Ostriker 1973, Illarionov, Syunyaev 1975).

In the reference system attached to a compact star, the specific angular momentum of a particle (per gram of mass) captured at a distance R is

$$k = [R \times v] = \{R \times [n_\Omega \Omega \times R + v_{\mathrm{w}}]\} \ ,$$

where n_Ω is the unit vector of the angular velocity of the binary system. Accordingly, the average angular momentum is equal to

$$k = \left[\frac{\int k\, dm}{\int dm} \right] .$$

Approximating the shock wave front as a cone with a height of the order of the capture radius, we obtain (Illarionov, Syunyaev 1975)

$$k \approx \tfrac{1}{4} \Omega R_{\mathrm{G}}^2 .$$

Henceforth, we shall use the following notation for the specific angular momentum:

$$k = \eta_k \Omega R_{\mathrm{G}}^2 . \tag{3.86}$$

This quantity is assumed to be positive if the direction of the angular momentum coincides with the orbital angular momentum of the binary system, and negative otherwise. It was shown by Wang (1981) that in the general case, η_k may differ strongly from unity, and may even reverse its sign due to the inhomogeneities in the stellar wind. Davies and Pringle (1980) attempted to find the gasdynamic solution of the problem, and showed that $\eta_k \ll 1$. However, it should be emphasized that the problem may be a lot more complicated in reality. For a considerably large angular momentum, the motion is not axially symmetric and the accretion axis is no longer a straight line. A basically new accretion regime appears, which must be investigated in detail. In order to illustrate the effects observed in this case, let us consider the two limiting positions of the shock wave cone. If $\beta = 0$, the angular momentum in the substance will be determined by the density gradient in the stellar wind. Indeed, the matter captured on the front has a higher density than the matter captured in the tail. Consequently, the captured angular momentum will be of the order of magnitude

$$k \approx \Omega R_{\mathrm{G}}^2$$

On the other hand, as $\beta \to \pi/2$, we have

$$k \to 0 .$$

The condition of formation of an accretion disk around a compact star can obviously be presented in the form (Illarionov, Syunyaev 1975)

$$\eta_k \Omega R_{\mathrm{G}}^2 \geqslant \sqrt{GM_x R_{\mathrm{min}}} , \tag{3.87}$$

where R_{min} is the minimum distance up to which free Keplerian motion is possible. In the case of a nonrotating black hole, $R_{\mathrm{min}} = 3R_{\mathrm{G}}$, while for a neutron star with a magnetic field, $R_{\mathrm{min}} = R_{\mathrm{st}}$ (R_{st} is the stopping radius).

From (3.87), we obtain the condition of disk formation as an inequality with the stellar wind velocity:

$$v_w(a) \leqslant v_{cr} \approx 320(4\eta)^{1/4} m^{3/8} T_{10}^{-1/4} R_8^{-1/8} (1+\tan^2 \beta)^{-1/2} \text{ km/s} \ . \qquad (3.88)$$

where $T_{10} = T/10$ days, $R_8 = R_{st}/10^8$ cm. If the stellar wind velocity is lower than the critical value, a quasi-spherical accretion regime sets in. Otherwise, an accretion disk is formed.

The total angular momentum captured by the accreting star is given by

$$K = k\dot{M}_c \sim R_G^4 \sim v_w^{-8} \ . \qquad (3.89)$$

The change in the angular momentum must lead to a strong fluctuation in the spin-up of the accreting star (Lipunov, Shakura 1976). The same factor may even change the accretion regime from disk-type to quasi-spherical type.

The effect of X-rays on the accretion from the stellar wind was considered by Syunyaev (1978). Calculations for the flow of stellar wind past an accreting star taking radiation into consideration were carried out by Krasnobaev and Syunyaev (1983). The formation of an accretion disk from a stellar wind was considered by Kolykhalov and Syunyaev (1979).

Let $\eta_k \Omega R_G^2$ be the specific angular momentum in the captured matter. We shall find the distance at which the angular momentum of the captured matter becomes equal to the Keplerian angular momentum:

$$\eta \Omega R_G^2 = \sqrt{GM_x R} \ .$$

This gives

$$R_1 = \eta^2 \frac{\Omega^2}{GM_x} R_G^4 \ . \qquad (3.90)$$

Using Kepler's third law, we obtain

$$R_1 = \eta^2 \left(\frac{1+q}{q}\right) \left(\frac{R_G}{a}\right)^3 R_G \ . \qquad (3.91)$$

In massive binary systems, $q \approx 0.1$ and $R_G \ll a/10$. Hence $R_1 \ll R_G$.

The following mechanism can be proposed for the formation of an accretion disk. In the interval $R_1 \leqslant R \leqslant R_G$, the motion of the captured matter is practically radial. However, at a distance $R \approx R_1$ (it is assumed that $R_1 \geqslant R_{st}$), the centrifugal forces become comparable with the gravitational forces. A nucleating ring is formed and begins to spread in the radial direction due to viscous forces. This leads to a two-flow pattern of accretion (Sect. 3.9, Fig. 3.1 d).

Equations of nonstationary disk accretion can be written in the following form by taking into account the accretion of matter from spherical flow:

$$\frac{\partial(2\pi\Sigma\omega R^3)}{\partial t} = -\frac{\partial}{\partial R}(2\pi\Sigma v_r\omega R^3 + 2\pi W_{r\varphi}R^2) + \frac{\partial(\dot{M}k)}{\partial r} \ , \tag{3.92}$$

$$\frac{\partial(2\pi\Sigma R)}{\partial t} = -\frac{\partial}{\partial R}(2\pi\Sigma v_r R) + \frac{\partial\dot{M}(R)}{\partial R} \ . \tag{3.93}$$

Equations (3.92) and (3.93) are, respectively, the equation of transport of angular momentum (equation of motion) and the continuity equation. Additional terms in both these equations describe the inflow of angular momentum and mass into the disk from top and bottom.

We shall consider two cases. Suppose that a substance is introduced into a narrow ring of radius $R = R_1$:

$$\frac{\partial\dot{M}}{\partial R} = \dot{M}_c\delta(R-R_1) \quad \text{and} \quad \frac{\partial(\dot{M}k)}{\partial R} = \dot{M}_c\sqrt{GMR_1}\,\delta(R-R_1) \ . \tag{3.94}$$

Assuming that the viscous forces disappear at the inner and outer edges of the disk ($W_{r\varphi}(R_d) = W_{r\varphi}(R_{out}) = 0$), and that the viscous stresses $W_{r\varphi}$ are continuous, we obtain for $R = R_1$ the following stationary solution:

$$\dot{M} = 2\pi\Sigma v_r R = -\dot{M}_1 < 0 \ ; \quad W_{r\varphi} = \frac{\dot{M}_1\omega}{2\pi}f_0(R) \quad \text{for} \quad R < R_1 \ ,$$

$$\dot{M} = \dot{M}_2 = \dot{M}_c - \dot{M}_1 > 0 \ ; \quad W_{r\varphi} = \frac{\dot{M}_1\omega}{2\pi}f_1(R) \quad \text{for} \quad R > R_1 \ , \tag{3.95}$$

where

$$\dot{M}_1 = \dot{M}_c\frac{1-(R_1/R_{out})^{1/2}}{1-(R_d/R_{out})^{1/2}} \approx \dot{M}_c\left[1-\left(\frac{R_d}{R_{out}}\right)^{1/2}\right] \ , \tag{3.96}$$

$$\dot{M}_2 = \dot{M}_c\frac{\left(\dfrac{R_1}{R_{out}}\right)^{1/2} - \left(\dfrac{R_d}{R_{out}}\right)^{1/2}}{1-\left(\dfrac{R_d}{R_{out}}\right)^{1/2}} \approx \dot{M}_c\left(\frac{R_d}{R_{out}}\right)^{1/2} \ . \tag{3.97}$$

When the plasma cools off rapidly for $R \approx R_G$, a conical shock wave is formed and it can be expected that matter will be deposited in a narrow sector on the disk. In this case,

$$\frac{\partial\dot{M}}{\partial R} = \frac{\dot{M}_c}{R_{out}-R_1} \ ;$$

$$\frac{\partial\dot{M}k}{\partial R} = \frac{\dot{M}_c k_1}{R_{out}-R_1} \ . \tag{3.98}$$

The solution of the system of Eqs. (3.92) and (3.93) in the zone $R > R_1$ assumes the form

$$\dot{M}(R) = -\dot{M}_1 + \dot{M}_c \frac{R - R_1}{R_{\text{out}} - R_1} \; , \qquad W_{r\varphi} = \frac{\dot{M}_1 \omega}{2\pi} f_2(R) \; . \tag{3.99}$$

For $R < R_1$, the solution coincides with the previous case.

There exists a zone in the disk where the radial velocity reverses its sign. For $R < R_{\text{cr}}$, matter flows out from the accreting star, while for $R < R_{\text{cr}}$, on the contrary, it flows to the star. Putting $\dot{M}(R_{\text{cr}}) = 0$, we obtain

$$\frac{R_{\text{cr}}}{R_{\text{out}}} \approx 1 - \left(\frac{R_d}{R_{\text{out}}} \right)^{1/2} \; .$$

The shape of the function $f_0(R)$ is the same as in the standard model

$$f_0(R) = 1 - \left(\frac{R_d}{R} \right)^{1/2} \; . \tag{3.100}$$

For $R > R_1$, the situation is quite different:

$$f_1(R) = \frac{\dot{M}_c}{\dot{M}_1} \left(\frac{R_1}{R} \right)^{1/2} - \frac{\dot{M}_2}{\dot{M}_1} - \left(\frac{R_d}{R} \right)^{1/2} \approx \left(\frac{R_1}{R} \right)^{1/2} \left[1 - \left(\frac{R}{R_{\text{out}}} \right)^{1/2} \right] \; ,$$

$$f_2(R) = \frac{\dot{M}_c}{\dot{M}_1} \left(\frac{R_1}{R} \right)^{1/2} \frac{R - R_1}{R_{\text{out}} - R_1} - \frac{\dot{M}_c}{\dot{M}_1} \left(\frac{R_{\text{out}} - R}{R_{\text{out}} - R_1} \right) \tag{3.101}$$

$$- \frac{\dot{M}_2}{\dot{M}_1} - \left(\frac{R_d}{R_1} \right)^{1/2} \approx 1 - \frac{R}{R_{\text{out}}} \; .$$

It should be emphasized that a mass flux $\dot{M}_1 \leqslant \dot{M}_c$ falls onto the accreting star. It can be seen from (3.96) that $\dot{M}_1$ practically coincides with $\dot{M}_c$.

Kolykhalov and Syunyaev (1979) also calculated the radial structure of the disk in the same way as in the standard model of disk accretion (Shakura, Syunyaev 1973). Apparently, the outer radius of the disk does not significantly exceed the gravitational capture radius R_G. Here, the matter flowing over the disk is blown by the stellar wind. In the case of a nonradiative shock wave for $R \approx R_G$, the problem concerning the outer edge of the disk is more complicated. It is possible that the radius of the outer edge becomes comparable with the size of the Roche lobe of a compact star: $R_{\text{out}} \approx a$. In this case, angular momentum may be transferred to the orbital motion of the binary system through tidal forces (Goldreich, Peal 1968).

3.10 Two-Stream Accretion

In a number of astrophysical situations, the motion of matter in the vicinity of a compact star can be presented as a sum of two flows, viz., the disk flow and the spherically symmetric flow (Lipunov 1980). We shall enumerate just a few cases of this type.

A. Suppose that a normal star fills its Roche lobe and overflows through the inner Lagrangian point. In other words, the most favorable conditions are realized for the creation of an accreting disk in accretion regime. Matter flowing in a gas jet has a considerable angular momentum and forms a disk flow around the compact star. At the same time, it is quite obvious that stellar wind may also flow from the normal star in this case. If the stellar wind velocity is much higher than the orbital velocity, the angular momentum of the captured matter will be small and it will move practically in the radial direction near the compact starr. Consequently, two independent (in the first approximation) flows of matter will be observed near the neutron star for $R \ll R_{\mathrm{G}}$ (R_{G} is the gravitational capture radius for the stellar wind).

B. A similar situation is observed in the case when a normal star rapidly rotates and sheds matter due to centrifugal and tidal forces along the equator of rotation. This matter has a low radial velocity and a large angular momentum, and can also form an accretion disk. On the other hand, as in the previous case, nothing can prevent the normal star from emitting a strong stellar wind, and thus a two-stream accretion pattern is observed once again. This situation may be realized in binary systems (with Be-stars).

C. Suppose that a normal star loses matter in the form of stellar wind. The parameters of the binary system and the wind are such that the accretion disk begins to be formed deep inside the capture radius and later spreads on to this radius. This picture was considered by Kolykhalov and Syunyaev (1979) (see Sect. 3.9). This also results in the formation of two flows.

D. Let us consider purely radial accretion onto a rapidly rotating magnetized star. It will be shown below (Chap. 7) that if the accreting star has a high rotational velocity, its magnetic field hampers the deposition of matter on its surface. Instead, matter accumulates around the star and takes away its angular momentum. Gradually, the shell starts approaching the equator of the rotating star and forms an accretion disk. This again leads to a two-stream accretion (which, however, is non-stationary).

It can be seen that two-stream accretion is quite prevalent and deserves a detailed investigation. Interaction of two flows can lead to observable effects. Firstly, a considerable amount of energy is released as a result of the collision of two flows. After the collision, the substance in the spherical flow is heated to a temperature corresponding to the free fall, which is much higher than the temperature of the accretion disk. Secondly, the dynamics of the disk changes. The equation of angular momentum transfer in the disk in this case is similar to that considered in Sect. 3.9.

3.11 Accretion of Magnetic Fields

It was mentioned in Sect. 3.2 that random magnetic fields in the accretion matter bind the motion of particles and ensure a gasdynamic accretion. But can the magnetic fields increase during the process of accretion to such an extent that they begin to affect the dynamics of the falling matter?

The answer to this question obviously depends on the mutual relation between the two competing mechanisms, viz., an enhancement in the magnetic fields associated with their freezing in the matter on the one hand, and the dissipation of the magnetic field due to a reclosure of the lines of force. As a rule, the ohmic losses are always small.

Shvartsman (1970b) formulated the theorem of equipartition of energy, which states that if an approximate equilibrium is established between the magnetic and gravitational energies in the falling matter at a certain distance from the accreting star, it will be preserved. Indeed, the gravitational energy increases as $\varepsilon_{gr} \propto R^{-5/2}$ when the star is approached [see (3.16)]. At the same time, the magnetic energy density of the "frozen" fields increases more rapidly: $\varepsilon_{m} = B^2/8\pi \propto R^{-4}$, so that the magnetic energy rapidly catches up with the gravitational energy. However, the inequality $\varepsilon_{m} \gg \varepsilon_{gr}$ is not possible since the magnetic field energy is supplied by the gravitational energy. Hence the annihilation of magnetic fields must lead to a uniform distribution:

$$\varepsilon_{gr} \approx \varepsilon_{m} \ . \tag{3.102}$$

For such a relation, the magnetic field does not strongly change accretion regime although it considerably increases the efficiency of energy release.

It was shown by Shvartsman (1970b) that in the case of accretion from the interstellar medium, a rough equality is established at the gravitational capture radius: the magnetic energy density in the interstellar medium is of the order of the thermal energy density, while the thermal energy at the capture radius is of the order of the gravitational energy.

An entirely different situation is realized in a binary system. Even in the case of accretion of the interstellar matter, magnetic and gravitational energies are equalized either later (to be more precise, when the shell is closer to the star), or may not be equalized at all. Certainly, this suppresses the role of the magnetic fields.

Let us consider once again the accretion of the interstellar medium, assuming that the square of the velocity of the matter constituting this medium is much larger than the square of the velocity of sound: $v_{\infty}^2 \gg a_{\infty}^2$. In this case, obviously, the ratio of the magnetic and gravitational energies at the capture radius is given by

$$\left(\frac{\varepsilon_{m}}{\varepsilon_{gr}}\right)_{R=R_{G}} \approx \left(\frac{a_{\infty}}{v_{\infty}}\right)^2 \ll 1 \ . \tag{3.103}$$

Accordingly, the distance from the accreting star at which a uniform distribution is established is estimated as

$$R_{\text{eq}} \approx \left(\frac{a_\infty}{v_\infty}\right)^{4/3} R_{\text{G}} \; . \tag{3.104}$$

In real conditions, $a_\infty/v_\infty \gtrsim 10^{-2}$ and hence

$$10^{-3} \lesssim R_{\text{eq}}/R_{\text{G}} \lesssim 1 \; .$$

It will be shown below that the intrinsic magnetic field for single neutron stars becomes effective at distances $R \approx 10^9 - 10^{10}$ cm, which is comparable with R_{eq}. Thus, uniform distribution is not attained at least in stars with standard magnetic fields. The role of accretion fields is not significant.

In binary systems, an additional effect is observed in the form of a magnetic evacuation. Let us consider the case of accretion from stellar wind. Suppose that a certain relation between the magnetic and kinetic energies is observed at the surface of a normal star (in the gravitational field of the normal star). It is logical to assume that the thermal energy at the surface of the normal star is of the order of the magnetic energy. In this case, we can write

$$\left(\frac{\varepsilon_{\text{m}}}{\varepsilon_{\text{k}}}\right)_{r=R_0} \approx \frac{1}{10}\left(\frac{a_0}{v_\infty}\right)^2 , \tag{3.105}$$

where v_∞ is the velocity of the stellar wind at infinity, a_0 is the velocity of sound at the surface and in the stellar wind (the stellar wind from hot stars is isothermal within the admissible errors). The factor $1/10$ takes into account the fact that the velocity of the stellar wind at infinity is about half an order of magnitude higher than the parabolic velocity. Obviously, at distances of the order of the semi-major axis, we can write

$$\left(\frac{\varepsilon_{\text{m}}}{\varepsilon_{\text{k}}}\right)_{r=a} = \left(\frac{\varepsilon_{\text{m}}}{\varepsilon_{\text{k}}}\right)_{r=R_0}\left(\frac{R_0}{a}\right)^2 \tag{3.106}$$

which indicates a sharp decline in the role of the magnetic field. This follows from the fact that the kinetic energy density decreases as r^{-2} for a constant velocity of the stellar wind, while the magnetic energy density decreases, as before, as r^{-4}. At the capture radius $\varepsilon_{\text{gr}} \approx \varepsilon_{\text{k}}$, and hence $\varepsilon_{\text{m}}/\varepsilon_{\text{gr}}$ is estimated at the capture radius with the help of (3.106). After the capture of the matter, the ratio $\varepsilon_{\text{m}}/\varepsilon_{\text{gr}}$ begins to rise again and, at a certain distance R from the accreting star, we can write

$$\frac{\varepsilon_{\text{m}}}{\varepsilon_{\text{gr}}} \approx \frac{1}{10}\left(\frac{a_0}{v_\infty}\right)^2 \left(\frac{R_0}{a}\right)^2 \left(\frac{R}{R_{\text{G}}}\right)^{-3/2} . \tag{3.107}$$

The distance at which the equipartition is established is given by

$$R_{\text{eq}} \approx 10^{-2/3} \left(\frac{a_0}{v_\infty}\right)^{4/3} \left(\frac{R_0}{a}\right)^{4/3} R_{\text{G}} \ . \tag{3.108}$$

It can be seen that in view of the additional "pumping out" of the magnetic energy in the stellar wind, the role of magnetic fields in the accretion matter becomes negligible.

The case of overflow through the inner Lagrangian point is taken into account by (3.107) for $R_{\text{G}} \approx a$.

4. Classification of Neutron Stars

The matter surrounding a neutron star is almost always in the form of a plasma at high temperature, and hence has a high electric conductivity. When captured by a star of mass on the order of the mass of the Sun, matter falls practically in nonradiative regime so that its temperature is close to that determined from the equality of gravitational and thermal energies (Sect. 3.1):

$$T_{\mathrm{ff}} = \frac{GM_x}{R_{\mathrm{u}} R} \approx 1.5 \times 10^{10}\, m_x R_8^{-1}\, \mathrm{K} \quad , \tag{4.1}$$

where $R_8 = R/10^8$ cm is the distance from the accreting star. The conductivity of a completely ionized plasma is estimated by the expression (Pikel'ner 1966)

$$\lambda_{\mathrm{c}} \approx 10^7\, T_{\mathrm{e}}^{3/2}\, \mathrm{s}^{-1} \quad , \tag{4.2}$$

where T_{e} is the electron temperature. For $T_{\mathrm{e}} \approx T_{\mathrm{ff}} \approx 10^8 - 10^{10}$ K, $\lambda_{\mathrm{e}} \approx 10^{19} - 10^{22}$ cm^{-1}, which is higher than the conductivity of copper.

The discovery of radio pulsars (Hewish et al. 1968) demonstrated that neutron stars possess powerful magnetic fields whose strength at the surface attains values of $10^{12} - 10^{13}$ Oe. This means that magnetized (rotating) neutron stars are surrounded by strong electromagnetic fields.

A good conducting accreting plasma must interact effectively with the magnetic field of a neutron star (Amnuel', Guseinov 1968). Hence the interaction of a neutron star with its surroundings cannot be treated as purely gravitational and consequently we cannot describe it by the purely gasdynamic accretion process considered in Chap. 3. In the general case, such an interaction should be described by the hydrodynamic equation as well as by Maxwell's equations. This makes the already complicated picture of the interaction of a neutron star with the surrounding medium even more complex.

The following classification of neutron stars will be based on characteristics of the interaction of the plasma surrounding them with their electromagnetic field. This approach was proposed by Shvartsman (1970) who isolated three stages of interaction of a rotating magnetized neutron star: the ejection stage, the "propeller" stage, which was later "rediscovered" by Illarionov and Syunyaev (1975) and named as such, and the accretion stage. Shvartsman (1971) was able to predict on the basis of this approach the phenomenon of an accreting X-ray pulsar in binary systems. New interaction regimes discovered in recent years have led to a general classification of neutron stars (Lipunov 1982a, 1984; Kornilov, Lipunov 1983a). Moreover, it was found that the new classification can be applied to any object which can be called a gravimagnetic rotator having a magnetic field, a gravitational field, and rotation (Lipunov 1987a). Among other things, these objects include white dwarfs, magnetic stars, and spinars.

It should be noted that the interaction of a neutron star with the surrounding plasma is yet to be understood completely. However, even the first approximation reveals a multitude of interaction modes. We shall take the first steps towards the solution of this problem in this chapter. To simplify the analysis, we assume the electromagnetic part of the interaction to be independent of accreting flux parameters, and vice versa.

Henceforth, we shall assume in almost all cases that the intrinsic magnetic field of a neutron star is a dipole field. This is not just a convenient mathematical simplification. It will be shown that the interaction between the plasma and the magnetic field takes place at a large distance from the surface of the neutron star, and the dipole moment makes the main contribution away from the surface. Moreover, it was mentioned above that the collapse of an ordinary star into a neutron star "cleanses" its field. According to the condition of conservation of magnetic flux, the ratio of the quadrupole magnetic moment q to the dipole moment μ decreases in direct proportion to the radius upon compression of the star:

$$q/\mu \propto R \ . \tag{4.3}$$

(It should be emphasized, however, that the contribution of the quadrupole component to the field strength at the surface remains unchanged.)

To begin, let us consider some properties of the dipole magnetic field which we shall be requiring for subsequent analysis.

4.1 Magnetic Dipole

First, we consider the static field of a non-rotating ($\omega = 0$) star. It is not necessary to assume that the dipole is concentrated at the origin of coordinates ($R = 0$). A dipole field is produced outside a sphere if a surface current varying according to $J \approx \sin \theta$ is passed over its surface. The magnetic field inside the sphere is uniform in this case. Outside the sphere, the magnetic field components are given by (Landau, Lifshitz 1973)

$$\boldsymbol{B}_{\mathrm{d}} = \frac{2\mu \sin \theta}{R^3} \boldsymbol{e}_r - \frac{\mu \cos \theta}{R^3} \boldsymbol{e}_\theta \ , \tag{4.4}$$

where $\boldsymbol{e}_r$ and $\boldsymbol{e}_\theta$ are unit vectors. The magnitude of the magnetic field strength is

$$B_{\mathrm{d}} = \frac{\mu}{R^3}(1 + 3\sin^2\theta)^{1/2} \ . \tag{4.5}$$

Among other things, these formulas indicate that the magnetic field strength B_0 at the magnetic poles is twice its value at the equator. Knowing the magnetic field strength at the poles and the radius of the star, we can find the dipole moment:

$$\mu = \frac{B_0 R_0^3}{2} \ . \tag{4.6}$$

Henceforth, we shall mainly use the quantity μ_{30} expressed in the units 10^{30} Oe cm^3. This is a convenient unit since the magnetic dipole moment of a neutron star with a magnetic field strength $B_0 = 2 \times 10^{12}$ Oe at the pole and a radius of 10 km is 10^{12} Oe $\times 10^{18}$ cm$^3 = 10^{30}$ Oe cm^3.

The dipole field is not a force field. The plasma cluster frozen in a magnetic field will not be deformed. However, if the lines of force are bent, a stress proportional to $B^2/4\pi$ will be produced. In the magnetohydrodynamic approximation, the plasma moves freely along the lines of force. In the gravitational field, the plasma cluster will creep to the magnetic pole (it is assumed that this cluster is quite light and does not deform the dipole field). This follows from the equation for the magnetic field lines:

$$\frac{1}{R}\frac{dR}{d\theta} = \tan\chi = \left(\frac{B_\theta}{B_r}\right)^{-1} = 2\tan\theta \ ; \tag{4.7}$$

where χ is the angle between the direction of vector B and the radius vector (Fig. 4.1). This leads to the following equation for a line of force:

$$R = R_{\mathrm{e}}\cos^2\theta \ , \tag{4.8}$$

where R_{e} is the distance from a given line of force at the magnetic equator. The line of force always lies within the circle $R = R_{\mathrm{e}}$ and hence the plasma cluster will indeed "roll" to one of the poles.

For a rotating dipole, we must also consider the electric component of the field. The exact solution describing the electromagnetic field of a dipole rotating in vacuum was given by Landau and Lifshitz (1973). For our purpose, it is sufficient to consider only the qualitative form of this solution. If the rotational axis does not coincide with the magnetic axis of a star, the dipole will emit electromagnetic waves at the rotational frequency, and we obtain the characteristic distance R_1 called the radius of the light cylinder:

$$R_1 = \frac{c}{\omega} \ . \tag{4.9}$$

At distances $R \ll R_1$, the electromagnetic field is static, and the order of magnitude of the electric field component E is given by

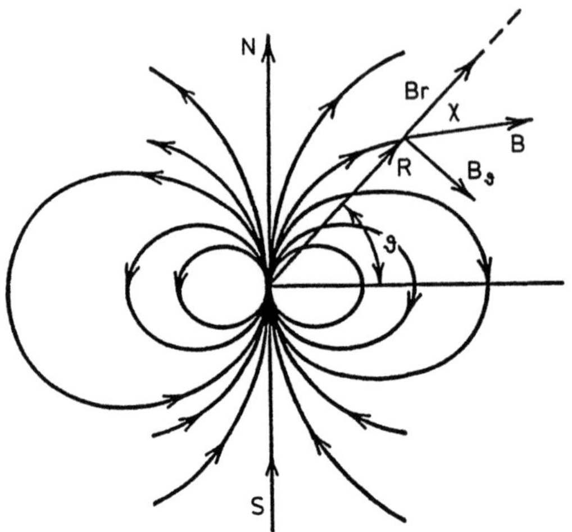

Fig. 4.1. Field lines of a magnetic dipole

$$E \approx \frac{v}{c} B = \frac{\omega R}{c} B = \frac{R}{R_1} B \; . \tag{4.10}$$

As R approaches the value R_1, the electric component E approaches B, as in the case of a free electromagnetic wave. The rotational velocity of the magnetic field lines becomes close to the velocity of light. The electromagnetic field is no longer static and becomes a freely propagating electromagnetic wave for $R \gg R_1$.

The energy carried away by the magnetic dipole radiation is

$$L_{\rm m} = \frac{2}{3} \frac{\mu^2 \omega^4}{c^3} \sin^2\beta \; , \tag{4.11}$$

where β is the angle between the rotational axis and the magnetic axis of the dipole. The radiation energy is derived from the rotational energy (the magnetic field does not dissipate). This results in a spin-down torque

$$\boldsymbol{K}_{\rm m} = -\left(\frac{2}{3}\right) \frac{\mu^2 \sin^2\beta}{c^3} \omega^3 \boldsymbol{e}_\omega \; ; \tag{4.12}$$

where $\boldsymbol{e}_\omega$ is a unit vector along the rotational axis. The spin-down torque arises from the φ-th component of the magnetic field at the surface of the neutron star.

Although the conditions near real neutron stars differ strongly from the idealized situation described above. [Because of the existence of plasma and relativistic particles all conventional models give energy losses close to magnetic dipole losses. Physically, this is due to the fact that in all models, the rotational energy is carried away by relativistic particles leaving the light cylinder (Zel'dovich, Novikov 1971).]

A formula of the type (4.11) can be derived easily from the following considerations. At distances of the order of the light cylinder radius, the electric and magnetic field strengths can be estimated from the dipole relation

$$E \approx B \approx \frac{\mu}{R_1^3} \; .$$

The total power carried away by the radiation is equal to the product of the Poynting's vector and the characteristic area:

$$L_{\mathrm{m}} \approx \frac{|E \times B|}{4\pi} \, c \, 4\pi R_1^2 \approx \frac{\mu^2}{R_1^4} c \approx \frac{\mu^2 \omega^4}{c^3} \; . \tag{4.13}$$

This expression is the same as the exact magnetic dipole relation (4.11) with the exception of a dimensionless factor of the order of unity. Note the form of the third term in (4.13). We shall use this form of notation most frequently.

4.2 Stopping Radius

We shall consider qualitatively the effect of the electromagnetic field of a rotating neutron star on the accreting plasma. Suppose that a neutron star has a magnetic dipole moment μ, rotational frequency ω, and mass M. The surrounding plasma has the following parameters at a distance $R \gg R_{\mathrm{G}}$: density ϱ_∞, velocity of sound a_∞ and/or velocity v_∞ relative to the star. Under the action of the gravitational field, the plasma will tend to accrete onto the neutron star. The electromagnetic field, however, will obstruct this process, and the accreting matter will come to a stop at a certain distance. This distance is the stopping radius which was already discussed in Chap. 3.

We shall successively consider two basically different cases: (1) when the interaction takes place outside the light cylinder: $R_{\mathrm{st}} > R_1$, and (2) when the interaction takes place inside the light cylinder: $R_{\mathrm{st}} < R_1$.

1) Suppose that the interaction takes place outside the light cylinder. In this case, which was first considered by Shvartsman (1970c), the neutron star is the generator of magnetic dipole radiation and relativistic particles. The form in which the major part of the neutron star energy is ejected is not important at this stage. What is important is that both the relativistic particles and the magnetic dipole radiation will give away their momentum and hence exert pressure on the accreting plasma. Indeed, random magnetic fields are always present in the accreting plasma. The Larmor radius of a particle with energy $\lesssim 10^{10}$ eV in the lowest magnetic field $\sim 10^{-6}$ Oe [the magnetic field of the interstellar medium (Kaplan, Pikel'ner 1979)] is found to be much smaller than the capture radius. The Larmor radius is given by (Landau, Lifshitz 1973)

$$R_L = \frac{E}{eB} \approx 10^{14} E_{10} B_{-6}^{-1} \text{ cm} \; , \tag{4.14}$$

where $E_{10} = E/10^{10}$ eV, $B_{-6} = B/10^{-6}$ Oe. For the characteristic velocity of sound $a_\infty = 10$ km/s or the velocity $v_\infty = 10$ km/s of the moving star, the capture radius is

$$R_G \approx 10^{14} m v_6^{-2} \text{ cm} \; . \tag{4.15}$$

The standard notation has been used here. In binary systems, the magnetic field in the stellar wind attains values of $10^{-2} - 10^{-4}$ Oe and $R_L \ll R_G$ always.

Thus the relativistic particles are "trapped" in the magnetic field of an accreting plasma and give away their momentum to it.

The same situation is observed for low-frequency electromagnetic radiation. It is well known that an electromagnetic wave may propagate in a plasma only if its frequency is higher than the plasma frequency v_p:

$$v_p = \sqrt{\frac{4\pi n_e e^2}{m_e}} \approx 9 \times 10^3 n_e^{1/2} \text{ Hz} \; . \tag{4.16}$$

Here, n_e is the number density of electrons. It can be seen that in almost all cases of interest, the magnetic dipole radiation does not pass through the plasma surrounding the neutron star [see, however, (Lipunov 1983a) and Sect. 8.4].

Thus, a relativistic stellar wind can effectively impede the accretion of matter. It was shown in Sect. 3.2 that even a low-power stellar wind is capable of stopping accretion. A cavern is formed around the neutron star, and the pressure of the ejected wind at its boundary balances the pressure of the surrounding plasma:

$$P_m = P_a|_b \; . \tag{4.17}$$

This equality defines a characteristic size, i.e., the Shvartsman radius R_{Sh}.

2) Let us now suppose that the accreting plasma has a pressure high enough to penetrate the light cylinder. We shall show that the accreting plasma behaves like a diamagnetic and expels (to be more precise, squeezes) the magnetic field of the neutron star. Let us compare the characteristic time t_r of plasma fall with the time t_d in which the magnetic field penetrates the plasma, determined by the ohmic losses (Pikel'ner 1966):

$$t_d \approx \frac{2\pi \lambda_c R^2}{c^2} \approx 10^{11} \frac{\lambda_c}{10^{15} \text{ s}^{-1}} R_8^2 \text{ s} \; . \tag{4.18}$$

The fall time is given by

$$t_r \approx R/v_r \approx 0.3 R_8^{3/2} \text{ s} \; , \tag{4.19}$$

where $R_8 = R/10^8$ cm. Comparing (4.18) and (4.19), we find that $t_d \gg t_r$ for all reasonable plasma parameters.

Since the magnetic field inside a light cylinder decreases according to the dipole law, the magnetic pressure is given by

$$P_{\mathrm{m}} = \frac{B^2}{8\pi} \approx \frac{\mu^2}{8\pi R^6} \ .$$

Substituting this expression into (4.17), we obtain the Alfvén radius R_{A}.

The magnetic pressure and the pressure of the relativistic wind can be presented in the following convenient form:

$$P_{\mathrm{m}} = \begin{cases} \dfrac{\mu^2}{8\pi R^6} \ , & R \leqslant R_1 \ , \\[2ex] \dfrac{L_{\mathrm{m}}}{4\pi R^2 c} \ , & R > R_1 \ . \end{cases}$$

We have intentionally replaced the approximate equality by an exact equality. We introduce a dimensionless factor κ_t such that the power of the ejected wind is

$$L_{\mathrm{m}} = \kappa_t \frac{\mu^2}{R_1^3} \omega \ .$$

In this case, the electromagnetic pressure is given by

$$P_{\mathrm{m}} = \begin{cases} \dfrac{\mu^2}{8\pi R^6} \ , & R \leqslant R_1 \ , \\[2ex] \dfrac{\kappa_t \mu^2}{4\pi R_1^4 R^2} \ , & R > R_1 \ . \end{cases} \tag{4.20}$$

Assuming that $\kappa_t = 1/2$, we obtain for $R = R_1$ a continuous function $P_{\mathrm{m}}(R)$ whose qualitative behavior is shown in Fig. 4.2. Keeping in view the rough estimates as before, we shall not bother about the multipliers.

The accretion pressure of a plasma outside the capture radius is nearly constant and hence gravitation does not affect the medium parameter significantly. In contrast, at distances smaller than the gravitational capture radius R_{G}, the matter falls almost freely and exerts pressure on the "wall" equal to the dynamic pressure [formula (4.16)]. Assuming the accretion to be spherically symmetric, we obtain

$$P_{\mathrm{a}} = \begin{cases} \dfrac{\dot{M}_{\mathrm{c}} v_\infty}{4\pi R_{\mathrm{G}}^2} \ , & R > R_{\mathrm{G}} \ , \\[3ex] \dfrac{\dot{M}_{\mathrm{c}} v_\infty}{4\pi R^2} \left(\dfrac{R_{\mathrm{G}}}{R} \right)^{1/2} \ , & R \leqslant R_{\mathrm{G}} \ . \end{cases} \tag{4.21}$$

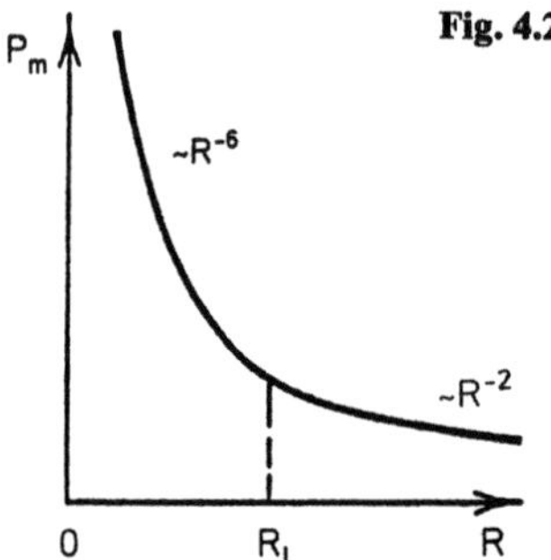

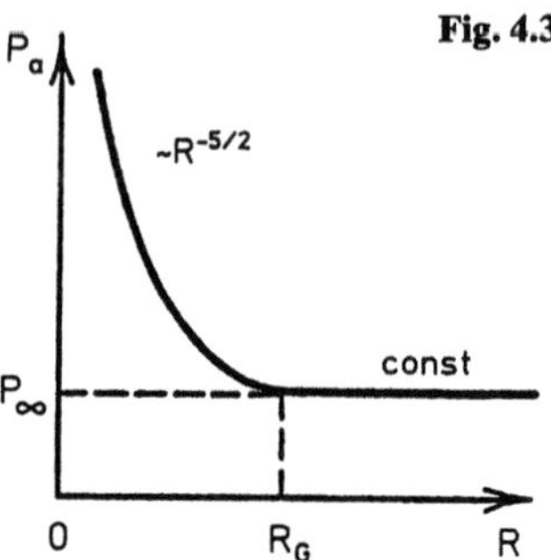

Fig. 4.2. Qualitative dependence of electromagnetic pressure on the distance from the star

Fig. 4.3. Dependence of dynamic plasma pressure on the distance from the star

Here, we have used the continuity equation $\dot{M}_c = 4\pi R_G^2 \varrho_\infty v_\infty$. When presented in this form, the pressure P_a is a continuous function of distance (Fig. 4.3).

Let us summarize the obtained results. If the pressure of the electromagnetic forces is due to a static magnetic field, the pressure balance equation (4.17) defines the Alfvén radius. If, however, the plasma is stopped by a relativistic wind, (4.17) defines the Shvartsman radius. Thus, the stopping radius is defined as

$$R_{st} = \begin{cases} R_A , R_{st} \leqslant R_1 , \\ R_{Sh}, R_{st} > R_1 \end{cases} . \tag{4.22}$$

Substituting (4.21) and (4.22) into (4.17), we obtain the following expression for the Alfvén radius:

$$R_A = \begin{cases} \left(\dfrac{2\mu^2 G^2 M^2}{\dot{M}_c v_\infty^5}\right)^{1/6} , & R_A > R_G , \\[4mm] \left(\dfrac{\mu^2}{2\dot{M}_c \sqrt{2GM}}\right)^{2/7} , & R_A \leqslant R_G . \end{cases} \tag{4.23}$$

The upper expression was first obtained by the geophysicists Zhigulev and Romishevskii (1959). It defines the characteristic distance from the Earth to the subsolar point of the Earth's magnetosphere.

The gravitational constant G does not generally appear in the expression for the Alfvén radius outside the capture radius (the gravitation is not significant). This can be easily verified by writing the expression for the accretion rate $\dot{M}_c$. However, we shall retain this form for the sake of convenience (see below). The second expression in (4.23) was first obtained by Lamb et al. (1973).

Let us now assume that the pressure is due to the relativistic wind. It follows from (4.20) that $P_{\mathrm{m}} \propto R^{-2}$, while the accretion pressure under the capture radius is $P_{\mathrm{a}} \propto R^{-5/2}$, i.e., increases more rapidly as we approach an accreting star. Consequently, a stable cavern can only be larger than the capture radius (Shvartsman 1970c):

$$R_{\mathrm{Sh}} = \left(\frac{8 \kappa_t \mu^2 (GM)^2 \omega^4}{\dot{M}_{\mathrm{c}} v_\infty^5 c^4} \right)^{1/2} , \quad R_{\mathrm{sh}} > R_{\mathrm{G}} . \tag{4.24}$$

As in the case of the Alfvén radius $R_{\mathrm{A}} > R_{\mathrm{G}}$, the gravitational constant appears only formally in this expression.

4.3 Stopping Radius in the Supercritical Case

Estimates for the stopping radius presented in Sect. 4.2 were obtained under the assumption that the energy liberated as a result of accretion does not exceed the Eddington limit (3.26). In other words, we neglected the reverse effect of radiation on the accretion flux parameters.

Following Lipunov (1982b), we shall now take this effect into consideration. Suppose that the accretion rate of matter captured by a neutron star is such that the luminosity at the stopping radius exceeds the Eddington luminosity:

$$\dot{M}_{\mathrm{c}} \frac{GM}{R_{\mathrm{st}}} \geqslant L_{\mathrm{Ed}} . \tag{4.25}$$

We shall base the analysis on the idea of self-regulation of the accretion rate, which was used to solve the problem of supercritical disk accretion (Sect. 3.7). We shall assume that the radiation sweeps away exactly that amount of matter so that the luminosity of the remaining matter at any radius is of the order of the Eddington energy:

$$\dot{M}(R) \frac{GM}{R} = L_{\mathrm{Ed}} .$$

This gives

$$\dot{M}(R) = \dot{M}_{\mathrm{c}} \frac{R}{R_{\mathrm{s}}} , \quad R_{\mathrm{s}} = \frac{\kappa}{4 \pi c} \dot{M}_{\mathrm{c}} .$$

Here, R_{s} has the same meaning as the spherization radius: the luminosity at this distance attains the Eddington limit for the first time. The dynamic pressure of the accretion plasma is found to depend on distance according to a new law:

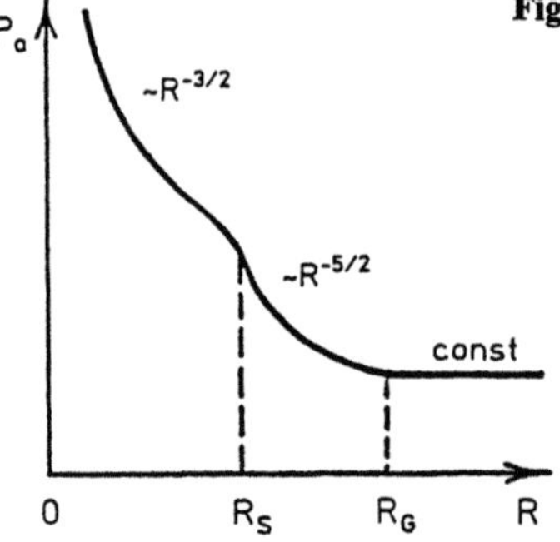

Fig. 4.4. Plasma pressure dependence at the supercritical stage

$$P_a \approx \varrho v^2 \approx \frac{\dot{M}(R)}{4\pi R^2} v = \frac{\dot{M}_c \sqrt{2GM_x}}{4\pi R_s} R^{-3/2} \; , \quad R \leqslant R_s \; . \tag{4.26}$$

It should be recalled that $P_a \propto R^{-5/2}$ in the subcritical regime. Figure 4.4 shows qualitatively the dependence of P_a on R. Using (4.20), we obtain from (4.17) the stopping radius in the supercritical case:

$$\left. \begin{aligned} R_A &= \left(\frac{\mu^2 \kappa}{8\pi c \sqrt{2GM_x}} \right)^{2/9} \\ R_{Sh} &= \left(\frac{\kappa_t \mu^2 \omega^4 \kappa}{4\pi c^5 \sqrt{2GM}} \right)^2 \end{aligned} \right\} \quad \dot{M}_c \geqslant \dot{M}_{cr} \; . \tag{4.27}$$

The critical accretion rate $\dot{M}_{cr}$ is defined by the boundary of the inequality (4.25):

$$\dot{M}_{cr} = \frac{4\pi c}{\kappa} R_{st} \; . \tag{4.28}$$

The dependence of the Alfvén radius on the accretion rate is quite informative. This dependence is such that the Alfvén radius (beyond the capture radius) decreases with increasing accretion rate as $\dot{M}^{-1/6}$, while it decreases as $\dot{M}^{-2/7}$ below the capture radius, and attains its lowest value (4.27) for the critical accretion rate $\dot{M}_c > \dot{M}_{cr}$, beyond which it is independent of the external conditions.

Let us also note that in the supercritical regime, the pressure of the accreting plasma increases more slowly (as $R^{-3/2}$) upon approaching the neutron star than the pressure of the relativistic wind (as R^{-2}) ejected by it. This means that in the supercritical case, a cavern may exist even below the capture radius. Of course, the estimates presented here are most suitable for the case of disk accretion. In actual practice, however, the supercritical regime seems to emerge mostly under these conditions. This statement can be simply explained as follows. The accretion rate is proportional to the square of the capture radius: $\dot{M}_c \propto R_G^2$. At the same time, an estimate of the angular momentum shows that this quantity is also proportional to R_G^2 [see (3.85)]. Hence the accretion disk is formed quite naturally under the conditions when the accretion rate is quite high.

4.4 The Effect of a Magnetic Field

Thus, we have obtained the characteristic distance at which the magnetic field pressure becomes comparable with the pressure of the gravitational forces. Now we know what we mean by a magnetized neutron star.

Apparently, the magnetic field of a star is significant only when the stopping radius exceeds the radius of the neutron star:

$$R_{st} > R_x \ . \tag{4.29}$$

We shall take the Alfvén radius R_A for R_{st}, since it is the smaller of the two quantities R_A and R_{Sh}. Using the above expressions (4.23) and (4.29) for the Alfvén radius, we can estimate the lowest magnetic field of a star which still influences the flow of matter:

$$\mu_{min} = \begin{cases} \left(\dfrac{\dot{M}_c v_\infty^5 R_x^6}{4 G^2 M_x^2} \right)^{1/2} , & R_A > R_G \\[2mm] (\dot{M}_c \sqrt{2 G M_x} R_x^{7/2})^{1/2} , & R_A \leqslant R_G \ . \end{cases} \left. \right\} \dot{M}_c \leqslant \dot{M}_{cr} \ , \\[4mm] \left(\dfrac{4 \pi c \sqrt{2 G M_x} R_x^{9/2}}{\kappa} \right)^{1/2} , \quad \dot{M}_c > \dot{M}_{cr} \ . \tag{4.30}$$

The case $R_A \leqslant R_G$, $\dot{M}_c \leqslant \dot{M}_{cr}$ is the one considered most frequently. The following numerical estimates have been obtained for the magnetic dipole moment and magnetic field strength at the surface of a star:

$$\mu_{min} \approx 10^{26} R_6^{7/4} \dot{M}_{17}^{1/2} m_x^{1/4} \ \mathrm{Oe \ cm}^3 \ , \tag{4.31}$$
$$B_{min} \approx 10^7 \dot{M}_{17}^{1/2} R_6^{5/4} m_x^{1/4} \ \mathrm{Oe} \ .$$

Most of the neutron stars being observed at present have magnetic fields $\sim 10^{12}$ Oe and dipole moments $\sim 10^{30}$ Oe cm^3. Hence it is obvious that we must take into consideration the magnetic fields of neutron stars.

4.5 Gravimagnetic Parameter

Glancing at the expression for the stopping radius in the subcritical regime ($\dot{M}_c \leqslant \dot{M}_{cr}$), we observe that the magnetic dipole moment μ and the accretion rate $\dot{M}_c$ always appear in the same combination. Let us express this combination through y:

$$y = \frac{\dot{M}_c}{\mu^2} \ . \tag{4.32}$$

The existence of such a universal ratio was first noticed by Davis and Pringle (1981). The parameter y characterizes the ratio between the gravitational and magnetic "properties" of a star, and will, therefore, be called the gravimagnetic parameter. Two neutron stars having quite different magnetic fields and external conditions but identical gravimagnetic parameters have similar magnetospheres. Of course, this statement is valid as long as the accretion rate is quite low ($\dot{M}_c \leq \dot{M}_{cr}$). Otherwise, the flux of matter near the stopping radius no longer depends on the accretion rate at a large distance.

Let us determine the qualitative dependence of the stopping radius on the gravimagnetic parameter [formulas (4.23) and (4.24)]. The Alfvén radius has the following dependence:

$$R_A \propto \begin{cases} y^{-1/6} & \text{for} \quad R_A > R_G \ , \\ y^{-2/7} & \text{for} \quad R_A \leqslant R_G \ , \end{cases} \tag{4.33}$$

while the Shvartsman radius behaves as

$$R_{Sh} \propto y^{-1/2} \ . \tag{4.34}$$

Qualitatively, this dependence can be easily explained. The gravimagnetic parameter has a large value when the accretion rate is high or the magnetic field is weak. As the gravimagnetic parameter increases, so does the pressure P_a of the accreting plasma, while the magnetic field pressure decreases. This leads to a decrease in the value of the stopping radius.

The gravimagnetic parameter appears in many relations determining the interaction of a magnetized star with the surrounding medium (see below). For example, the condition (4.29) which stipulates whether magnetic fields must be considered assumes an especially simple form if we write it as an inequality in the gravimagnetic parameter. The magnetic fields are important if the parameter y is smaller than a certain critical value (subcritical regime)

$$y \leqslant y_{max} = \begin{cases} \dfrac{2GM_x}{R_x^3 v_\infty^{5/2}} \ , & y < y_G \ , \\[2ex] (\sqrt{2GM_x R_x^{7/2}})^{-1} \ , & y \geqslant y_G \ , \end{cases} \tag{4.35}$$

where y_G is defined by the condition $R_A = R_G$:

$$y_G = \frac{v_\infty^7}{(2GM_x)^4} \ . \tag{4.36}$$

Accordingly, we arrive at the critical value of the gravimagnetic parameter when the luminosity at the stopping radius becomes equal to the Eddington limit. Substituting (4.27) into (4.28), we obtain

$$y'_{cr} = \frac{4\pi c}{\kappa} \frac{R_A}{\mu^2} \;,$$

$$y''_{cr} = \frac{4\pi c}{\kappa} \frac{R_{sh}}{\mu^2} \;.$$

(4.37)

For $y < y'_{cr}$ and $y < y''_{cr}$, the luminosity at the Alfvén radius and Shvartsman radius, respectively, is lower than the Eddington limit.

4.6 Corotation Radius

The corotation radius is an important characteristic of a rotating star. Suppose that an accreting plasma penetrates the light cylinder and is stopped by the magnetic field at a certain distance R_{st} given by the balance between the static magnetic field pressure and the plasma pressure. What happens next? The answer to this question depends to a large extent on the rotational velocity of the neutron star.

Suppose that the plasma penetrates and is "frozen" in the magnetic field of the neutron star. The magnetic field will drag the plasma and force it to rotate as a solid with the angular velocity of the star. Will the plasma fall onto the surface of the star? Obviously, matter will fall on the star surface only if its rotational velocity is smaller than the Keplerian velocity at the given distance R_{st}:

$$\omega R_{st} < \sqrt{GM_x/R_{st}} \;.$$

If this in equality is not satisfied, a centrifugal barrier emerges and the rapidly rotating magnetic field impedes the accretion of matter (Shvartsman 1970a; Pringle, Rees 1972; Lamb et al. 1973; Davidson, Ostriker 1973; Illarionov, Syunyaev 1975). The last authors assume that if $\omega R_{st} \gg \sqrt{GM/R_{st}}$, the magnetic field throws back the plasma beyond the capture radius. They called this effect the "propeller" regime. In fact, matter may not be shed away (Lipunov 1980) but it is important to note that stationary accretion is also not possible.

Thus, two basically different situations separated by the equality

$$\omega R = \sqrt{\frac{GM_x}{R}} \;,$$

(4.38)

may be realized, whence we obtain an expression for the corotation radius R_c:

$$R_c = (GM_x/\omega^2)^{1/3} \approx 2.8 \times 10^8 \, m_x^{1/3} p^{2/3} \; \text{cm} \;,$$

(4.39)

p being the period of rotation of the star in seconds.

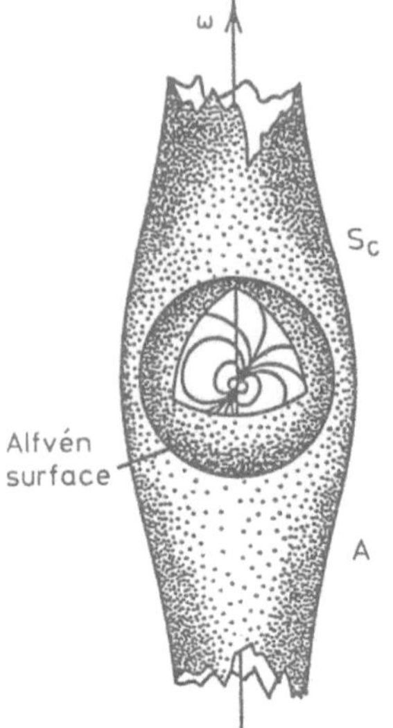

Fig. 4.5. Alfvén surface inside the surface S_c; rotation does not impede accretion

If $R_{st} < R_c$, rotation does not significantly affect the accretion possibility. Conversely, stationary accretion is not possible for $R_{st} > R_c$. There exists a surface S_c beyond which the velocity of particles is too high to let them fall on the surface of the star:

$$\omega R \cos \theta = \sqrt{\frac{GM_x}{R}} \qquad (4.40)$$

(Fig. 4.5). Using the definition of the corotation radius, we obtain the equation for the generator of the rotational surface (4.40):

$$R = R_c \cos^{-2/3} \theta \; . \qquad (4.41)$$

In order to illustrate how a rotating magnetic field locks accretion, let us consider the following idealized situation. Suppose that the surface at which an accreting plasma is stopped has the form of a sphere of radius R_{st}. We assume that the particles of the accretion flow do not interact with one another and, upon approaching the stopping surface (magnetosphere boundary), get stuck to it and acquire the rotational velocity of the solid. The particles are incident purely radially. In this case, the fraction of particles whose further fall is forbidden is proportional to the area $\Delta S'$ of the spherical segment confined within the surface S_c (Fig. 4.5):

$$\dot{M}_{out} = \left(1 - \frac{2\Delta S}{4\pi R_{st}^2} \right) \dot{M}_c \equiv \dot{M}_c - \dot{M} \; . \qquad (4.42)$$

The stopping sphere and the surface S_c intersect at

$$R_{st} = R_c \cos^{-2/3} \theta_{cr} \; .$$

Simple geometrical considerations lead to the following relations:

$$\dot{M}_{out} = \begin{cases} \dot{M}_c\sqrt{1-(R_c/R_{st})^3} \; , & R_c \leqslant R_{st} \; , \\ 0 \; , & R_c > R_{st} \; . \end{cases} \qquad (4.43)$$

We introduce the critical period p_A such that for $p = p_A$ the stopping radius is equal to the corotation radius: $R_{st} = R_c$. In this case, obviously, we can write

$$\dot{M}_{out} = \begin{cases} \dot{M}_c\sqrt{1-(p/p_A)^2} \; , & p \leqslant p_A \; , \\ 0 \; , & p > p_A \; . \end{cases} \qquad (4.44)$$

The amount of substance falling on the surface of the magnetized star is given by (Lipunov, Shakura 1976)

$$\dot{M} = \begin{cases} \dot{M}_c(1-\sqrt{1-(p/p_A)^2}) \; , & p \leqslant p_A \; , \\ \dot{M}_c \; , & p > p_A \; . \end{cases} \qquad (4.45)$$

In order to remain within the realms of reality, let us estimate the characteristics introduced by us for the case of the X-ray pulsar Her X-1. The period of its pulsation is $p = 1.24$ s. The luminosity of the pulsar is $L_x \simeq 10^{37}$ erg/s. For an accretion efficiency $\eta \simeq 0.1$, we obtain the estimate $\dot{M} \simeq 10^{17}$ g/s for the accretion rate. Observations of the binary system show that the mass of the neutron star is $\simeq 1.5\,M_\odot$. We assume that its magnetic dipole moment has the "standard" value $\mu = 10^{30}$ Oe cm^3. The flow of matter is this system occurs through the inner Lagrangian point, and hence the capture radius is of the order of the semi-major axis of the binary system: $R_G \approx a \simeq 10^{11}$ cm. It can easily be verified that the stopping radius is smaller than the capture radius. Hence it is defined by the second expression for the Alfvén radius in (4.23). Table 4.1 shows the characteristic quantities for the Her X-1 pulsar. The Alfvén radius for this pulsar is $\sim 10^8$ cm. It can be seen from (4.39) that the corota-

Table 4.1. Estimated dimensions of the X-ray pulsar Her X-1

Parameter	Notation	Estimate [cm]
Radius of light cylinder	R_l	$\sim 6 \times 10^9$
Capture radius	R_G	$\sim 10^{11}$
Alfvén radius	R_A	$\sim 10^8$
Corotation radius	R_c	$\sim 3 \times 10^8$
Radius of the star	R_x	$\sim 10^6$

tion radius is 3×10^8 cm. Hence for the pulsar Her X-1, $R_{st} \approx R_c$ and $p \approx p_A$. This astonishing coincidence, which will be explained later, shows that the rotational effect is indeed significant. It should also be emphasized that the Alfvén radius is hundreds of times larger than the radius of the star. Long before the matter falls on the surface of a neutron star, its motion begins to be controlled completely by the magnetic field.

4.7 Nomenclature

The interaction of a magnetized neutron star with the surrounding plasma depends to a large extent on the relation between four characteristic distances, that is, the stopping radius R_{st}, the gravitational capture radius R_G, the light cylinder radius R_l, and the corotation radius R_c. The difference in interaction regimes is so significant that neutron stars behave entirely differently in different regimes. Hence the classification of interaction regimes may well mean the classification of neutron stars. The classification, notation, and terminology described below and summarized in Table 4.2 and based on the works of Lipunov (1982a, 1984a, 1987a), and Kornilov and Lipunov (1983a).

Naturally, not all combinations of the characteristic parameters can be realized. For example, the inequality $R_l > R_c$ is not possible in principle. On the other hand, some combinations require anomalously large or small parameters of neutron stars and are not realistic. For the same internal and external conditions, the same neutron star may gradually pass through several interaction regimes. Such a process will be termed the evolution of a neutron star.

Let us describe the classification by considering an idealized scenario of the evolution of a neutron star. Suppose that the parameters ϱ_∞, v_∞, and $\dot{M}_c$ of

Table 4.2. Classification of neutron stars

Notation	Name	Relation between characteristic distance s	Accretion rate
E	Ejector	$R_{st} > \max\{R_G, R\}$	$\dot{M}_c \leqslant \dot{M}_{cr}$
P	Propeller	$R_c < R_{st} \leqslant \max\{R_G, R\}$	$\dot{M}_c \leqslant \dot{M}_{cr}$
A	Accretor	$R_{st} \leqslant \min\{R_c, R_G\}$	$\dot{M}_c \leqslant \dot{M}_{cr}$
G	Georotator	$R_G < R_{st} \leqslant R_c$	$\dot{M}_c \leqslant \dot{M}_{cr}$
M	Magnetor	$R_{st} > a, R_c > a$	$\dot{M}_c \leqslant \dot{M}_{cr}$
SE	Superejector	$R_{st} > R_l$	$\dot{M}_c > \dot{M}_{cr}$
SP	Superpropeller	$R_c < R_{st} \leqslant R_l$	$\dot{M}_c > \dot{M}_{cr}$
SA	Superaccretor	$R_{st} \leqslant R_c, R_{st} \leqslant R_G$	$\dot{M}_c > \dot{M}_{cr}$
BH	Black hole	$M_x > M_{OV}$	Arbitrary

the surrounding medium remain unchanged. We shall also assume that the magnetic moment of the neutron star remains constant. Let the potential accretion rate $\dot{M}_c$ at the beginning be not too high, so that the reverse effect of radiation pressure can be neglected: $\dot{M}_c \leq \dot{M}_{cr}$. Moreover, we assume that the star initially rotates at such a high speed that it is a powerful source of relativistic wind.

Ejecting Neutron Stars (Ejectors). We shall call a neutron star an ejecting star (or simply an ejector E) if the pressure of the electromagnetic radiation and ejected relativistic particles is so high that the surrounding matter is swept away beyond the capture radius or the radius of the light cylinder (if $R_l > R_G$). Figure 4.6 shows the qualitative configuration of the light cylinder and the stopping surface (cavern) of the ejector. An illustrative example of an ejector is a radiopulsar which is a single neutron star (Chap. 8).

In the ejection regime, a neutron star is spun down and the power of the relativistic wind is reduced: $L_m \propto \omega^4 \propto p^{-4}$. The Shvartsman radius decreases and the cavern collapses at a certain instant. The condition of collapse depends on the relation between the gravitational capture radius R_G and the radius R_l of the light cylinder. The situation when $R_G > R_l$ is astrophysically more plausible. However, the situation with $R_G < R_l$ also cannot be ruled out. In this case, the gravitation is not significant and the condition for the termination of the ejector stage is naturally $R_{st} < R_l$, when a new stage sets in.

Propeller (P) Regime. After the termination of the ejector stage, the propeller stage sets in under quite general conditions, when the accretion of matter at the surface of a magnetized star is hampered by a rapidly rotating magnetic field. It will be shown below that for a sufficiently large value of the gravimagnetic parameter the propeller regime may not be realized at all. In this regime, the stopping (Alfvén) surface contains in it the surface $S_c : R_{st} < R_c$ (Fig. 4.7). Owing to a finite magnetic viscosity, the rotational angular momentum is transferred to the accreting matter and the neutron star is spun down. The necessary (but, apparently, not sufficient) condition for the onset of accretion is that the velocity at the Alfvén surface be lower than the Keplerian velocity: $R_{st} < R_c$. So far, the propeller stage remains one of the least investigated phenomena. It has been borne out by a number of investigations that the condition $R_{st} < R_c$ is not a sufficient condition for the onset of accretion (Davis, Pringle 1981). However, it is clear that sooner or later the neutron star is retarded to such an extent that the rotational effects are no longer important and the accretion stage sets in.

Accretors. At the accretion stage, the stopping radius is estimated to high accuracy by (4.23). It must be smaller than the corotation radius (Fig. 4.5). Such magnetized neutron stars in the accretion stage will be called accretors. This is the most thoroughly investigated regime of the interaction of a neutron star with accreting plasma. Examples of such systems are the X-ray pulsars and X-ray bursters.

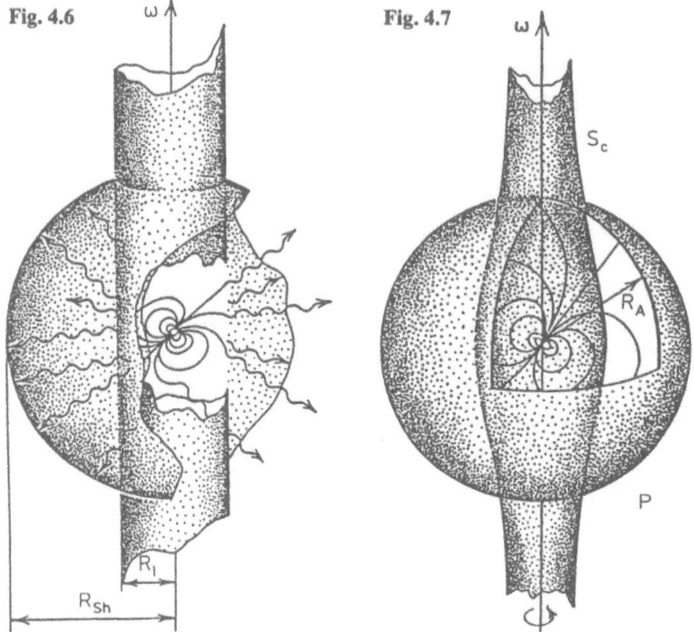

Fig. 4.6. Stopping surface embracing the light cylinder (ejector)

Fig. 4.7. The propeller regime

Georotators. Suppose that a star begins to rotate so slowly that it cannot impede the accretion of plasma, i.e., all the conditions mentioned in the last paragraph are satisfied. Still, matter cannot fall on the surfce of the neutron star if the Alfvén radius is larger than the capture radius (Illarionov, Syunyaev 1975; Lipunov 1982c). This means that the attractive force of the star at the Alfvén surface is not significant. A similar situation is observed in the interaction of solar wind with the magnetic field of the Earth. It should be recalled that the velocity of the solar wind is ~ 300 km/s, while the parabolic velocity (escape velocity) for the Earth is 11 km/s. The gravitation of the Earth has almost no effect on the plasma in the solar wind, $\simeq (11/300)^2 \simeq 10^{-5}$. The plasma mainly flows around the Earth's magnetosphere and recedes to infinity (Fig. 4.8). This analogy explains the term "georotator" used for this stage.

Consequently, the condition supplementary to weak rotation will be $R_A > R_G$. This leads to the relation [see (4.36)]

$$y < y_G = \frac{v_\infty^7}{(2GM)^4} \tag{4.46}$$

or

$$\mu > \mu_G \approx 3 \times 10^{33} \, m^2 \, v_8^{-7/2} \dot{M}_{17}^{1/2} \ . \tag{4.47}$$

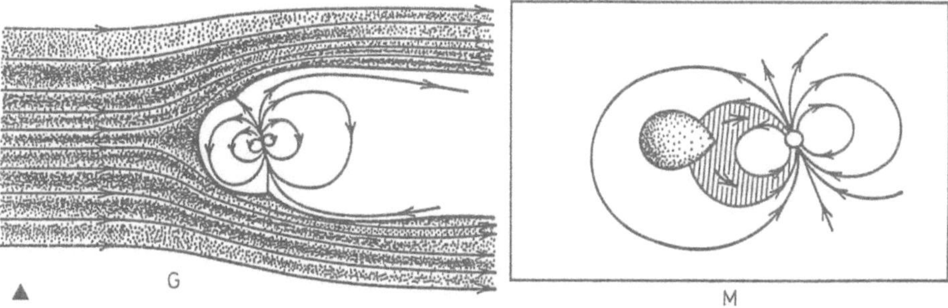

Fig. 4.8. Georotator

Fig. 4.9. A binary magnetic system (magnetor)

Clearly, a georotator must either have a high magnetic field, or must be located in a highly rarefied medium.

Magnetic Binary Systems (Magnetors). So far, we have considered only an isolated neutron star. To be more precise, the above regimes were analyzed irrespective of whether or not the neutron star forms a part of a binary system. However, binary systems also have another characteristic distance, the semi-major axis a of the system. It may so happen that the Alfrén radius, which is formally defined by (4.23), is larger than a: $R_A > a$. The normal companion star will then lie within the Alfrén surface of the magnetized star (Fig. 4.9). Such magnetized stars in magnetic binary systems are called magnetors (M). Such a regime was first considered by Mitrofanov et al. (1977) for white dwarfs in close binary systems called polars. For neutron stars, this type, M, may be realized only under extreme conditions when there is practically no matter in binary systems.

Supercritical Interaction Regimes. So far, we have assumed that the luminosity at the stopping surface is lower than the Eddington limit. This is fully justified for G and M regimes since gravitation is not important for them. For the types, E, P, and A, however, this is not always true. The critical accretion rate at which the Eddington limit is reached does not differ in any way from other accretion rates (3.39):

$$\dot{M}_{cr} \approx 10^{18} R_6 \, \text{g/s} \approx 1.5 \times 10^{-8} \, R_6 M_\odot / \text{year} \ .$$

However, in the case of mass exchange in massive binary systems the accretion rate may assume values tens of thousands times larger than $\dot{M}_{cr}$.

It was shown in Chap. 3 that the main part of matter in the dynamic model of supercritical accretion forms an outflowing flux covering the neutron star by a completely opaque shell. The astrophysical manifestations of the neutron star in such a regime are quite extraordinary (Chap. 9). Hence the following three additional types are distinguished, depending on the relation between

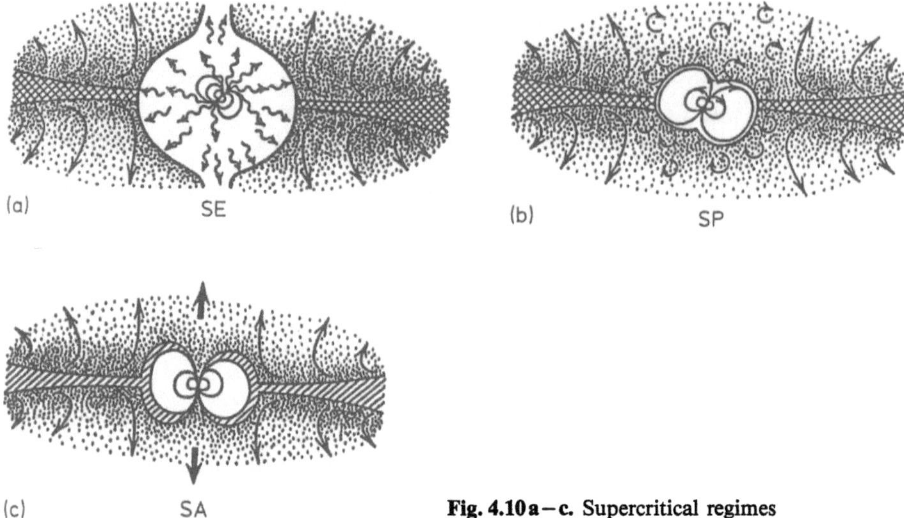

Fig. 4.10a–c. Supercritical regimes

characteristic dimensions (Table 4.2): superejector (SE) (Fig. 4.10a), super-propeller (SP) (Fig. 4.10b), and superaccretor (SA) (Fig. 4.10c).

Collapse into a Black Hole (BH). In the process of accretion onto a neutron star, its mass increases monotonically (especially rapidly in the supercritical accretion regime). If the mass of the neutron star exceeds the Oppenheimer-Volkoff limit M_{OV}, the star will inevitably collapse into a black hole. We believe that this is the undisputable and fully realistic way of creation of black holes in the Galaxy (Kornilov, Lipunov 1983b).

4.8 Critical Periods. The $p-y$ and $p-L$ Diagrams

The above classification was based on the relation between characteristic radii, i.e., quantities which cannot be observed directly. This drawback can be removed if we note that the characteristic values of the light cylinder radius R_l, Shvartsman radius R_{Sh} and corotation radius R_c are functions of a quantity that can be measured quite accurately in observations, viz., the rotational frequency or the rotational period p. Hence the above classification can be reformulated in the form of inequalities in terms of the rotational period of a magnetized star.

We introduce two critical periods p_E and p_A such that each of the following inequalities describes a certain type of neutron star:

$$p < p_E \rightarrow E \text{ or SE} ;$$
$$p_E \leqslant p \leqslant p_A \rightarrow P \text{ or SP} ; \tag{4.48}$$
$$p \geqslant \dot{p}_A \rightarrow A, \text{ SA, G or M} .$$

The values of p_E and p_A can be determined in the first approximation with the help of inequalities given in Table 4.2 by using the above estimates of the characteristic radii. Naturally, this will be just a first approximation which does not take into account the reverse effect of emergence of the stopping surface on the parameters of the electromagnetic field of a neutron star and the accreting flux.

We put $R_{Sh} = \max\{R_G, R_l\}$ for $\dot{M}_c \leqslant \dot{M}_{cr}$ and $R_{st} = R_l$ for $\dot{M}_c > \dot{M}_{cr}$. Using formulas (4.27) and (4.24), we obtain

$$p_E = \begin{cases} 2\pi \left(\dfrac{4\kappa_t}{v_\infty c^4} \right)^{1/4} (\mu^2/\dot{M}_c)^{1/4} & \text{for} \quad \begin{cases} \dot{M}_c \leqslant \dot{M}_{cr}\,, \\ R_G < R_l\,, \end{cases} \\[4mm] 2\pi \left[\dfrac{8\kappa_t(GM_x)^2}{v_\infty^5 c^6} \right]^{1/6} (\mu^2/\dot{M}_c)^{1/6} & \text{for} \quad \begin{cases} \dot{M}_c \leqslant \dot{M}_{cr}\,, \\ R_G \geqslant R_l \end{cases} \\[4mm] 2\pi \left(\dfrac{\kappa_t \kappa}{4\pi^2 c^{11} GM_x} \right)^{1/9} \mu^{4/9} & \text{for} \quad \dot{M}_c > \dot{M}_{cr}\,. \end{cases} \qquad (4.49)$$

Putting $R_A = R_c$, and using formulas (4.27) and (4.25), we find that

$$p_A = \begin{cases} 2\pi \left(\dfrac{2}{v_\infty^5} \right)^{1/4} \left(\dfrac{\mu^2}{\dot{M}_c} \right)^{1/4} & \text{for} \quad \begin{cases} R_A > R_G\,, \\ \dot{M}_c \leqslant \dot{M}_{cr}\,, \end{cases} \\[4mm] 2^{11/14}\pi(GM_x)^{-5/7} \left(\dfrac{\mu^2}{\dot{M}_c} \right)^{3/7} & \text{for} \quad \begin{cases} R_A \leqslant R_G\,, \\ \dot{M}_c \leqslant \dot{M}_{cr}\,, \end{cases} \\[4mm] 2\pi \left[\dfrac{\kappa}{8\sqrt{2}\pi c(GM_x)^2} \right]^{1/3} \mu^{2/3} & \text{for} \quad \dot{M}_c > \dot{M}_{cr}\,. \end{cases} \qquad (4.50)$$

These formulas show that the critical periods p_E and p_A mainly depend on two parameters, viz., the accretion rate $\dot{M}_c$ of matter potentially falling under the capture radius, and the magnetic dipole moment μ.

Hence the interaction of a neutron star with the surrounding medium is mainly determined by three parameters, two of which (p and μ) characterize the electrodynamic interaction while the third ($\dot{M}_c$) describes the gravitational interaction. Instead of $\dot{M}_c$, we introduce the potential luminosity

$$L \equiv \dot{M}_c \frac{GM_x}{R_x}\,. \qquad (4.51)$$

The physical meaning of the potential luminosity L is quite simple: the accreting star would have this luminosity if the matter formally falling on the gravitational capture cross section were to fall on the star surface.

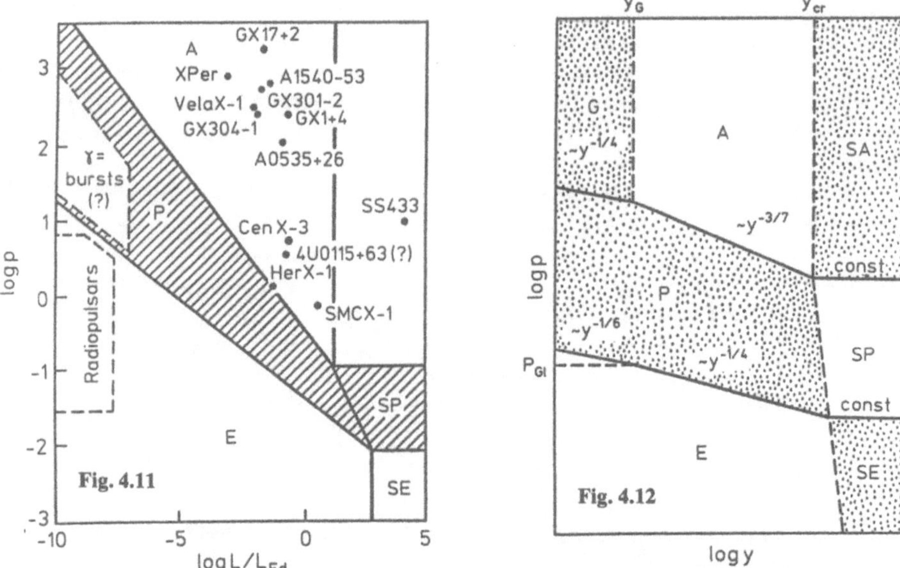

Fig. 4.11. $p-L$ diagram for neutron stars. The boundary lines are drawn for a magnetic dipole moment $10^{30}\,\mathrm{Oe\,cm^3}$

Fig. 4.12. $p-y$ diagram

Treating the magnetic dipole moment as a parameter, we find that an overwhelming majority of the neutron star states can be shown on a "$p-L$" diagram (Lipunov 1982a). The quantity L is also convenient because it can be observed at the accretion stage.

To verify the correctness of these arguments, we construct the $p-L$ diagram for neutron stars having standard parameters: $\mu_{30} = m = v_7 = 1$, where $v_7 = v_\infty/10^7\,\mathrm{cm/s}$. We plot two types of observed manifestations of neutron stars on this diagram, X-ray pulsars and radiopulsars (Fig. 4.11).

We can write (4.49, 50) in a form convenient for making estimates:

$$p_E \approx \begin{cases} 0.42 \times v_7^{-1/4}\mu_{30}^{1/2}L_{38}^{-1/4}\,\mathrm{s}\;, & \dot{M}_c \leqslant \dot{M}_{cr} \;\; \text{and} \;\; p > p_{Gl}\;, \\ 1.8 \times v_7^{-5/6}m^{1/3}\mu_{30}^{1/3}L_{38}^{-1/6}\,\mathrm{s}\;, & \dot{M}_c \leqslant \dot{M}_{cr} \;\; \text{and} \;\; p \leqslant p_{Gl}\;, \\ 1.4 \times 10^{-2}m^{-1/9}\mu_{30}^{4/9}\,\mathrm{s}\;, & \dot{M}_c > \dot{M}_{cr}\;. \end{cases} \qquad (4.52)$$

$$p_A \approx \begin{cases} 4 \times 10^2 \times v_7^{-5/4}\mu_{30}^{1/2}L_{38}^{-1/4}\,\mathrm{s} & \text{for} \;\; R_A > R_G \;\; \text{and} \;\; \dot{M}_c \leqslant \dot{M}_{cr}\;, \\ 1.2 \times m^{-5/7}\mu_{30}^{6/7}L_{38}^{-3/7}\,\mathrm{s} & \text{for} \;\; R_A \leqslant R_G \;\; \text{and} \;\; \dot{M}_c \leqslant \dot{M}_{cr}\;, \\ 0.17 \times m^{-2/3}\mu_{30}^{2/3}\,\mathrm{s} & \text{for} \;\; \dot{M}_c > \dot{M}_{cr}\;. \end{cases} \qquad (4.53)$$

Here we had to introduce a new critical period p_{Gl} from the condition $R_G = R_l$:

$$p_{Gl} = \frac{4\pi G M_x}{v_\infty^2 c} \approx 500\, m_x v_7^{-2}\, \text{s} \; . \tag{4.54}$$

The positions of X-ray pulsars on the $p-L$ diagram are given according to the data of Table 6.1. The quantities L cannot be observed directly for radiopulsars, but we can estimate their value. Radiopulsars are single neutron stars. The potential accretion rate onto a single neutron star cannot be much higher than the value estimated by using the Bondi-Hoyle-Littleton formula (3.39) for parameters $\xi = 1$, $\varrho_\infty = 10^{-24}$ g/cm^3 and $v_\infty^2 + a_\infty^2 = 10^6$ cm/s, $\dot{M}_c \lesssim 7 \times 10^{10}$ g/s, and $L \lesssim 7 \times 10^{30}$ erg/s.

As expected, radiopulsars fall in the range of ejectors (E) and X-ray pulsars in the range of accretors (A). Note that Fig. 4.11 does not show the position of the superfast X-ray pulsar A 0538-66 which has a period $p \simeq 0.067$ s. As a matter of fact, the superfast pulsar has a magnetic field much weaker than 10^{12} Oe (Chap. 6). This example shows that it would be good to eliminate the superfluous parameter μ. This can be done, but only in the subcritical regime.

We use the gravimagnetic parameter introduced by Davies and Pringle (1981): $y = \dot{M}_c/\mu^2$ (Sect. 4.5). Figure 4.12 shows the $p-y$ diagram for neutron stars not accounting for their magnetic field [see (4.49) and (4.50)].

Looking at the $p-L$ and $p-y$ diagrams, we observe that as the luminosity or the gravimagnetic parameter increases, the critical lines converge and the propeller region vanishes above a critical value of p, obtained from the condition $p_E = p_A$.

5. Boundaries.
Magnetospheres of Slowly Rotating Neutron Stars

In Chap. 4, we singled out the main types of interaction of a magnetized neutron star with an accreting plasma. The effect of magnetic fields begins to be felt at distances from the neutron star that are large in comparison with its radius. The magnetic field stops the accreting plasma at a position where the pressures of the field and the plasma become of the same order. An expression was obtained for the characteristic distance, the stopping radius R_{st}, at which the plasma flow is stopped. For slowly rotating stars, this quantity is estimated by the Alfvén radius R_{A}.

We shall now endeavor to take into account the reverse effect of the accreting flux on the magnetic field structure, as well as the effect of the magnetic field on the structure of the accreting flux. In this chapter, we shall consider the simplest case in which the rotation of the neutron star can be neglected. It was shown in Chap. 4, that according to the condition of a "slow" rotation, the velocity in the Alfvén zone corresponding to a rigid body must be much lower than the Keplerian velocity. In terms of characteristic radii, this condition can be written $R_{\mathrm{A}} \ll R_{\mathrm{c}}$.

The nature of accretion in the Alfvén zone has been determined more precisely as follows (we shall use the same sequence of analysis in this chapter): the structural variation of the magnetic field is obtained first and then the solution is used for analyzing the reverse effect of magnetic field on an accreting plasma.

For the present, we shall not concern ourselves with the stability of the plasma-field boundary. These questions will be considered in Chap. 6 (and the Appendix).

The shape of the Alfvén surface of a neutron star was first investigated by Amnuel' and Guseinov (1972), while a qualitative analysis was carried out by Shvartsman (1970a). First detailed calculations (which were, however, partly erroneous) were made by Inoue and Hoshi (1975). A correct solution was obtained by Arons and Lea (1976), Elsner and Lamb (1976), and Basko (1977). These first works were basically devoted to the shape of the magnetosphere of a magnetized star. Similar problems were considered earlier in the beginning of the sixties by geophysicists (Akasofu, Chapman 1974). A general analysis of the problem of interaction of an ideal conducting flow with a magnetic field was first carried out by Maxwell (1873).

5.1 Physical Conditions in the Alfvén Zone

Before simplifying the problem to a rigorous mathematical formulation, we must study the physical situation near a magnetized neutron star. Let us estimate the parameters of accreting flow at a distance of the order of Alfvén radius (4.23):

$$R_A \approx 10^8 \mu_{30}^{4/7} \dot{M}_{18}^{-2/7} m_x^{1/7} \text{ cm} \ . \tag{5.1}$$

We used for estimates the formulas of radial accretion in the gasdynamic approximation. It was shown in Sect. 3.2 that at distance of the order of the capture radius, weak magnetic fields make the plasma a continuous medium.

Let us consider some examples of a radially accreting plasma. Suppose that $R_A \ll R_G$. The density of matter in the Alfvén zone is

$$\varrho \approx \varrho_{ff} = \frac{\dot{M}}{4\pi R^2 v_{ff}} \approx 5 \times 10^{-8} R_8^{-3/2} \dot{M}_{18} m_x^{-1/2} \text{ g/cm}^3 \ , \tag{5.2}$$

where $R_8 = R_A/10^8$ cm, and v_{ff} is the free-fall velocity:

$$v_{ff} = \sqrt{2GM_x/R} \approx 1.6 \times 10^9 R_8^{-1/2} m_x^{1/2} \text{ cm/s} \ . \tag{5.3}$$

The characteristic time scale of radial fall is

$$t_r = \frac{R}{v_{ff}} \approx 6 \times 10^{-2} R_8^{3/2} m_x^{-1/2} \text{ s.} \tag{5.4}$$

The accreting flow is stopped by the magnetic field of the neutron star. This process is accompanied by the formation of a shock wave (most probably, a collisionless wave). Since the magnetic pressure increases more rapidly than the pressure of the accreting plasma, we can expect the formation of a magnetosphere, that is, a surface separating the plasma from the magnetic field. Obviously, the characteristic size of the magnetosphere is $R_m \approx R_A$. Assuming that the entire energy of ordered motion is transformed into heat when the accreting matter is stopped, the temperature will be close to the value T_{ff}:

$$T_{ff} = \frac{GMm_i}{kR} \approx 1.6 \times 10^{10} A R_8^{-1} m_x \text{ K} \ , \tag{5.5}$$

where A is the atomic mass.

The physical parameters characterizing the plasma and the magnetic field near the magnetosphere for a neutron star with $\mu_{30} = 1$ and $\dot{M}_{18} = 0.1$, presented in Table 5.1, have been taken from Aly (1985). In the case of disk accretion, we assume the ratio $H/R = 0.01$ for the ratio of the disk thickness to its radius. The density and temperature can be roughly estimated from the formulas

Table 5.1. Physical conditions in the Alfvén zone

Parameter	Notation	Magnetopause	Disk	Accretion flux
Distance from the star	$R\,(\mathrm{cm})$	10^8	10^8	10^8
Transition layer thickness	$\delta_\mathrm{m}\,(\mathrm{cm})$	10^7	10^6	10^7
Radial fall time	$t_r\,(\mathrm{s})$	4×10^{-2}	4×10^{-2}	3×10^{-2}
Density	$n\,(\mathrm{cm}^{-3})$	10^{15}	10^{20}	10^{16}
Temperature	$T\,(\mathrm{K})$	10^9	10^6	10^6
Field strength	$B\,(\mathrm{GS})$	10^6	10^6	10^6
Debye radius	$l_\mathrm{D}\,(\mathrm{cm})$	1.2×10^{-2}	1.2×10^{-6}	1.2×10^{-4}
ζ	$\zeta = 24\pi n l_\mathrm{D}^3$	1.3×10^{11}	1.3×10^4	1.3×10^6
Mean free path	$l_\mathrm{ee}\,(\mathrm{cm})$	10^7	4×10^{-4}	3
Plasma frequency	$\omega_\mathrm{p} = (4\pi n e^2/m_\mathrm{e})^{1/2}\left(\dfrac{\mathrm{rad}}{\mathrm{s}}\right)$	2×10^{11}	6×10^{14}	6×10^{12}
Cyclotron frequency of electrons	$\omega_\mathrm{ce} = eB/m_\mathrm{e}c\,(\mathrm{s}^{-1})$	1.8×10^{13}	2×10^{13}	2×10^{13}
Cyclotron frequency of protons	$\omega_\mathrm{cp} = eB/m_\mathrm{p}c\,(\mathrm{s}^{-1})$	10^{10}	10^{10}	10^{10}
$e-p$ collision frequency	$\nu_\mathrm{ep} = 6\omega_\mathrm{p}\left(\dfrac{\log\zeta}{\zeta}\right)(\mathrm{s}^{-1})$	2×10^3	2×10^{12}	4×10^8
$p-p$ collision frequency	$\nu_\mathrm{pp}\,(\mathrm{s}^{-1})$	35	4×10^{10}	6×10^{10}
Electrical conductivity	$\lambda_\mathrm{e} = \omega_\mathrm{p}^2/4\pi\nu_\mathrm{ep}\,(\mathrm{s}^{-1})$	10^{18}	10^{16}	8×10^{15}
Magnetic field diffusion time	$t_\mathrm{B} = 4\pi\lambda_\mathrm{e}\delta_\mathrm{m}^2/c^2\,(\mathrm{s})$	2×10^{12}	10^8	10^{10}
Kinematic viscosity	$\nu = (3k_\mathrm{B}T)^{5/2}e^4 m_\mathrm{p}^{1/2}\log\zeta$	10^{15}	10^3	10^7

(CGS)

$$v_r \approx \alpha \left(\frac{H}{R}\right)^2 v_\varphi \; ,$$

$$a \approx \frac{H}{R} v_\varphi \; , \tag{5.6}$$

whence we obtain the following expression for the density of matter in a disk:

$$\varrho_{\mathrm{d}} \approx \alpha^{-1} \frac{R}{H} \varrho_{\mathrm{ff}} \; . \tag{5.7}$$

The temperature in the disk is

$$T_{\mathrm{d}} \approx \left(\frac{H}{R}\right)^2 T_{\mathrm{ff}} \; . \tag{5.8}$$

The structure of the unperturbed magnetosphere (Fig. 5.1) looks like this: outside the magnetosphere, the neutron star does not have a magnetic field, and there is no plasma inside the magnetosphere. In the transition layer separating the plasma and the magnetosphere, electric currents flow and screen the magnetic field of the neutron star. If we neglect the magnetohydrodynamic instabilities which may facilitate the entry of the plasma into the magnetosphere, the thickness of the transition layer is found to be of the order of the Larmor radius of the ion. It can be seen from Table 5.1 that in this approximation, its thickness is much smaller than the size of the magnetosphere:

$$\delta_{\mathrm{m}} \ll R_{\mathrm{A}} \approx R_{\mathrm{m}} \; .$$

In this case, the condition of equilibrium of the magnetosphere boundary can be written in the form

$$\left(\frac{B^2}{8\pi}\right)_{\mathrm{in}} + P_{\mathrm{in}} = \left(\frac{B^2}{8\pi}\right)_{\mathrm{out}} + P_{\mathrm{out}} \; . \tag{5.9}$$

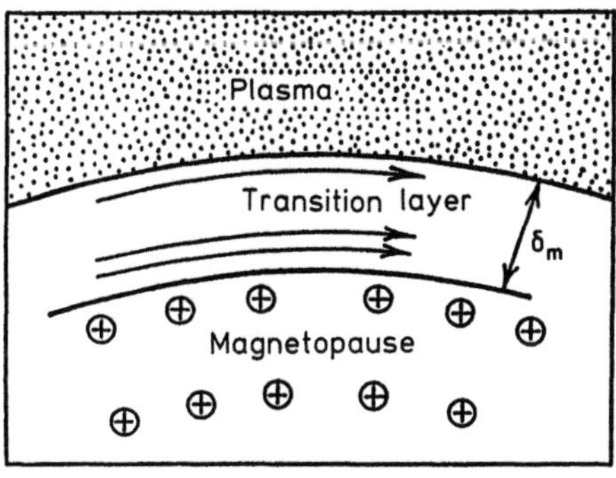

Fig. 5.1. Structure of magnetosphere near its boundary

The subscripts "in" and "out" stand for "inside" and "outside" the boundary layer.

We shall now present certain quantities in a form convenient for estimates. The mean free path l_i for ion-ion collisions and l_e for electron-electron collisions can be written as

$$l_i \approx 1.8 \times 10^9 \left(\frac{T_i}{10^{10}\,\mathrm{K}}\right)^2 \left(\frac{n_e \ln \Lambda}{10^{16}\,\mathrm{cm}^{-3}}\right)^{-1} \mathrm{cm}\ ,$$

$$l_e \approx 1.8 \times 10^5 \left(\frac{T_e}{10^{10}\,\mathrm{K}}\right)^2 \left(\frac{n_e \ln \Lambda}{10^{16}\,\mathrm{cm}^{-3}}\right) \mathrm{cm}\ . \tag{5.10}$$

The Larmor radius $R_L(i)$ for ions and $R_L(e)$ for electrons is written in the form

$$R_L(i) \approx 0.13\, T_{10}^{1/2} A^{1/2} z^{-1} B_6^{-1}\ \mathrm{cm}\ ,$$

$$R_L(e) \approx 3.1 \times 10^{-4} T_8^{1/2} B_6^{-1}\ \mathrm{cm}\ . \tag{5.11}$$

It follows from (5.10) that, in any case the mean free path of ions is comparable with, or larger than, the characteristic size of the magnetosphere. Consequently, the transition layer is collisionless. However, the Debye radius

$$l_D \approx 2.2 \times 10^{-3} T_8^{1/2} n_{15}^{-1/2}\ \mathrm{cm}\ , \tag{5.12}$$

is much smaller than the layer thickness, which means that the plasma can be treated as electrically neutral.

The system of magnetohydrodynamic equations for an ideally conducting inviscid fluid has the form

$$\frac{\partial \varrho}{\partial t} + \mathrm{div}\,\varrho v = 0\ , \tag{5.13}$$

$$\frac{dv}{dt} = -\frac{1}{\varrho}\nabla P + g + \frac{1}{\varrho c}(j \times B)\ , \tag{5.14}$$

$$\frac{\partial B}{\partial t} = \nabla \times (v \times B)\ , \tag{5.15}$$

$$\mathrm{curl}\,B = \frac{4\pi}{c}j\ , \tag{5.16}$$

$$\mathrm{div}\,B = 0\ , \tag{5.17}$$

$$P = A\varrho^\gamma\ , \tag{5.18}$$

where j is the electric current density. Equation (5.15) is a combination of Ohm's law and Maxwell's equation

$$\text{curl } E = \frac{1}{c} \frac{\partial B}{\partial t} \; .$$

The last term in the equation of motion (5.14) describes the magnetic force, and can be presented in the form

$$\frac{(j \times B)}{c} = -\nabla \frac{B^2}{8\pi} + k \frac{B^2}{4\pi} n_s \; , \tag{5.19}$$

where n_s is the normal to the magnetic line of force, and k is its curvature. This notation illustrates the anisotropic nature of the Lorentz force. The first term can be associated with pressure, but the second term acts in an anisotropic manner and the magnetic field lines tend to straighten out.

5.2 Formulation of the Problem

We shall find the shape of the magnetosphere, assuming that the current layer is infinitely thin. As mentioned above, the magnetosphere boundary divides the space into two regions: outside the magnetosphere, the magnetic field is equal to zero, and inside it there is no plasma. The magnetosphere is static. Hence (5.13−18) assume the form

$$\text{curl } B = \frac{4\pi}{c} j \; , \qquad \nabla P = \frac{1}{c} (j \times B) \; ,$$

$$\text{div } B = 0 \; , \qquad \text{div } \varrho v = 0 \; , \tag{5.20}$$

$$P = A \varrho^\gamma \; .$$

Combining the first and third equations, we obtain

$$\frac{B^2}{8\pi} = P \bigg|_b \tag{5.21}$$

and

$$\frac{1}{c} j = \frac{B}{B^2} \times \nabla P \; . \tag{5.22}$$

We introduce the surface current density:

$$J = j \, \delta_m \; . \tag{5.23}$$

In this case, we obtain from (5.21, 22)

$$J = \frac{c}{\sqrt{2\pi}} P^{1/2} \; .$$
(5.24)

The last expression illustrates the fact that the jump in the magnetic field strength is proportional to the surface current. In the first approximation, we can directly obtain the value of the surface current by specifying the force of pressure P (without solving the corresponding gasdynamic problem). In this case, the problem can be formulated as

$$\text{curl } \boldsymbol{B} = 0 \; ,$$
$$\text{div } \boldsymbol{B} = 0$$
(5.25)

(inside the magnetosphere).

The magnetic field at the origin must tend to the dipole field:

$$\boldsymbol{B}(R \to 0) \to \boldsymbol{B}_{\mathrm{d}} \; .$$
(5.26)

At the boundary, we have

$$\left. \begin{array}{l} J = \dfrac{c}{\sqrt{2\pi}} P^{1/2} \\[2ex] \dfrac{B^2}{8\pi} = P \\[2ex] (\boldsymbol{B} \times \boldsymbol{n}_\mathrm{s}) = 0 \end{array} \right| \; \mathrm{b}$$
(5.27)

where $\boldsymbol{n}_\mathrm{s}$ is the normal to the magnetosphere boundary. Finally, there is no magnetic field outside the magnetosphere:

$$\boldsymbol{B} = 0 \; .$$
(5.28)

5.3 Simple Configurations

Let us consider some simple examples of the distortion of a dipole magnetic field under the action of an ideally conducting medium. We shall specify the form of the "medium-field" boundary arbitrarily.

The simplest case is that of a conducting plane and a dipole (Fig. 5.2). The current distribution was obtained (for a moving plane having a finite conductivity) by Maxwell (1873). This problem acquired "astronomical" proportions in connection with magnetic storms. Chapman and Ferraro (1931) calculated the magnetic field structure shown in Fig. 5.2. The complete structure of the

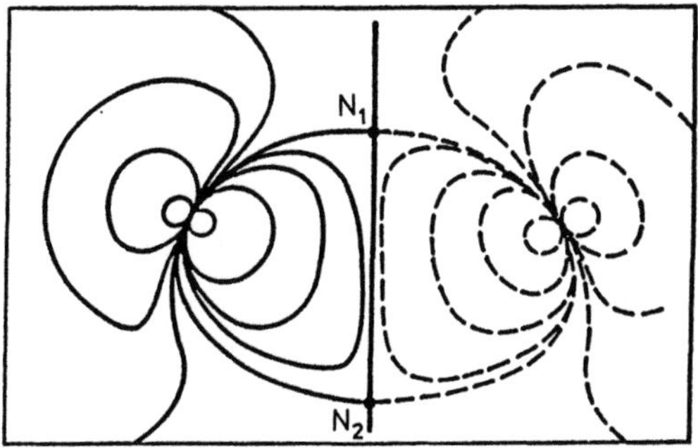

Fig. 5.2. A dipole near an ideally conducting plane

magnetic field can be obtained by the specular mapping method: the resulting field is equal to the superposition of the fields of the real and "reflected" dipoles. Since there are no currents to the left of the plane, the field is potential:

$$\boldsymbol{B} = -\nabla V_{\mathrm{m}} \; , \tag{5.29}$$

where V_{m} is the scalar potential. There are two singular points N_1 and N_2 on the plane, called neutral points, at which the magnetic field is equal to zero. There is no field to the right of the plane. At the boundary, the field lines are parallel to the plane.

If the direction of the dipole moment is parallel to the plane, the field is doubled at the line of intersection of the magnetic equator and the plane:

$$B = 2B_{\mathrm{d}} \; , \tag{5.30}$$

where B_{d} is the strength of the unperturbed dipole (Sect. 4.1). This means that when the magnetic dipole is confined by an ideally conducting plane, the magnetic field pressure increases fourfold on this line.

Let us now replace the ideally conducting plane by the plane-parallel plasma flow of non-interacting particles moving with a velocity v_∞. The pressure exerted by such a flow on the plane is described by the Newton function (Akasofu, Chapman 1972):

$$P = 2m_i n_i v_\infty^2 \cos^2 \chi = 2\varrho v^2 \cos^2 \chi \; , \tag{5.31}$$

where m_i and n_i are the mass and concentration of particles, and χ is the angle between the direction of the flow and the normal to the magnetic field. The factor 2 appears because a charged particle entering a field is twisted by the Lorentz force and reflected specularly.

Obviously, the magnetosphere boundary will not be planar in this case. Indeed, for a planar boundary the pressure is constant and does not vanish anywhere. The magnetic field, however, is variable in a plane and even equal to zero at the neutral points. For the equilibrium equation

Fig. 5.3. A dipole inside an ideally conducting sphere

$$P = \frac{B^2}{8\pi} \tag{5.32}$$

to hold, the plane must be deformed. The nature of such a deformation can be predicted qualitatively. Suppose that an ideally conducting plane suddenly loses its rigidity. Obviously, it starts bending at the edges and near the neutral points.

The next important case is that of a dipole inside an ideally conducting sphere (Fig. 5.3). The field at any point of the space is the sum of the dipole field and the field of currents flowing over the sphere:

$$B = B_J + B_{\mathrm{d}} \ . \tag{5.33}$$

Outside the sphere, $B = 0$ and hence the currents flowing over the sphere must also create outside the sphere a dipole field equal in magnitude to the initial field but opposite in direction:

$$B_J = -B_{\mathrm{d}} \tag{5.34}$$

(outside the sphere).

It is well known that the circular currents flowing over the sphere and varying from the equator according to the law

$$J = \dot{J}_0 \cos\theta \ , \tag{5.35}$$

produce a dipole magnetic field outside the sphere. This can be easily verified by calculating the multipole moments of such currents. With the exception of the dipole moment, all moments vanish. The dipole moment is given by (Jackson 1965)

$$e_\mu \mu = \frac{1}{2} \int [R \times J]\, dS \ ,$$

$$\mu \ = \frac{1}{2} J_0 R_0^3 \int\limits_0^{2\pi} \int\limits_0^{\pi/2} \cos^2\theta\, d\theta\, d\varphi = \frac{\pi^2 J_0 R_0^3}{4} \ , \tag{5.36}$$

where R_0 is the radius of the sphere, and e_μ is a unit vector along the magnetic axis. The conditions of the problem will be satisfied if we put

$$e_\mu \mu_J = e_\mu \mu \ . \tag{5.37}$$

Since the surface current at the boundary is connected with the field through the relation

$$J = \frac{1}{4\pi} [n_s \times B] \ , \tag{5.38}$$

we find that the strength of the resulting magnetic field at the boundary is

$$B = -\frac{3\mu}{R_0^3} \cos(\theta) e_\theta \ . \tag{5.39}$$

Two neutral points N_1 and N_2 are located at the poles of the sphere $\{B[\theta = \pm \pi/2] = 0\}$. Comparing (5.39) with the dipole field strength, we observe that the magnetic field strength at the equator becomes equal to three times the unperturbed value.

An interesting situation was observed by Kornilov (private communication). A sphere is an exact solution of the problem if plasma particles arrive axially symmetrically with respect to the dipole axis with constant velocities. Indeed, the pressure of noninteracting particles is defined by Newton's formula (5.31) in which $\chi = \theta$:

$$P = 2\varrho v_\infty^2 \cos^2 \theta \ .$$

From the equilibrium condition (5.32) we obtain

$$|B| = \sqrt{16\pi\varrho_\infty v_\infty^2} \cos \theta \ . \tag{5.40}$$

Comparing (5.39) and (5.40), we get the radius of the magnetosphere:

$$R_{\rm m} = \left(\frac{9\mu^2}{16\pi\varrho_\infty v_\infty^2}\right)^{1/6} \ . \tag{5.41}$$

Introducing in accordance with Sect. 4.3 the Alfvén radius and the condition

$$\frac{\mu^2}{8\pi R_{\rm A}^6} = \varrho_\infty v_\infty^2 \ ,$$

we obtain

$$R_{\rm m} = \left(\frac{9}{2}\right)^{1/6} R_{\rm A} \approx 1.28\, R_{\rm A} \ . \tag{5.42}$$

It can be seen that the magnetosphere radius differs from the rough estimate of the Alfvén radius by just 22%.

5.4 Magnetosphere in Spherically Symmetric Accretion

In the next two sections we shall consider accretion onto a magnetized star having a dipole field. The shape of the magnetosphere boundary is not specified in this case, but must be obtained in the course of the solution of the problem. We can forecast a qualitative solution by using the problem of a dipole inside a sphere. It was shown in Sect. 5.3 that there are two neutral points on a sphere at which the magnetic field pressure is equal to zero. Hence an external pressure acting on a sphere will "depress" it at the poles. Consequently, the polar radius of the magnetosphere will become smaller than the equatorial radius.

The problem will be solved specifically for two cases, viz. the case of a continuous and the collisionless state. To begin, we assume the accreting matter to be collisionless and the pressure anisotropic. We shall consider radial accretion in the approximation of noninteracting particles. Suppose that the size of the magnetosphere is much smaller than the capture radius: $R_A \ll R_G$. This problem was formulated by Inoue and Hoshi (1975). We assume that the accretion rate $\dot{M}$ is constant and that the particles fall freely, giving away their momentum at the magnetosphere boundary. In this case, the pressure created by the accreting plasma is

$$P = \varrho v^2 \cos^2 \chi = \frac{\dot{M}\sqrt{2GM_x}}{2\pi} R^{-5/2} \cos^2 \chi \, , \tag{5.43}$$

where, as before, χ is the angle between the normal to the boundary and the radial direction. The Alfvén radius in the problem under consideration is

$$R_A = \left(\frac{\mu^2}{2\dot{M}\sqrt{2GM_x}} \right)^{2/7} . \tag{5.44}$$

Numerical methods must be applied to obtain an exact solution of the three-dimensional problem. In the general case, the problem is reduced to the solution of an integral equation. However, before embarking on a numerical solution, it is interesting to comprehend the results to be obtained.

We use Beard's approximate method (Mead, Beard 1964) which reduces the problem of finding the shape of the magnetosphere to the solution of an ordinary differential equation. This method (to be more precise, its first approximation) uses a peculiar principle of local specular reflection. Each small region of the magnetosphere can be approximately treated as plane. We know (Sect. 5.3) that the perturbed field is doubled near the plane. Hence it is approximated that the strength of the distorted magnetic field is nearly equal to double the dipole field at the same point:

$$B = 2B_d \ . \tag{5.45}$$

Substituting this approximate equality into the equilibrium condition (5.32), we can write it in vector form:

$$|n_s \times B| = -\sqrt{8\pi P}\, n_s \cdot n_v = \sqrt{8\pi P}\cos\chi \ ; \tag{5.46}$$

where n_v is a unit vector in the direction of the velocity. Obviously, the unit normal vector n_s can be written in the form

$$n_s = \left(e_r - \frac{1}{R}\frac{dR}{d\theta}e_\theta\right)\Big/\sqrt{1+\left(\frac{dR}{R\,d\theta}\right)^2} \ . \tag{5.47}$$

Introducing the notation

$$r \equiv \frac{R}{R_A} \ ,$$

we obtain an approximate differential equation in the form

$$\cos\theta - 2\sin\theta\,\frac{dr}{r\,d\theta} = \frac{1}{2}r^{7/4} \ . \tag{5.48}$$

While deriving this equation, we have used the obvious relation

$$\cos\chi = \left[1+\left(\frac{dr}{r\,d\theta}\right)^2\right]^{-1/2} \ . \tag{5.49}$$

Equation (5.48) allows us to directly obtain the equatorial radius of the magnetosphere. Indeed, it follows from symmetry considerations that $dr/d\theta$ must be equal to zero for $\theta = 0$. This gives

$$R_e = 2^{4/7}R_A \approx 1.49\,R_A \ . \tag{5.50}$$

We introduce the enhancement factor k_m which shows the extent to which the magnetic field is distorted by the accreting plasma:

$$k_m = \left(\frac{B}{B_d}\right) \quad \text{(on the equator)} \ . \tag{5.51}$$

For a plane, $k_m = 2$, while for a sphere $k_m = 3$. In the approximate solution considered here, $k_m = 2^{12/7} \approx 3.28$. This is quite natural since the curvature of the surface is larger than for a sphere. The solution of the equation is shown in Fig. 5.4.

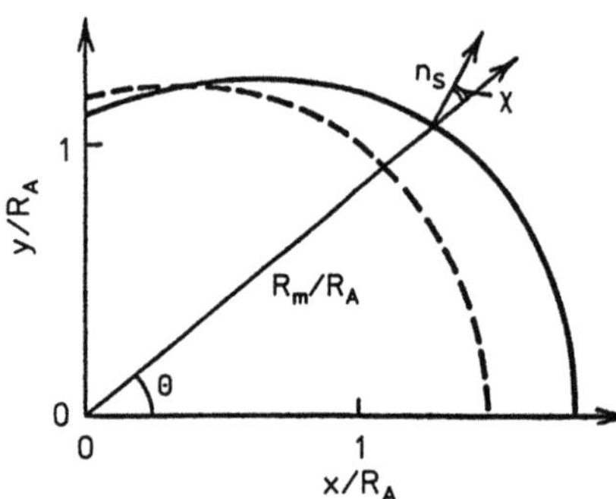

Fig. 5.4. The shape of a magnetosphere calculated by Beard's method (*dashed line*) and variational method for radial accretion with Newton's pressure law (collisionless gas)

To solve the exact problem of finding the shape of the Earth's magnetosphere, a method of moments was proposed by Medgley and Davis (1962). This method is based on the following considerations. By specifying the pressure at the magnetosphere boundary in a functional form $P(r,\chi)$, we thereby specify, in accordance with (5.27), the value of the surface currents in the form $J(r,\chi)$, i.e., as a function of the shape of the surface. Next, we must select such a form of the surface that all multipole moments (higher than dipole) vanish and the dipole moment is equal to the dipole moment of the star with the opposite sign. As a result, we obtain a system of integral equations for which we must specify the shape of the surface in the form of a series. The stumbling block in this case is the sagging of the polar cusp. In principle, it can be assumed that the magnetosphere sags along the magnetic axis to the dipole itself. In this case, we could "transmit" the accreting matter directly to the magnetic poles along polar channels. Such a shape of the magnetosphere was proposed in the work of Inoue and Hoshi (1975). Since the polar region makes a small contribution to the lower-order multipole moments, the numerical problem converges. However, such a topology was found to be erroneous. This is proved on the one hand by the exact two-dimensional solution (see below), and on the other hand by the solution of the problem by using a variational technique.

In order to avoid any arbitrariness in the choice of the form of the series approximating the shape of the surface, we developed the variational technique. The idea behind this method was put forth by Maguire and Carovillano (1966), who showed that the energy required to be spent in order to confine a dipole by any closed surface is

$$U = -\tfrac{1}{2}\boldsymbol{\mu}\cdot\boldsymbol{B}(0)\cdot\boldsymbol{e}_\mu \ , \tag{5.52}$$

where $B(0)$ is the magnetic field strength due to currents flowing over the magnetosphere at the dipole point. In the case of accretion, this work on the dipole confinement is done by the gravitational field.

Obviously, the stable surface corresponds to the minimum energy U. Varying the expression (5.52), we obtain the corresponding Euler equation which is the differential equation for the magnetosphere boundary:

$$\frac{\partial \Phi}{\partial r}+\frac{d}{d\theta}\frac{\partial \Phi}{\partial \dot{r}}=0 \ . \tag{5.53}$$

The multiple moments created by the currents satisfy the conditions

$$I_1 = 4 \ ,$$
$$I_j = 0 \ , \quad \text{where} \quad j = 2i+1, i>0 \ . \tag{5.54}$$

In these equations, $\dot{r} \equiv dr/d\theta$, and I_j are the multipole moments of currents flowing over the magnetosphere. Only odd moments are taken since the even moments automatically vanish in view of the symmetry of magnetosphere relative to the magnetic equator. For the pressure under consideration in the form (5.43), the function Φ has the form

$$\Phi = \tfrac{1}{2} r^{-5/4} \cos^2 \theta + \sum_{n=1}^{\infty} \lambda_n r^{n+3/4} P_n^1(\sin \theta) \ , \tag{5.55}$$

where λ_n is the Lagrangian indeterminate multiplier and P_n^1 are Legendre's adjoint polynomials (Korn, Korn 1968). Since $\partial \Phi/\partial \dot{r} = 0$, Euler's equation assumes the form

$$\frac{\partial \Phi}{\partial r}=0 \ . \tag{5.56}$$

Solving this equation, we obtain the exact equation for the magnetosphere boundary in the form

$$\frac{8}{5}\sum_{n=1}^{\infty} \lambda_n \left(n+\frac{3}{4} \right) r^{n+2} P_n^1(\sin \theta) = \cos \theta \ , \tag{5.57}$$

$$I_i = \int_{-\pi/2}^{\pi/2} r^{i+3/4} P_i^1(\sin \theta) \cos \theta \, d\theta \ . \tag{5.58}$$

Moreover, we put

$$I_1 = 4 \ ,$$
$$I_j = 0 \ , \quad \text{where} \quad j = 2i+1, i>0 \ . \tag{5.59}$$

We have solved the system of equations (5.57–59) numerically and the results are presented in Fig. 5.4.

Let us write down a number of useful relations. We multiply (5.57) by $(1/2)r^{-5/4} \cos \theta$ and integrate with respect to θ:

$$\frac{1}{2}\int_{-\pi/2}^{\pi/2} r^{-5/4} \cos^2 \theta \, d\theta = \frac{4}{5}\sum_{n=1}^{\infty} \lambda_n \left(n+\frac{3}{4} \right) \int_{-\pi/2}^{\pi/2} r^{n+3/4} P_n^1(\sin \theta) \cos \theta \, d\theta \ . \tag{5.60}$$

The left-hand side is just the strength of the magnetic field produced by currents flowing over the magnetosphere at the dipole [in units of $b(0) = B(0)/(\mu^2/R_A^3)$]. The integral on the right-hand side is equal to the multipole moment I_n. Consequently, we obtain

$$b(0) = \frac{4}{5} \sum_{n=1}^{\infty} \lambda_n \left(n + \frac{3}{4} \right) I_n \; .$$

Since $I_n = 0$ for $n > 1$ by hypothesis and $I_1 = 4$, the strength of the field produced by external currents at the dipole point is

$$B(0) = \frac{28}{5} \lambda_1 \frac{\mu^2}{R_A^3} \; .$$

In principle, the variational method can be used successively. In the first approximation, we simply require that $I_1 = 4$, without imposing any restrictions on the higher-order multipole moments. This is equivalent to the condition $\lambda_i = 0$ for $i > 1$. In this case, we find from (5.57) that in the first approximation, the magnetosphere is a sphere of radius

$$r_m = \left(\frac{8}{\pi} \right)^{4/7} \approx 1.72 \; .$$

This is very close to the value of the equatorial radius in the exact solution $r_e \simeq 1.78$ (Fig. 5.4).

The physical meaning of Lagrangian multipliers is that they characterize the rate of decrease of energy U or $b(0)$ upon a variation of the multipole moment I_i (Zel'dovich, Myshkis 1972). The obtained solution shows that a "hole" is not formed in the magnetosphere.

5.5 Pascal's Pressure Law

Let us now assume that the pressure of plasma particles near the magnetosphere becomes isotropic as a result of collisions and/or random magnetic fields, and that it obeys Pascal's law. In this case, we are left only with the dependence of pressure on the radius, which can be presented as a power law

$$P = P_0 \left(\frac{R}{R_0} \right)^{-n} \; . \tag{5.61}$$

Let us consider some physical situations in which the power law is observed. One of the most important cases was first considered by Cole and Huth (1959), namely, the case $n = 0$ in which a dipole is immersed in a homogeneous plasma with pressure $P = P_0 = \text{const}$. According to the classification of

Chap. 4, this case corresponds to a georotator with $R_G < R_A$. The velocity v_∞ of the star relative to the medium is much lower than the velocity of sound in it: $v_\infty \ll a_\infty$. By the "Alfvén radius", we shall mean the quantity

$$R_A = \left(\frac{\mu^2}{8\pi P_0 R_0^n}\right)^{1/(6-n)} .$$

(5.62)

If the Alfvén radius $R_A \lesssim R_G$ and there is spherically symmetric accretion, a collision with the magnetosphere leads to the formation of a shock wave, and the pressure behind the wavefront is of the order of the dynamic pressure (3.16) in the plasma:

$$P = \varrho v^2 = \frac{\dot{M}\sqrt{2GM_x}}{4\pi} R^{-5/2} .$$

(5.63)

This expression is the same as (5.61) for $n = -5/2$. It is interesting to note that a similar dependence of pressure on distance is also observed when a slowly settling or a static plasma atmosphere prevails around the magnetosphere. We neglect the term $v\nabla v$ in the Euler equation (3.2). This gives

$$\frac{1}{\varrho}\frac{dP}{dR} = -\frac{GM_x}{R^2} .$$

Substituting into this relation the equation of state $P = P_0(\varrho/\varrho_0)^\gamma$, we obtain

$$P = P_0\left(\frac{R}{R_0}\right)^{-\gamma/\gamma-1}$$

(5.64)

where

$$R_0 = GM_x\varrho_0(\gamma-1)/\gamma P_0 .$$

Consequently, the pressure obeys a power dependence in atmosphere with an exponent n given by

$$n = \frac{\gamma}{\gamma-1} .$$

(5.65)

For an adiabatic atmosphere, $\gamma = 5/3$ and $n = 5/2$, for an isothermal atmosphere $n = \infty$, while for $\gamma = 4/3$, $n = 4$.

We shall solve the problem concerning the shape of the magnetosphere boundary by considering model two-dimensional problems first.

5.5.1 Two-Dimensional Solutions

As a rule, the solution of a three-dimensional problem requires numerical computations on a computer. The obtained solution is not always exact, for exam-

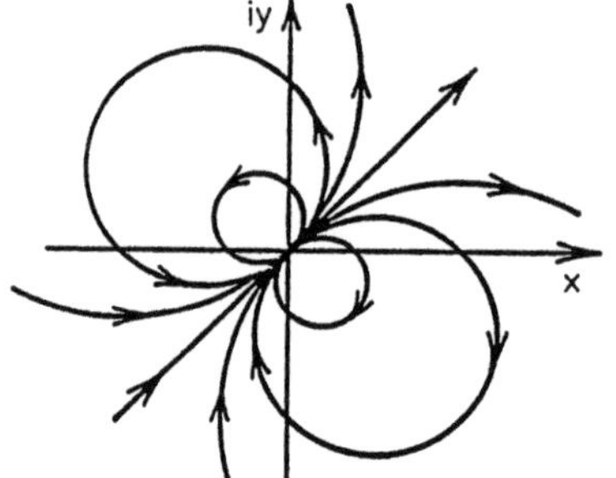

Fig. 5.5. A two-dimensional dipole in a complex plane

ple, at singular points. On the other hand, the two-dimensional analogs of three-dimensional problems qualitatively retain all the features of the three-dimensional solution and can be obtained in an analytical form.

The conformal mapping method is found to be especially effective. It allows us to obtain the solution in an analytical form convenient for analysis, and does not differ from the numerical three-dimensional solution by more than $10-20\%$.

The conformal mapping method is widely used in hydrodynamics, electrostatics, and strain theory (Lavrent'ev, Shabat 1973). However, the problems considered here differ somewhat conceptually from the standard problems, say, in fluid dynamics. Generally the formulation of a problem in fluid dynamics is that for a given flow boundary, the flow field has to be determined. However, when the magnetosphere structure is to be found, its boundary is not specified (it must be determined), although some functional condition on the boundary is given (in the form of the equality of the magnetic and external pressures). This, of course, complicates the solution of the problem.

Let us consider a complex plane $z = x+iy$. A two-dimensional dipole with a magnetic dipole moment d is placed at the origin (Fig. 5.5). It should be recalled that the dipole field strength decreases quadratically with distance, $|B| \propto R^{-2}$. Let the dipole be surrounded by a plasma with an isotropic pressure (5.61). This leads to a two-dimensional analog of the Alfvén radius:

$$R_A^{(2)} = \left(\frac{d_2}{8\pi P_0 R_0^n} \right)^{1/4-n} .\tag{5.66}$$

The solution of this problem is reduced to finding the shape of the conformal map of the region with a known field structure onto the required region. Obviously, we should seek the solution only for $n \leqslant 4$, otherwise the magnetosphere will be globally unstable as the external pressure increases more rapidly upon approaching the magnetic dipole than the magnetic pressure.

We shall write down a number of solutions. The case $n = 0$ corresponding to a uniform pressure leads to the formation of a boundary described by the following parametric system of equations (Cole, Huth 1959):

$$R \cos \theta = R_A^{(2)} \left[\sin \frac{\varphi}{2} - \frac{1}{3} \sin \frac{3\varphi}{2} \right] ,$$

$$R \sin \theta = R_A^{(2)} \left[\cos \frac{\varphi}{2} - \frac{1}{3} \cos \frac{3\varphi}{2} \right] . \tag{5.67}$$

The solution is sought for the first quadrant, and then extended to other quadrants by symmetric mapping.

It was shown by Elsner and Lamb (1976) that the case $n = 2$ is the closest to the three-dimensional case $n = 5/2$. The solution has the form:

$$R = R_A^{(2)} \exp \left(\frac{\cos 2\varphi}{2} \right) ,$$

$$\theta = \varphi + \frac{1}{2} \sin 2\varphi . \tag{5.68}$$

In the first quadrant, $0 \leqslant \varphi \leqslant \pi/2$. It is obvious that the polar radius does not vanish

$$R_p = \frac{R_A}{\sqrt{e}} ,$$

and is larger than the equatorial radius by a factor of e.

The polar region of the magnetosphere, called the polar cusp, is of special interest. The cusp is a singular point of the Legendre differential equation describing the scalar potential of a two-dimensional magnetic field. The polar line of force branches into two at the crusp (Fig. 5.6). Obviously, the behavior of the magnetic field near the cusp and its form are independent of the type of magnetosphere (i.e., independent of n). This is quite clear since the quest for the shape of a cusp on any magnetosphere boils down to the problem of finding the magnetic field and the shape of the "plasma-field" boundary near the branching point of the initial uniform field located in a homogeneous

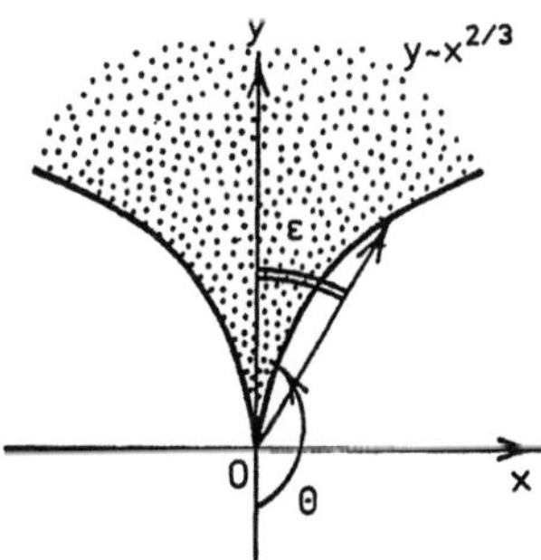

Fig. 5.6. Cusp, the branching point of lines of force

plasma. (The plasma pressure in the small neighborhood of a given point can always be treated as constant.)

The shape of the cusp can be determined, say, from the solution (5.68) for the case $n = 2$. Expanding the solution near the branching point (Fig. 5.6), we obtain (Morozov, Solov'ev 1963):

$$\varphi = \frac{R_p^{-5/4}}{4B_p}\,\varepsilon^2 \ , \quad y = \text{const} \cdot x^{2/3} \ , \tag{5.69}$$

here B_p is the magnetic field strength at the branching point:

$$B_p = P_0 \left(\frac{R_p}{R_0}\right)^{-2} .$$

The field lines converge at the branching point with an infinite derivative $(dy/dx \to \infty$ as $x \to 0)$. This circumstance is decisive. If the field lines were to converge with a finite angle, the magnetic field on them would be equal to zero and the singular point would be neutral. However, the existence of a neutral point would violate the equilibrium condition (5.27) at the magnetosphere boundary.

The field is not equal to zero in the vicinity of the cusp. It was shown by Morozov and Solov'ev (1963) that the cusp has the same shape in the three-dimensional case also.

5.5.2 Three-Dimensional Solutions

The magnetosphere boundary in the three-dimensional case, was simulated numerically by Arons and Lea (1976a) for the case when the dynamic pressure varies according to a power law with $n = 5/2$. Away from the cusps, the equation of the surface was sought in the form of a series

$$R(\theta) = R_e \left[1 + \sum_{k=1}^{N} c_k \left(\frac{2\theta}{\pi}\right)^{2k}\right] \ , \tag{5.70}$$

while the solution near the cusps was joined with the solution obtained by Morozov and Solov'ev (1966):

$$R = R(\theta) + S \left(1 - 2\frac{|\theta|}{\pi}\right)^{2/3} \ , \tag{5.71}$$

where S is a constant. The results of calculation of the coefficients c_k were published by Arons and Lea (1976a), while the shape of the surface boundary is shown in Fig. 5.7. The equatorial and polar radii were found to have the values

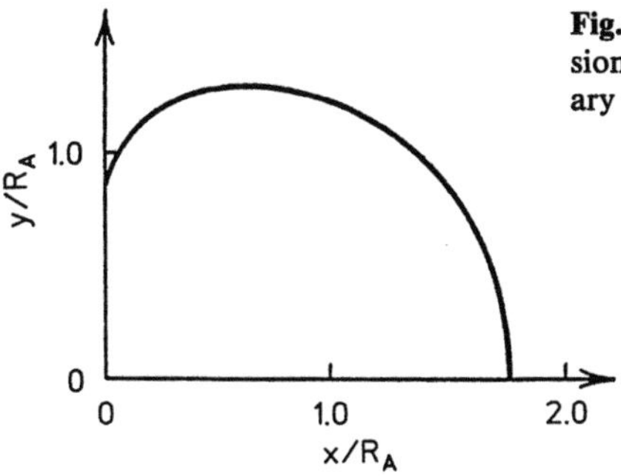

Fig. 5.7. Numerical computation of the three-dimensional problem on the shape of a magnetosphere boundary for Pascal's pressure law in a plasma

$$R_e \approx 1.78 \, R_A \ ,$$

$$R_p \approx R_A \ .$$

A comparison with the two-dimensional solution (5.68) obtained by Elsner and Lamb (in this solution, $R_e = \sqrt{e}\,R_A \approx 1.65\,R_A$) shows that the difference is not large (but the difference in the efforts involved in obtaining these solutions is very large!). In a series of works, Michel (1977 a, b) considered the problem analytically and obtained an integral equation whose numerical solution gives the shape of the boundary for any $n \leqslant 6$. Michel (1977) obtained a simple analytical solution for the case $n = 6$:

$$R = R_e \cos \theta \ . \tag{5.72}$$

The magnetosphere was found to sag to a dipole. It should be emphasized that, firstly, the case $n = 6$ is degenerate with the plasma pressure increasing in the same manner as the magnetic field pressure. Secondly, this case corresponds to a quite specific equation of state with $\gamma = 6/5$ [see (5.65)].

Finally, the numerical calculations for the case $n = 0$ carried out by Medgley and Davis (1962) lead to a solution that is not valid near the cusp since it does not take into account the fact that the expansion of the vector potential near the branching point contains Legendre's integral as well as half-integral polynomials.

5.6 A Dipole Confined by an Ideally Conducting Disk

Let us now consider the other limiting case, disk accretion onto a magnetized neutron star. Pringle and Rees (1972) were the first to consider such an accretion. The interaction of a disk with the magnetic field of a star was considered qualitatively and it was observed that the disk must be destroyed by magnetic forces at distances of the order of the Alfvén radius. The analysis of the magnetic field structure in the case of disk accretion was first carried out by Lipunov (1978 a, b), Sharleman (1978), and Ghosh and Lamb (1978).

Sharleman considered a model in which an accretion disk was replaced by a circular current, and paid special attention to the problem of determining the role of magnetohydrodynamic instabilities. Ghosh and Lamb studied an idealized model in which the magnetic axis of the star, the rotational axis and the disk axis coincide. They also tried to take into account the magnetohydrodynamic interaction of the star and the disk. Lipunov used a model in which the disk axis and the magnetic axis of the star do not coincide. He found the global structure of the magnetic field, assuming the disk to be an ideal conductor. Anzer and Börner (1980) studied the case when the dipole axis lies in the plane of the accreting disk.

The standard model of disk accretion (Chap. 3) shows that in the subcritical regime the disk thickness is much smaller than the disk radius:

$$\frac{H}{R} \approx \frac{a_{\rm s}}{v_\varphi} \ll 1 \ .$$

It is intuitively clear that the disk thickness should not be significant for finding the structure and shape of the magnetosphere (the characteristic scale along the z-axis vanishes). Hence, in the first approximation, the problem can be reduced to a dipole confined by an ideally conducting infinitely thin disk (Lipunov 1978 a, b).

5.6.1 Two-Dimensional Model

Following Lipunov (1978 a, b), let us determine the magnetic field structure for the two-dimensional analog of disk accretion. We shall assume that (a) the disk is infinitely thin and ideally conducting, and (b) the inner boundary of the disk lies at a distance a from the dipole. In Fig. 5.8, the disk is shown in the form of two cuts in the complex plane $z = x + iy$ along the real axis $|x| \geqslant a$. A dipole with moment d, directed at angle ψ to the magnetic axis, is placed at the origin. This problem is much simpler than the preceding ones since the magnetosphere boundary is specified beforehand. Hence it is not difficult to map the given region onto a simple domain in which the solution of the problem is known (Lavrent'ev and Shabat 1973). For the domain with a known solution, we choose a unit circle in the complex plane $\zeta = u + iv$, in which the direction of the dipole coincides with that of the magnetic axis. The conformal mapping transforming the region with discarded segments into the inner part of a unit circle has the form

$$z = \frac{2\zeta a e^{-i\psi}}{1 + \zeta^2 e^{-2i\psi}} \ . \tag{5.73}$$

Figure 5.8 shows the correspondence of points in the conformal mapping.

The magnetic field strength at any point is

$$\boldsymbol{B} = B_x + iB_y = \frac{dW(\zeta)}{d\zeta} \left(\frac{dz}{d\zeta}\right)^{-1} \ . \tag{5.74}$$

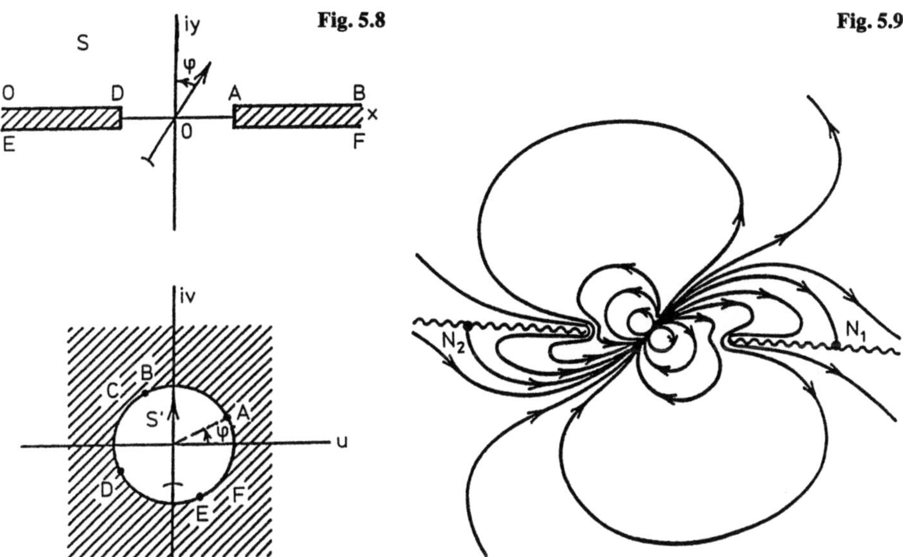

Fig. 5.8. A two-dimensional disk model and its mapping onto the inner surface of a unit circle. The plane S is mapped onto the inside of the circle S'. The symbols A, B, C, D, E and F show the correspondence of points in conformal mapping. The *arrow* indicates the dipole moment vector

Fig. 5.9. A dipole confined by an ideally conducting disk. Results of the two-dimensional solution (Lipunov 1978a, b)

Along the disk (cuts), we have

$$B = B_x = \pm \frac{d}{a^2} \left[-\frac{a^3}{x^3 \sqrt{1-(a/x)^2}} \cos \psi \pm \left(\frac{a}{x} \right)^2 \sin \psi \right] . \tag{5.75}$$

This expression corresponds to the "right" cut (Fig. 5.8) and the signs "+" and "−" to the magnetic field strength above and below the cut. The meaning of the obtained expression is clear: the first term is the field of currents flowing over the disk, and the second term describes the dipole field. The qualitative pattern of field lines for the obtained solution is shown in Fig. 5.9.

The global structure of the magnetic field has the following important features:

1) Near the inner edge of the disk ($x = \pm a$), the magnetic field has a singularity (which is well known in electrodynamics; see, for example, Landau and Lifshitz 1982, p. 44):

$$|B(x \to a)| \sim \Delta^{-1/2} , \tag{5.76}$$

were Δ is the distance from the inner edge.

2) When the dipole axis is directed along the disk axis ($\psi = 0$), the magnetic field acquires a quasiquadrupole form:

$$|\boldsymbol{B}| \sim \frac{1}{x^4}\,, \quad x \to \infty\,, \tag{5.77}$$

3) Conversely, if the dipole lies in the plane of the disk, the dipole field is not distorted at all.

4) The disk has two neutral points N_1 and N_2 at which the polar field lines enter and the magnetic field vanishes. From the condition $B_x = 0$ we find that the distance between the dipole and the neutral points is

$$x_{1,2} = \pm \frac{a}{\sin \psi}\,. \tag{5.78}$$

5) A torque, tending to turn the disk along the magnetic axis of the dipole, is exerted by the magnetic field on the disk. This follows from the calculation of the torque with the help of Chaplygin-Blazius formulas (Lipunov 1978a), which leads to the following expression:

$$K = \frac{d^2}{2a^2} \sin 2\psi\,. \tag{5.79}$$

This result becomes qualitatively clear from energy considerations. Although the torque is equal to zero in both positions $\psi = 0$ and $\psi = \pi/2$, the latter position results in a lower energy state: the magnetic field remains practically undistorted in this case and hence the energy expenditure in creating such a situation is minimum (the disk passes like "a warm knife through butter").

An interesting comparison can be made in this case with an ideal incompressible liquid (Shakura, private communication). It is well known (Lavrent'ev, Shabat 1973) that two-dimensional fluid mechanics is analogous to magnetostatics. The magnetic field lines can be compared with the stream lines in a fluid. However, the following example clearly demonstrates the important difference between fluid mechanics and magnetostatics.

Let us consider two plates in a fluid flow and a magnetic flow (Fig. 5.10), both inclined at an angle to the direction of the flows. The structures of the field lines and stream lines are exactly identical in both cases (the plates are ideally conducting and frictionless). The sum of the forces applied to both plates is equal to zero (D'Alembert's paradox), but the torques in the two cases have different signs. The plate turns across the flow of the fluid (this can be easily verified experimentally at home), while the plate in the magnetic flow turns along it. The difference is described by Bernoulli's integral in fluid mechanics (the pressure has its highest value where the velocity is minimum). The magnetic field pressure is proportional to its strength ($\propto B^2/8\pi$). At the neutral points N_1 and N_2 (Fig. 5.10), the velocity and field strength vanish. However, the pressure becomes zero only in a magnetic field, while its value is maximum in a fluid. The direction in which the plate will turn can be easily determined.

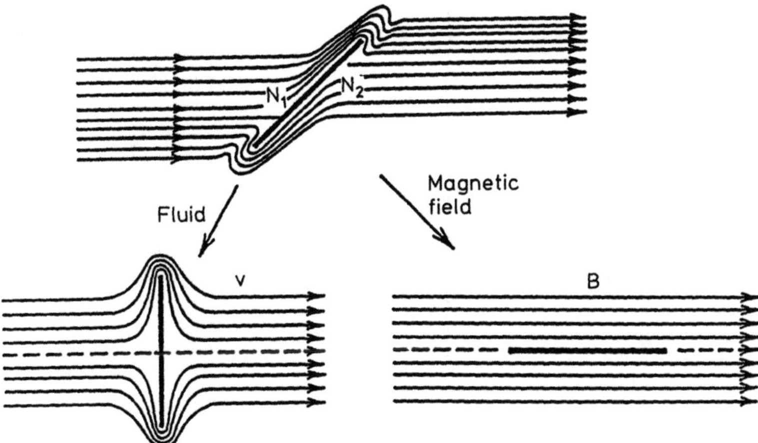

Fig. 5.10. A plate behaves in different ways when placed in the flow of an ideal fluid or a magnetic field

5.6.2 Three-Dimensional Problem

The results obtained in the last subsection can be easily generalized to the three-dimensional case (Lipunov 1978a, b). Suppose that an ideally conducting disk has an inner boundary of radius R_d. In this case, a generalization to the three-dimensional case is made by the substitution $d \to \mu$, $a \to R_d$, and by changing the exponents according to a dimensional analysis. The following properties of the three-dimensional solution can be predicted.

1) Near the inner edge of the disk, the nature of the singularity remains unchanged:

$$|B(R \to R_d)| \propto B_d \left(\frac{R_d}{\Delta} \right)^{1/2} , \tag{5.80}$$

where B_d is the strength of the dipole field at the same point.

2) At large distances from the inner edge, the field structure is reminiscent of a quadrupole:

$$B(R \gg R_d) \propto R^{-4} . \tag{5.81}$$

3) If the magnetic axis of the dipole lies in the plane of the disk, the structure of the dipole field is not distorted.

4) The distance of the neutral points on the disk from the dipole is approximately given by

$$R_{1,2} \approx \frac{R_d}{\sin \psi} . \tag{5.82}$$

5) The magnetic field tends to turn the disk along the magnetic axis, and the torque can be estimated by using the formula

$$K = \frac{\mu^2}{2R_{\mathrm{d}}^3} \sin 2\psi \; . \tag{5.83}$$

The singularity at the inner edge of the disk is a consequence of the assumption that its thickness is infinitely small. In the approximation of a finite thickness H, we obtain from (5.80) the following estimate:

$$B(R = R_{\mathrm{d}}) \doteq B_{\mathrm{d}} \left(\frac{R_{\mathrm{d}}}{H} \right)^{1/2} \; . \tag{5.84}$$

This singularity is not so strong, and hence the total energy required for holding the disk around the dipole is finite:

$$U \sim \int_V \left(\frac{B^2}{8\pi} - \frac{B_{\mathrm{d}}^2}{8\pi} \right) dV \; .$$

Indeed, the edge of the disk makes a small contribution to this energy:

$$\int \frac{B^2}{8\pi} R_{\mathrm{d}} \Delta \, d\Delta \sim \int \frac{B_{\mathrm{d}}^2}{8\pi} \left(\frac{R_{\mathrm{d}}}{\Delta} \right) \Delta \, d\Delta \sim \Delta \; .$$

It is clear that the force expelling the accretion disk is finite. In this respect, the model under consideration differs strongly from Sharleman's simplified model (1978) in which the disk was replaced by a circular current. In the latter case, the field at the inner edge has a much stronger singularity: $B \propto \Delta^{-1}$.

The results obtained by generalizing the two-dimensional solution were fully confirmed by the exact calculations of the three-dimensional problem (Aly 1980) and by the correction of certain inaccuracies in the three-dimensional solution by Kundt and Robnik (1980). The solution of this problem is quite tedious and we shall describe only the main results here. Figure 5.11 shows the lines of force for different angles of inclination of the dipole axis to the disk plane.

The position of neutral points on the disk, described by the condition $B = 0$, is determined from the expression

$$R_{1,2} = R_{\mathrm{d}} \sqrt{\left[1 + \left(\frac{2}{\pi} \cot \psi \right)^2 \right]} \; , \tag{5.85}$$

which becomes identical with (5.82) if we replace $2/\pi$ by 1. The total torque exerted on the disk by the magnetic field is

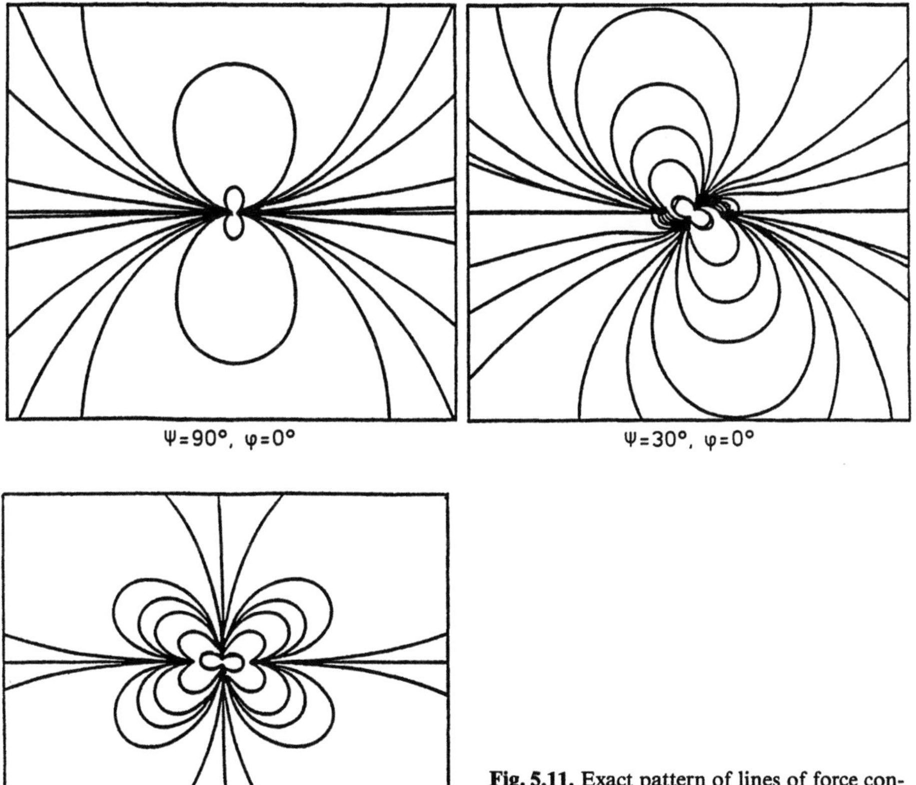

Fig. 5.11. Exact pattern of lines of force constructed from the results of the solution of the three-dimensional problem (Kundt, Robnik 1980)

$$K = \frac{4}{3\pi}\frac{\mu^2}{R_d^3}\sin 2\psi \; . \tag{5.86}$$

The factor $4/3\pi \simeq 0.42$ is close to 0.5 in (5.83). A comparison of both solutions was carried out by Lipunov and Shakura (1980), and Kundt and Robnik (1980).

5.6.3 Dipole Rotation

In the approximation of an ideally conducting disk considered here, we can easily take into account the rotation of the dipole in the region $R \ll R_l$ (as before, R_l is the radius of the light cylinder). Deep inside the light cylinder, the electric fields are weak and the resulting pattern is obtained by turning the magnetostatic pattern about the rotational axis. An averaged pattern is more important for processes occurring more slowly than the rotation of the neutron star. It follows from the theory of disk accretion (Chap. 3) that the longest characteristic time is the time of radial motion of matter in the disk:

$$t_{\rm r} \approx \frac{R}{v_r} \approx \frac{1}{\alpha} t_{\rm ff} \left(\frac{R}{H}\right)^2 \approx \frac{1}{\alpha} \left(\frac{R}{R_{\rm c}}\right)^{3/2} \left(\frac{R}{H}\right)^2 p \, , \tag{5.87}$$

where α is the turbulence parameter in the disk, $R_{\rm c}$ the corotation radius, and p the period of rotation of the neutron star. It can be seen that if $R \lesssim R_{\rm c}$, $t_r \gg p$. Consequently, the disk will acquire a magnetic field pattern averaged over the period of rotation.

Following Lipunov and Shakura (1980), let us average the torque exerted on the disk by the magnetic field over a period of revolution. The torque applied to an annular disk of thickness dR at a distance R can be presented in the exact three-dimensional solution in the form

$$dK = \frac{4\mu^2 dR}{\pi R^4 (R^2/R_{\rm d}^2 - 1)} (n_{\rm m} \cdot n_{\rm d}) [n_{\rm m} \times n_{\rm d}] \, , \tag{5.88}$$

where $n_{\rm m}$ and $n_{\rm d}$ are unit vectors directed along the dipole and along the disk axis, respectively.

We shall consider the most general case in which all the three axes, that is, the dipole axis, the axis of rotation, and the disk axis do not coincide (Fig. 5.12). Let α_0 be the angle between the rotational and the disk axes, and β the angle between the rotational axis and the magnetic axis of the star. We introduce the following unit vectors: n_ω along the direction of rotation and $n_\perp$ at right angles to the rotational axis. We assume that the rotational axis lies in the Z0Y plane. In this case, we can present the vector $n_{\rm m}$ in the form

$$n_{\rm m} = n_\omega \cos\beta + n_\perp \sin\beta \, . \tag{5.89}$$

Hence

$$(n_\omega \cdot n_{\rm d})[n_{\rm m} \times n_{\rm d}] = (n_{\rm d} \cdot n_\omega) \cdot [n_\omega \times n_{\rm d}] \cos^2\beta$$

$$+ \sin\beta \cos\beta (n_\omega \cdot n_{\rm d})[n_\perp \times n_{\rm d}] + \sin\beta \cos\beta (n_\perp \cdot n_{\rm d})[n_\omega \times n_{\rm d}]$$

$$+ \sin^2\beta (n_{\rm d} \cdot n_\perp)[n_\perp \times n_{\rm d}] \, .$$

The first term is constant while the second and third terms vanish when averaged over the period of rotation of the star. In order to average the last term, we introduce the axis Y' at right angles to the axis n_ω in the plane Z0Y. Writing the vector $n_{\rm d}$ in the form

$$n_{\rm d} = n_\omega \cos\alpha_0 - n_{Y'} \sin\alpha_0$$

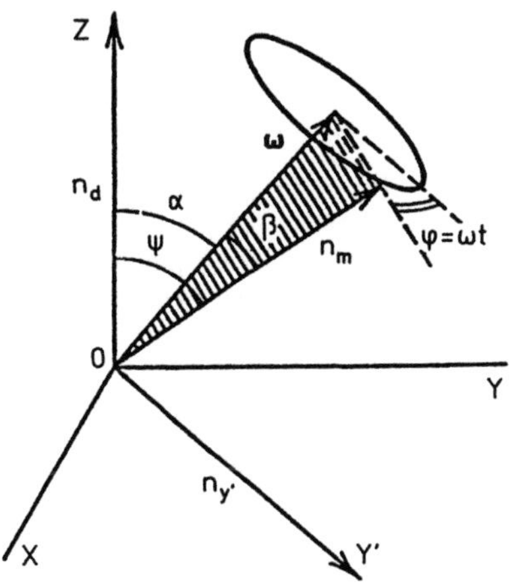

Fig. 5.12. The dipole axis $n_{\rm m}$ does not coincide with the disk axis $n_{\rm d}$ and the rotational axis n_ω of the neutron star

and making simple computations, we find that the average over a period is

$$\langle (n_{\rm m} \cdot n_{\rm d})\,[n_{\rm m} \times n_{\rm d}] \rangle = \frac{\cos \alpha_0}{2}\,(3 \cos^2 \beta - 1)\,[n_\omega \times n_{\rm d}] \ .$$

Substituting this expression into (5.88), we get

$$\langle d\boldsymbol{K} \rangle = \frac{2\mu^2 dR \cos \alpha_0}{\pi R^4 (R^2/R_{\rm d}^2 - 1)^{1/2}}\,(3 \cos^2 \beta - 1)\,[n_\omega \times n_{\rm d}] \ . \tag{5.90}$$

Integrating this expression over the entire disk, we obtain the total torque applied to the disk and averaged over a period (Lipunov, Shakura 1980; Kundt, Robnik 1980):

$$\boldsymbol{K}_{\rm m} = \frac{4\mu^2}{3\pi R_{\rm d}^3}\,\cos \alpha_0 (3 \cos^2 \beta - 1)\,[n_\omega \times n_{\rm d}] \ . \tag{5.91}$$

Obviously an equal and opposite torque is exerted on the dipole by the disk.

It should be remarked that the torque $\boldsymbol{K}_{\rm m}$ vanishes not only in trivial cases $\alpha_0 = 0$ and $\alpha_0 = \pi/2$, but also for a certain finite value of the angle between the rotational axis and the magnetic axis of the dipole:

$$\beta_0 = \arccos(\sqrt{3}/3) \approx 54°\,44' \ . \tag{5.92}$$

The averaged energy of magnetostatic interaction is

$$U_{\rm m} = \frac{4\mu^2}{3\pi R_{\rm d}^3}\,[2 \cos^2 \beta + (1 - 3 \cos^2 \beta) \sin^2 \alpha_0] \ . \tag{5.93}$$

5.7 Magnetosphere in a Plane-parallel Plasma Flow

Let us consider a situation in which the gravitational capture radius $R_G \ll R_A$. The case when the velocity of an accreting star is much lower than the velocity of sound in the plasma can be reduced to the problem with constant isotropic pressure discussed in Sect. 5.5. Hence we shall now consider the opposite case $v_\infty \gg a_\infty$, which is close to "terrestrial" conditions. In this case, the pressure of the oncoming flow is described by Newton's formula

$$P = P_0 \cos^2 \chi \ . \tag{5.94}$$

This case has been investigated in the greatest detail, and hence we refer the reader to monographs (for example, Akasofu, Chapman 1972).

The solution to the problem of finding the shape of the magnetosphere boundary will be described within the simplest approximation in which the plasma does not contain any intrinsic magnetic fields and, conversely, when there is no plasma inside the magnetosphere. For accreting compact stars, the first condition is justified by the "magnetic pumping" effect (Sect. 3.10).

On the magnetosphere, the ratio of the magnetic energy in accreting matter to the intrinsic energy of the field of the neutron star (for $R_G \ll R_A$) always satisfies the condition $B_{\rm out}^2 / B_{\rm in}^2 \simeq (a_\infty / v_\infty)^2 \ll 1$ [cf. (3.102)].

5.7.1 Two-Dimensional Solution

An exact analytical solution of this problem was obtained by Zhigulev and Romishevskiĭ (1959) with the help of the conformal mapping technique. The shape of the surface in x, y coordinates is described parametrically by

$$x(\varphi) = \frac{2}{\pi} \ln \frac{1}{2[1 - \cos(\varphi + \psi)]} + \frac{2}{\pi} \sum_{n=2}^{\infty} \frac{1 + (-1)^n}{n(n^2 - 1)} \cos n\varphi \ , \tag{5.95}$$

and

1) $y(\varphi) = 1 + \cos\varphi - 2\psi/\pi$ for $|\psi| \leqslant \varphi \leqslant \pi$;

2) $y(\varphi) = -1 - \cos\varphi - 2\psi/\pi$ for $\pi \leqslant \varphi \leqslant 2\pi - |\psi|$;

3) for $\psi > 0$, relation (1) is also valid in the region $0 \leqslant \varphi \leqslant \psi$;

4) $y(\varphi) = 3 - \cos\varphi - 2\psi/\pi$ for $2\pi - \psi \leqslant \varphi \leqslant 2\pi$;

5) for $\psi < 0$, relation (2) is also valid in the region $2\pi - |\psi| \leqslant \varphi \leqslant 2\pi$;
    and

6) $y(\varphi) = -3 + \cos\varphi - 2\psi/\pi$ for $0 \leqslant \varphi \leqslant |\psi|$.

The x and y coordinates are measured in units of Alfvén radius. The angle $0 \leqslant |\psi| \leqslant \pi$ is the angle between the direction of the flow and the direction of the dipole. In order to calculate the magnetic field strength at any point inside the magnetosphere, we must use (5.74) for conformal mapping. The mapping of the obtained magnetosphere onto a unit inner circle in the plane $\xi = u + iv$ is described by Schwarz' integral (Lavrent'ev, Shabat 1973)

$$z = -\frac{1}{2\pi i} \int_0^{2\pi} y(\varphi) \frac{e^{i\varphi} + \xi}{e^{i\varphi} - \xi} \, d\varphi \; . \tag{5.96}$$

The magnetosphere is not symmetric and a force

$$F = F_x + iF_y = F_0 \left(\frac{5\pi}{48} + \frac{\pi}{60} \cos 2\psi - \frac{\pi}{432} \cos 4\psi \right)$$

$$+ iF_0 \left(\frac{\pi}{60} \sin 2\psi - \frac{\pi}{432} \sin 4\psi \right) \tag{5.97}$$

is exerted by the oncoming flow. The torque turning the dipole is

$$K_\mathrm{m} = \frac{d^2}{R_\mathrm{A}^2} \sin 2\psi \; , \tag{5.98}$$

where R_A is the two-dimensional analog of the Alfvén radius:

$$R_\mathrm{A} = \left(\frac{d^2}{8\pi P_0} \right)^{1/4} \; .$$

It should be observed that the torque depends on the angle ψ in exactly the same way as for a dipole confined by an ideally conducting disk. This can be explained as follows. The torque applied to a dipole placed in an external magnetic field is given by (Landau, Lifshitz 1982)

$$K = [e_\mu \times B(0)]\mu \; , \tag{5.99}$$

where $B(0)$ is the magnetic field produced by external currents at the dipole point. Apparently, the direction of $B(0)$ is a preferred direction in the accretion flow. In the case of a disk this direction is that of the disk axis, while in the present case, it is the direction of flow. Consequently, we find from (5.99) that $|K| = \mu |B(0)| \sin \psi$. The dependence on ψ in (5.98) appears due to the fact that $B(0) \propto \cos \psi$.

Fig. 5.13. A geosimilar magnetosphere in a binary system (qualitative pattern)

5.7.2 Three-Dimensional Solution

The three-dimensional confinement of a dipole by the plane-parallel flow of a plasma was solved numerically by several authors (Akasofu, Chapman 1972). These results are applicable to both single neutron stars and neutron stars in binary systems if the Alfvén radius is much smaller than the semi-major axis of the binary system: $R_A \ll a$. For most real cases, this inequality is satisfied by a large margin. However, it may so happen that the size of the magnetosphere is comparable to the semi-major axis.[1]

If matter is ejected by a normal star in the form of a spherically symmetric wind, a new problem emerges concerning the shape of the magnetosphere (Fig. 5.13) when the plasma pressure obeys

$$P = \frac{\dot{M} v_{\mathrm{w}}}{4\pi r^2} \cos^2 \chi \; ,$$

where r is the distance from the center of the mass-losing star to the magnetosphere boundary, v_{w} the velocity of the stellar wind at a distance r, and χ has the same meaning as before, that is, the angle between the normal to the magnetosphere surface and the radial direction to the neighboring star.

5.8 Two-Stream Accretion

It was mentioned in Chap. 3 that the real accretion regime of a neutron star in a close binary system cannot be reduced either to a disk regime or to a radial regime. In many cases, nonradial accretion can be reduced to two-stream accretion (Lipunov 1980). In the two-stream model of nonradial accretion, it is assumed that there are two flows which accrete independently onto the neutron star − a disk flow described by the accretion rate $\dot{M}_{\mathrm{d}}$, and a spherically symmetric flow with an accretion rate $\dot{M}_{\mathrm{s}}$ (Fig. 3.1).

---

[1] Such a situation is most likely in white dwarfs.

Let us consider the structure of the magnetosphere in the vacuum approximation in the case of a two-stream accretion (Lipunov 1980, 1982). The radius R_d of the inner disk flow boundary must be obtained by solving the self-consistent problem whose solution has not been obtained so far. Hence we shall assume for the sake of precision that R_d is known. As before, we shall neglect the effect of plasma penetration into the magnetosphere.

We consider two limiting cases. In the first case, the magnetic axis of the neutron star is perpendicular to the plane of the disk flow ($\psi = 0$), while in the second case the dipole axis lies in the plane of the disk flow ($\psi = \pi/2$).

1) $\psi = 0$. An analysis of the solution for a dipole confined by a thin diamagnetic disk shows that the magnetic field structure of "disk + dipole" can be replaced by a quadrupole. For a two-dimensional analog, for example, the corresponding quadrupole moment is equal to $a \cdot d$ [see (5.75)]. An analysis of the three-dimensional solution shows that the quadrupole moment is (Lipunov 1980a)

$$q = \frac{8}{3\pi} \mu R_d \ .$$
(5.100)

Next, replacing the two-dimensional system "disk + dipole" by a two-dimensional quadrupole with moment q, we shall consider the radial accretion onto the quadrupole. For the case $n = 2$ [see (5.61)], the following solution was obtained (Lipunov 1978b) by the method of conformal mapping:

$$\begin{cases} R_m = R_A(q) \exp\left[-\tfrac{1}{4}\cos(4\varphi)\right] \ , \\ \theta = \varphi - \tfrac{1}{4}\sin(4\varphi) \ , \end{cases}$$
(5.101)

where $0 \leqslant \varphi \leqslant \pi/2$ in the first quadrant, and the solution in the remaining quadrants is obtained by specular reflection. Here, $R_A(q)$ is the two-dimensional analog of the Alfvén radius of the quadrupole. This solution can be generalized as follows to the three-dimensional case: we replace the Alfvén radius in (5.101) by the Alfvén radius of a three-dimensional quadrupole (Lipunov 1978b; 1980):

$$R_A(q) = \left(\frac{32\mu^2 R_d^2}{9\pi^2 \dot{M}_s \sqrt{2GM_x}}\right)^{2/11} \ .$$
(5.102)

Turning the obtained curve about the disk axis, we find the three-dimensional surface (Fig. 5.14a). The three-dimensional magnetosphere obtained in this way reflects all the characteristic features of the exact solution (which has not been obtained so far), with only one exception: the magnetosphere of a three-dimensional dipole in two-stream accretion for $\psi = 0$ will be extended more strongly along the magnetic axis since the field of a three-dimensional dipole is not isotropic and increases towards the pole for a constant R. A characteristic feature of this magnetosphere is the existence of three cusps of which two

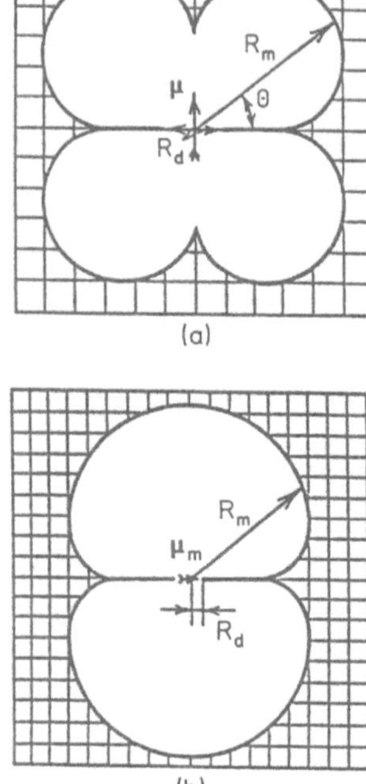

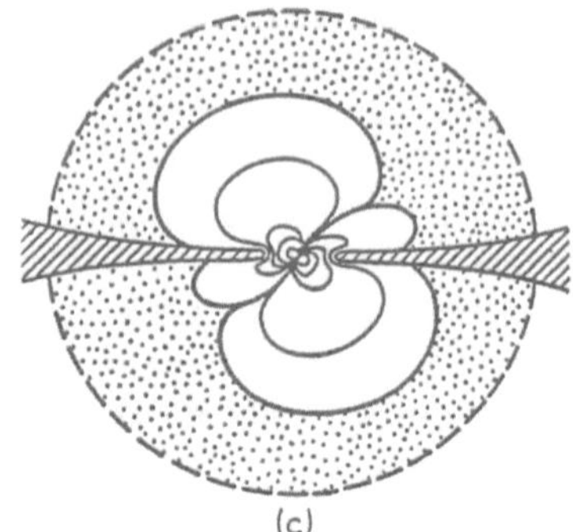

Fig. 5.14a–c. A magnetosphere in a two-stream accretion for different angles of inclination of the dipole axis to the disk flow

are polar and have identical structure for all magnetospheres, while the third cusp is circular and lies in the plane of the magnetic equator of the neutron star.

2) Let us now consider the other limiting case in which the dipole lies in the plane of the disk flow ($\psi = \pi/2$). In this case, a thin disk flow does not change the structure of the dipole magnetic field at all, and all the results obtained for the radial accretion onto a dipole remain valid (Sects. 5.4, 5). For analytic solutions, we can use the two-dimensional solution (5.67) (Fig. 5.14b).

An analysis of these two cases makes it possible to construct the (qualitative) model of the magnetosphere for any inclination ψ of the dipole axis to the disk (Fig. 5.14c). The part of the magnetosphere closed by a spherical flow will be called the external magnetosphere.

6. Accreting Neutron Stars

In this chapter we shall describe the theory, observational manifestations, and interpretation of the observational data on accreting neutron stars.

Accreting neutron stars emit X-rays. The brightest stars have a luminosity close to 10^{39} erg/s. X-ray stars can be divided into two categories: (a) those emitting pulsed X-ray radiation (X-ray pulsars) and usually belonging to a comparatively younger stellar population in the Galaxy (plane component), and (b) the sources of (not periodically) varying radiation (e.g., X-ray bursters) belonging to the galactic bulge, a quasi-spherical subsystem with a radius on the order of 5 kpc. All these stars are undoubtedly binary systems. It should be emphasized that four pulsating sources are among the galactic bulge sources.

The existence of pulsations can be attributed to very powerful magnetic fields. It was shown in Sect. 4.4 that the magnetic field starts exerting considerable influence on the accretion of stars whose magnetic moments are $\geqslant 10^{26} - 10^{27}$ Oe cm^3 ($B_0 \gtrsim 10^8 - 10^9$ Oe at the surface). Apparently, the X-ray pulsar fields at the surface are not weaker than 10^{12} Oe (see below).

A star with such a strong magnetic field has a magnetosphere which is hundreds of times the size of the star itself. In Chap. 5, we mainly confined ourselves to the shape and size of the magnetosphere boundaries. What we must understand above all is the manner in which accreting matter, which is a practically ideally conducting plasma, can penetrate the magnetosphere and reach the surface of a neutron star after traversing enormous layers of magnetic fields.

In the preceding chapter, we studied the structure of the magnetosphere of a neutron star in the vacuum approximation, i.e., under the assumption that matter does not penetrate the magnetosphere. Such magnetospheres can indeed exist for some neutron stars in the propeller or georotator regimes.

However, it is clear that in some powerful X-ray sources (like X-ray pulsars), matter penetrates to the surface across thousands of kilometers of magnetic field strata. This problem is much more complicated than the one considered in Chap. 5. At the existing level of theoretical and observational data, we can simply try to find out (at least qualitatively) the main processes occurring in the magnetospheres of accreting neutron stars. After all, it is the interaction of plasma and the magnetic field in the magnetosphere that is responsible for a number of observational facts. We shall enumerate just some of these effects:

1) liberation of energy of the accreting matter on the surface and its emission (anisotropic in a powerful magnetic field): circumpolar region;

2) exchange of angular momentum between a star and accreting matter, leading to a change in the rotational period of the neutron star (Alfvén zone – transition layer);

3) determination of the time evolution of a source by a valve at the magnetosphere boundary.

We begin the analysis of these questions with the most important one, that is, the passage of the plasma across the magnetosphere boundary. The first ideas on this subject were put forth by Shvartsman (1970a). Plasma may pass through the magnetosphere boundary due to magnetohydrodynamic instabilities (e.g., groove instability). This question was investigated in greater detail after the discovery of X-ray pulsars (Arons, Lea 1976a,b; Elsner, Lamb 1977; Sharleman 1978; Lipunov 1978b,c; Börner 1979).

The significant role of the gravitational force of the star, which is typical of accreting neutron stars, was pointed out by Arons and Lea (1976a). It should be recalled that in the case of the magnetosphere around the Earth, for example, the gravitational force does not play any role since the solar wind velocity is tens of times larger than the parabolic velocity at the magnetosphere boundary.

At present, a large number of different kinds of plasma instabilities are known to exist (Mikhailovskii 1975). Hence, astrophysicists are faced with the difficult task of isolating the main types of instability leading to the penetration of matter into the magnetosphere. It was natural that such investigations were started with the classical magnetohydrodynamic instabilities (see Appendix).

The main types of magnetohydrodynamic instability at the plasma-field boundary have their analogs in fluid mechanics. This is because the magnetic field is elastic and can be treated as a sort of fluid. The main difference between the field and a fluid is that the pressure in the former does not obey Pascal's law but depends on the structure of the field. In very slow processes in which displacement currents are not produced, a uniform magnetic field interacts with an ideally conducting plasma like an incompressible fluid during its motion across the field (Northrop 1956). However, any movement along the field occurs as if there were no magnetic field at all.

6.1 Boundary Stability

Let us analyze the stability of the magnetosphere boundaries considered in Chap. 5.

6.1.1 Spherically Symmetric Accretion

The stability of the magnetosphere boundary under conditions of spherically symmetric accretion was first considered by Arons and Lea (1976), Elsner and

Lamb (1977), and later in greater detail by Arons and Lea (1980). We shall briefly describe the results of these investigations. First of all, we assume that neutron stars rotate very slowly (see Sect. 5.9 for the definition of slow rotation). In order to investigate the boundary stability, we use the energy principle (Bernstein et al. 1958). The instability condition can be presented in the form

$$n_s \nabla P_0 - n_s \nabla (B_0^2 / 8\pi) > 0 \ . \tag{6.1}$$

We assume that outside the magnetosphere the plasma moves slowly (in the "settling" plasma regime, Sect. 5.5) so that the hydrostatic equilibrium condition is satisfied to a high degree of accuracy:

$$\nabla P_0 = \varrho_0 g = - \varrho_0 \frac{GM_x}{R^2} e_r \ . \tag{6.2}$$

Considering that

$$(\nabla \times B_0) \times B_0 = 0 \ , \tag{6.3}$$
$$n_s (B_0 \nabla) B_0 = k_c B_0^2 \ ,$$

where k_c is, as before, the curvature of the lines of force on the magnetosphere, the boundary instability condition can be written

$$g_{\text{eff}} = \frac{GM}{R_m^2} \cos \chi - k_c \frac{B_{\text{in}}^2}{4\pi \varrho_{\text{out}}} > 0 \ . \tag{6.4}$$

It should be recalled that χ is the angle between the radius vector and the outward normal to the magnetosphere boundary. The first term in (6.4) is the destabilizing gravitational force, while the second describes stabilization by convex magnetic field lines. Hence we are dealing with a Rayleigh-Taylor (RT) instability.

At the magnetosphere boundary, the equilibrium condition

$$\frac{B_{\text{in}}^2}{8\pi} = n_i k_B (T_i + z T_e) \tag{6.5}$$

is satisfied.

In this case, the boundary instability condition assumes the form

$$(T_i + z T_e)_{\text{out}} < T_{\text{cr}}(\theta) = \alpha(\theta) T_{\text{ff}} \ , \tag{6.6}$$

where

$$\alpha(\theta) = \frac{\cos \chi}{2 k_c R_m} \tag{6.7}$$

and T_{ff} is defined by (5.5). The meaning of condition (6.6) is quite clear: for the Rayleigh-Taylor instability to hold, the plasma must cool down sufficiently and become "heavier".

In a static atmosphere, the plasma atmosphere temperature is given by

$$T_{\text{out}} = \frac{\gamma-1}{\gamma} T_{\text{ff}} \; . \tag{6.8}$$

In this case, the instability condition can be presented in the following convenient form:

$$\alpha(\theta) > \frac{\gamma-1}{\gamma} \; . \tag{6.9}$$

Figure 6.1 shows the behavior of the function $\alpha(\theta)$ for different types of magnetospheres considered in Chap. 5 for spherically symmetric accretion. The criteria obtained by us show that

1) the cusp region is absolutely stable to perturbations (the component of gravitational force normal to the boundary is equal to zero);

2) the function $\alpha(\theta)$ is quite smooth everywhere except in the neighborhood of the cusp. Hence if the magnetosphere becomes unstable, the instability extends over a larger part of the magnetosphere;

3) for the unstable magnetosphere, the condition $\gamma < 5/3$ is satisfied, i.e., the plasma is cooled considerably. If, on the contrary, there is an additional source of energy liberation at the magnetosphere boundary (for example, due to rotation of the neutron star), the magnetosphere boundary will be stable. It should be emphasized that the above analysis is linear and hence even an

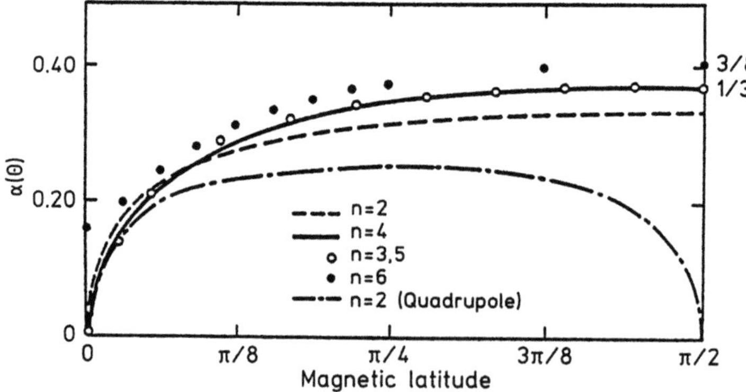

Fig. 6.1. The $\alpha(\theta)$ function for different types of magnetospheres: $n = 2$ (*dashed line*) and $n = 4$ are two-dimensional solutions, $n = 3.5$ and $n = 6$ are three-dimensional solutions, while $n = 2$ (quadrupole) (dash-dot line) is the two-dimensional solution. Pascal's law for plasma pressure is obeyed everywhere

instability in the linear approximation cannot guarantee an instability at the nonlinear stage. Hence the obtained condition is necessary but not sufficient.

Equation (6.6) indicates that the instability will be effective if the plasma temperature is $\lesssim (0.1-0.3)\,T_{\mathrm{ff}}$, i.e., if there is an effective cooling of matter. The main cooling mechanisms include the thermal (free-free) emission of plasma and the inverse Compton effect for the colder X-ray quanta emitted by the neutron star. Apart from these processes, synchrotron (to be more precise, cyclotron) radiation at the magnetosphere boundary may also be important. The most important heating process is the Compton effect (if the radiation temperature of the neutron star is higher than the electron temperature of the plasma surrounding the magnetosphere). If there is no accretion, heating due to the dissipation of rotational energy of the neutron star at the magnetosphere boundary may play a more important role.

Depending on the accretion rate, angular velocity, and the magnetic field strength, the relations between the rates of the above processes may differ considerably. This results in a variety of regimes so that, for example, the stationary inflow of matter into the magnetosphere is rather the exception than the rule.

Let us consider the characteristic times of certain processes. First, in order to estimate the importance of any process, we must compare its rate with the characteristic dynamic time in the accretion flow:

$$t_r \approx t_{\mathrm{ff}} \approx 0.06\, R_8^{3/2} m_x^{-1/2}\ \mathrm{s}\ . \tag{6.10}$$

The plasma cooling time as a result of free-free transitions is estimated as

$$t_{\mathrm{br}} \approx 1.4\,(z\bar{g})^{-1} R_8 m_x L_{37}^{-1}\ \mathrm{s} \tag{6.11}$$

where $\bar{g}$ is the Gaunt-factor. In deriving the above formula, we have used the continuity equation (3.8) and the relation between the luminosity due to accretion and the accretion rate. Comparing this time with the characteristic time of radial motion, we find that cooling due to thermal radiation will be effective if

$$L_x > 2.3 \times 10^{38} \frac{m^{3/2}}{R_8^{1/2}(z\bar{g})}\ . \tag{6.12}$$

It should be recalled that for most of the stationary X-ray sources this relation is not satisfied (even if $R_8 = 10$). Hence Bremsstrahlung cannot ensure a steady flow of matter into the magnetosphere. Either the accretion regime in these X-ray sources differs significantly from the spherical symmetry, or there exist more effective cooling mechanisms.

The inverse Compton effect may be one such mechanism (Elsner, Lamb 1977). Indeed, the ion temperature may attain values of about 1 MeV (Table 5.1), which is much higher than the spectral temperature of typical X-ray pulsars (5 – 20 keV). If electron-ion collisions are frequent in the plasma, the electron temperature becomes close to the ion temperature. The electron and ion temperatures' equalization time $(T_i \gg T_e)$ is given by (Zel'dovich, Raizer 1966)

$$t_{ei} \approx 2.5 \times 10^{-2} \frac{A}{z^2} \left(\frac{n_e \ln \Lambda}{10^{16}} \right)^{-1} \left(\frac{T_e}{10^8} \right)^{3/2} \text{s} \ . \tag{6.13}$$

The time-scale for plasma cooling due to the inverse Compton effect is

$$t_c = \frac{3 m_e c}{8 \sigma_T \varepsilon_r} \text{s} \ , \tag{6.14}$$

where ε_r is the radiation density. For an optically thin plasma in the layer above the magnetosphere we can assume that

$$\varepsilon_r = \frac{L}{4 \pi R_m^2 c} \tag{6.15}$$

while the characteristic time scale is given by

$$t_c \approx 6 \times 10^{-3} L_{37}^{-1} R_8^2 \text{s} \ . \tag{6.16}$$

Hence we can assume the following stationary accretion regime in the magnetosphere: The temperatures are assumed to follow the relation $T_i \gg T_e \gg T_r$, where T_r is the radiation temperature.

In a shock wave, the kinetic energy of the incident plasma is transformed into the thermal energy of ions which heat the electrons as a result of Coulomb collisions so that their temperature becomes higher than that of the radiation. The energy is consequently transferred to photons due to the inverse Compton effect. The ion temperature drops to $0.1 T_{ff}$. The magnetosphere is always open. Apart from the observance of all conditions ensuring the temperature balance, steady accretion also requires that the continuity condition must be satisfied, i.e., the amount of matter penetrating the magnetosphere must be equal to the amount arriving at the boundary. It is difficult to present a more detailed pattern of motion in such a regime, since the behavior of the RT-instability in the nonlinear regime is not clear. Apparently, matter must penetrate the magnetosphere in the form of individual clusters with a characteristic size $\lambda \approx 0.1 R_A$ (Arons, Lea 1976) (Fig. 6.2). The question concerning the fate of these clusters is discussed by Arons and Lea (1980) who describe the problem in all its complexity. At this stage, we simply observe that the fate of an individual cluster depends on several factors. The fragmentation of clusters is facilitated by the Kelvin-Helmholtz instability associated with their motion relative to the magnetic field. In principle, after disintegrating into a fine spray, these clusters could "freeze" in the magnetic lines of force (on account of resistive losses) at distances on the order of R_A. In this case, matter will ultimately flow to the poles of the neutron star. However, it is also quite possible that individual clusters move the magnetic field lines apart and fall radially

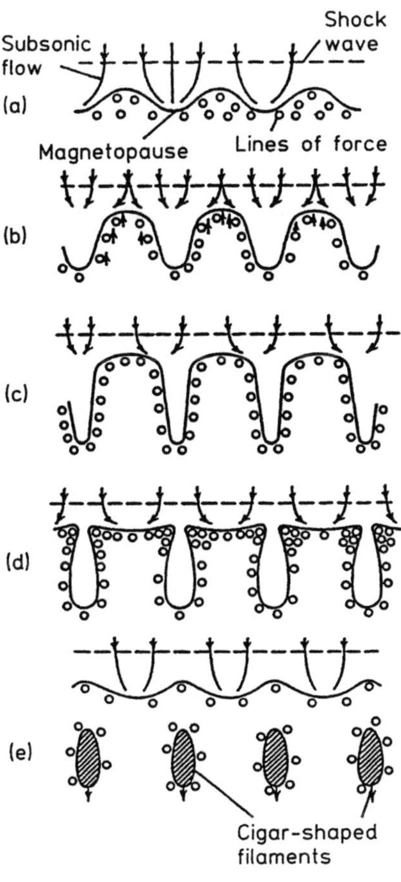

Fig. 6.2a–e. The time evolution of the Rayleigh-Taylor instability

and isotropically all over the surface of the neutron star. In this case, one can hardly expect the emergence of pulsed radiation (Arons, Lea 1980).

We believe that a steady (distant) spherically symmetric accretion onto a neutron star is hardly possible. There are several factors stabilizing the magnetosphere boundary.

1) It was mentioned above that inverse Compton effect may be a strong plasma cooling mechanism. However, the pulsar regime is directional and this effect can be observed only on a part of the magnetosphere. Moreover, the cooling time t_c is proportional to R^2, i.e., it increases more rapidly than the free-fall time $t_r \propto R^{3/2}$. For $R_A \approx 10^9$ cm, $t_c \gtrsim t_r$ and hence the plasma has no time to cool down.

2) Since real neutron stars rotate, the energy of their rotation at the boundary must dissipate into heat, that is, cause an additional heating of the plasma. If thermal radiation does not manage to remove heat, the plasma temperature will be on the order of T_{ff} and the magnetosphere will be stable (Davies,

Pringle 1981). It can be shown (Chap. 7) that the energy dissipation in the regime $R_A < R_c$ can be estimated by the expression

$$L_{rot} \approx \frac{1}{3} \frac{\mu^2}{R_c^3} \omega \approx \frac{1}{3} \times 10^{36} \mu_{30}^2 m^{-1} p^{-3} \text{ erg/s} \ . \tag{6.17}$$

Equating this energy to the energy of thermal radiation of the plasma, we obtain the critical period of rotation of a neutron star, during which the plasma does not get time to cool down (Davies, Pringle 1981):

$$p_{br} \approx 60 \, \mu_{30}^{16/21} \, \dot{M}_{15}^{-5/7} \, m^{-4/21} \text{ s} \ . \tag{6.18}$$

For $p < p_{br}$, accretion is not possible.

3) In principle, thermal energy could be removed through electronic heat conduction in the atmosphere. However, even negligible random magnetic fields ($\sim 1 \, \text{Oe}$) in the accretion flux can render this process ineffective. It should be recalled that the intrinsic field of a neutron star in the vicinity of the Alfvén radius has a strength of $\sim 10^6 \, \text{Oe}$!

All this makes the process of spherically symmetric accretion highly unfavorable for a steadily accreting neutron star. It has been mentioned repeatedly (Arons, Lea 1976; Elsner, Lamb 1977) that the rotation of a neutron star must lead to an instability of the magnetosphere on account of the Kelvin-Helmholtz (KH) instability. However, it is not clear so far whether the KH instability, which grows effectively only for small perturbation wavelengths, can ensure a global hydrodynamic flow of matter. Obviously, for an arbitrary inclination of the rotational axis, it is necessary that the characteristic time-scale of the KH instability be smaller than the period of rotation, i.e.,

$$\Gamma_{KH} \gg \omega \ . \tag{6.19}$$

Substituting the linear increment of the KH instability into this expression, we find that only those perturbations can grow whose wavelengths satisfy the inequality

$$\lambda \ll \frac{v_A}{c} R_A \ , \tag{6.20}$$

where $v_A = \omega R_A$. In real conditions, $v_A/c \simeq \sqrt{R_g/R_A} \simeq 1/30$. This means that clusters with a wavelength $\lambda \approx (1/30) R_A$, i.e., those having a fairly large size, manage to grow during one revolution. However, there are many factors stabilizing the KH instability and hence this question must be investigated further.

6.1.2 Disk Accretion onto a Magnetized Neutron Star

The magnetic field structure described in Sect. 5.6 was obtained under the assumption that plasma does not penetrate the magnetic field. Naturally, this assumption is not true in the case of accreting stars considered in this chapter. Plasma does penetrate the magnetosphere and reaches the magnetic poles of the neutron star. Let us consider the mechanism of this process.

We consider the idealized model of a magnetic dipole confined by an ideally conducting thin disk. It was first suggested by Sharleman (1978) and Lipunov (1978b) that the inner edge of the disk is the most unstable. Indeed, if the angle between the magnetic axis of a star and the rotational axis satisfies the condition

$$\tan \beta \lesssim \left(\frac{R}{H}\right)^{1/2} , \tag{6.21}$$

the field is enhanced at the inner edge by a factor of $(R_d/H)^{1/2}$ as compared to the dipole. We assume that the internal parts of the disk rotate in the equator plane of the star (it will be shown below that the combined action of magnetic and viscous forces results in a rotation of the disk in the equator plane).

The magnetic field strength decreases with distance from the inner edge of the disk. Consequently, the inner edge is not stable relative to the commutative instability. In contrast to the spherically symmetric accretion considered in the preceding section, the Rayleigh-Taylor instability does not play a significant role in the present case since matter rotating in Keplerian orbits is weightless.

The Kelvin-Helmholtz instability may turn out to be quite significant (Sharleman 1978; Lipunov 1978c). Such an instability can affect the entire surface of the disk and not just its inner edge. It was mentioned by Anzer and Börner (1980), however, that the Kelvin-Helmholtz instability is suppressed strongly due to a high relative velocity between the plasma and field. With the exception of a narrow ring at a distance equal to the corotation radius, the velocity is everywhere considerably higher than the velocity of sound.

This leads to the conclusion that the commutation instability is the principal channel responsible for the influx of mass into the magnetosphere (Fig. 6.3). The Kelvin-Helmholtz instability may just serve as an initiator. Matter is injected into the surface layers of the disk owing to the Kelvin-Helmholtz instability and is then entrained by turbulent diffusion (Ghosh, Lamb 1979). In this case, the plasma injection into a magnetic field depends considerably on the nature of the turbulence distribution along the z-coordinate. In some cases, turbulence may obstruct the penetration of the magnetic field by pushing it out. This phenomenon is called turbulence diamagnetism (Zel'dovich 1956). The field inside a highly turbulized medium is effectively suppressed (Zel'dovich, Ruzmaikin 1982):

$$B_{in} = R_{mR}^{-1/2} B_{out} , \tag{6.22}$$

where R_{mR} is Reynolds' magnetic number. For real accretion disks, $R_{mR} \approx 10^6 - 10^7$. It should be emphasized, however, that in order to estimate the role of turbulence diamagnetism, we must know the distribution of param-

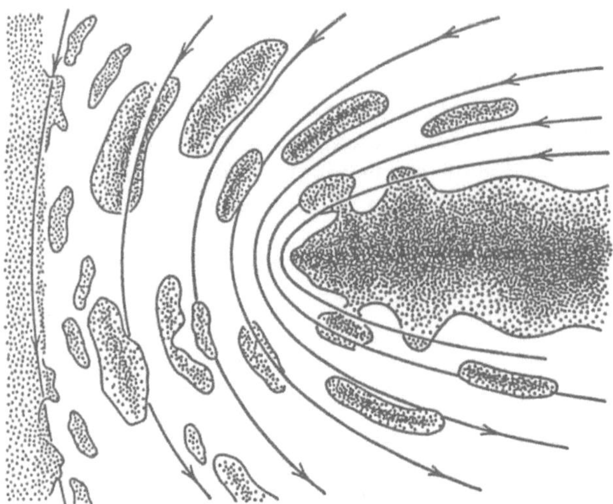

Fig. 6.3. Commutative instability at the inner edge of the disk

eters characterizing the turbulence along the z-axis. At present, these dependences are not known so well (Shakura et al. 1978).

After penetrating the magnetosphere from the inner edge, matter is crushed and "frozen" in the magnetic field. The difference in the field and plasma rotation velocities results in the appearance of a φ-component of the magnetic field which decelerates the matter (Fig. 6.4).

Ghosh and Lamb (1978; 1979) where the first to try to calculate this deceleration. For this purpose, the system of magnetohydrodynamic equations (5.13–18) must be used. Let us transform the equation of motion

$$(v\nabla)v = -\frac{1}{\varrho}\nabla P - \nabla\Phi + \frac{1}{4\pi\varrho}(\nabla\times B)\times B \ . \tag{6.23}$$

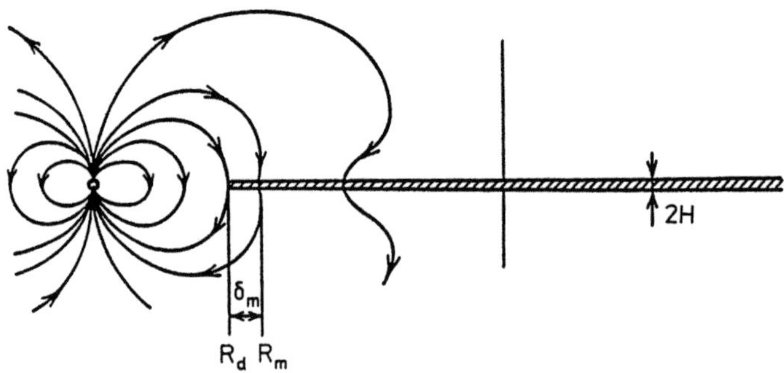

Fig. 6.4. Structure of the magnetosphere for disk accretion (Ghosh, Lamb 1978). R_d and R_m are the radii of the disk and the magnetosphere, δ_m is the thickness of the transient layer, and $2H$ is the disk thickness

We write this equation in cylindrical coordinates R, φ, z with the help of the following vector equalities:

$$[(\nabla \times B) \times B]_\varphi = \frac{1}{R} B_r \frac{\partial}{\partial R} (R B_\varphi) + B_z \frac{\partial B_\varphi}{\partial z} \; ,$$

$$[(\nabla \times v) \times v]\Big|_\varphi = \frac{1}{R} v_r \frac{\partial}{\partial R} (R v_\varphi) + v_z \frac{\partial v_\varphi}{\partial z} \; ,$$

$$\varrho(v\nabla)v = \varrho \left[\frac{1}{2} \nabla v^2 + (\nabla \times v) \times v \right] \; .$$

This system of equations must be supplemented by Maxwell's equation

$$\text{div } B = \frac{1}{R} \frac{\partial}{\partial R} (R B_r) + \frac{\partial B_z}{\partial z} = 0 \; .$$

Assuming that equilibrium prevails along the z-coordinate, i.e., $v_z = 0$, we obtain from the φ-component in (6.23) the following equation for the transport of angular momentum:

$$\varrho v_r R \frac{\partial}{\partial R} v_\varphi R = \frac{1}{4\pi} \left[\frac{\partial}{\partial R} R^2 B_r B_\varphi + R^2 \frac{\partial}{\partial z} B_z B_\varphi \right] \; . \tag{6.24}$$

Integration over the thickness of the disk gives

$$\dot{M} \frac{d\omega R^2}{dR} = R^2 B_z B_\varphi \; . \tag{6.25}$$

This expression is analogous to (3.45) for the case when the angular momentum is taken away by the magnetic field. We can now write down the general expression for the case when both turbulent viscosity and magnetic field are in action:

$$\dot{M} \frac{d\omega R^2}{dR} = 2\pi \frac{d}{dR} W_{r\varphi} R^2 + R^2 B_z B_\varphi \; . \tag{6.26}$$

Let us return to (6.25), keeping in mind that the magnetic field is the main reason behind the removal of the angular momentum. We assume that

$$B_z = B_\varphi = \sqrt{\kappa_t} B_d \; . \tag{6.27}$$

Since the disk is situated in the magnetic equator plane, we can write (6.25) in the form

$$\dot{M}\frac{d\omega R^2}{dR} = \kappa_t \frac{\mu^2}{R^4} \; .$$

(6.28)

We also write down the expression for the Alfvén radius:

$$\frac{\mu^2}{8\pi R_A^6} = \frac{\dot{M}}{4\pi R_A^2}\sqrt{\frac{2GM_x}{R_A}} \; .$$

Introducing the notation $r = R/R_A$, we obtain

$$\frac{dk}{dr} = 2\sqrt{2}\kappa_t r^{-4} k_A \; , \quad k = \omega r^2 \; , \quad k_A = \omega r^2 (R = R_A) \; .$$

(6.29)

We can now estimate the thickness of the transition layer:

$$\delta_m \approx \frac{r^4}{2\sqrt{2}\kappa_t} R_A \; .$$

(6.30)

It can be seen that the layer thickness δ_m over which matter is stopped depends strongly on the quantities κ_t and r_d. Ghosh and Lamb obtained the estimate $\delta_m \approx 2H$. In the standard model (Sect. 3.5), the disk thickness in the B zone is given by

$$\frac{H}{R} \approx 2\times 10^{-3} \left(\frac{R}{R_g}\right)^{1/20} \alpha^{-1/10} m_x^{-3/10} \dot{M}_{17}^{1/5} \; .$$

(6.31)

The deceleration zone is found to be very narrow.

On the basis of the numerical model under assumptions considerably simplifying the computations, Ghosh and Lamb (1979) obtained the torque applied to a neutron star. Qualitatively and quantitatively similar results were obtained by the author of this book (1982c) by using an analytical model. This model will be described in Sect. 6.4.

6.1.3 Torsion of an Accretion Disk by Magnetic Forces

If the accreting matter is stopped over a region $\delta_m \ll R_d$ which is quite narrow, the magnetic field structure away from the inner edge is close to that described in Sect. 5.6. An analysis of the magnetic field of a dipole confined by an ideally conducting disk shows that a torque (5.90) is applied to the accreting disk and tends to turn it. Under the action of this torque, matter in the disk begins to precess and the disk is no longer planar. The torsion of accreting disks was first considered by Bardeen and Petterson (1975) in connection with the Lense-Thirring effect (Landau, Lifshitz 1973) emerging in the case of accretion to a rotating black hole. Later, Petterson (1977) considered the precession of disks under the action of tidal, radiative and other torques.

The torsion of a disk by magnetic forces was considered by Lipunov and Shakura (1980) and by Lipunov et al. (1981). Time-independent solutions were found for the shape of a disk under the assumption that the magnetic field has the same structure as in the case of a dipole confined by an ideally conducting disk. In the steady state, matter precesses and the disk profile does not change in time.

We shall assume that the torque applied to an annular disk is defined by (5.90) obtained by averaging the magnetostatic profile over a period of rotation of the star. Let us consider the conditions under which such an averaging is justified. The frequency of precession of the ring under the action of the torque (5.90) is obviously given by

$$\Omega_r = \left| \frac{[dK(dI) \times e_k \omega_k]}{(dI)^2 \omega_k^2} \right| = \frac{\mu^2 \sin 2\alpha_0 (3 \cos^2 \beta - 1)}{2\pi^2 \Sigma R^7 (R^2/R_d^2 - 1)^{1/2} \omega_k} \, , \tag{6.32}$$

where e_k is a unit vector along the rotational axis of the ring of the disk, α_0 is the angle between the disk axis and the rotational axis of the star (at large distances), β is the angle between the rotational axis and the magnetic axis of the star, ω_k is the Keplerian velocity, $dI = 2\pi R^3 \Sigma dR$ is the moment of inertia of the ring and Σ the surface density on the ring of the disk. In the stationary case, we can use the continuity equation $\dot{M} = 2\pi R \Sigma v_r$, where v_r is the radial velocity in the disk. In this case, the velocity of precession is defined as

$$\Omega_r = \frac{2\sqrt{2}\, v_r \sin(2\alpha_0)(3\cos^2 \beta - 1)}{\pi R (R^2/R_d^2 - 1)^{1/2}} \left(\frac{R_A}{R} \right)^4 . \tag{6.33}$$

Formulas (6.32, 33) are applicable under the conditions

$$\Omega_r \ll \omega_k \, , \quad \Omega_r \ll \omega \, . \tag{6.34}$$

It should be recalled that ω is the angular velocity of the rotating star. It follows from (6.33) that $\Omega_r/\omega_k \approx (v_r/v_\varphi)(R_A/R)^4 \approx \alpha(H/R)^2(R_A/R)^4 \ll 1$ over a wide range of R, and hence the first of the conditions (6.34) is always satisfied. Similarly, we obtain $\Omega_r/\omega \approx (v_r/v_\varphi)(\omega_k/\omega)$. Substituting this estimate into (6.34), we find that the second requirement is equivalent to the inequality

$$p \ll \alpha^{-1} \left(\frac{R}{H} \right)^2 \frac{2\pi}{\omega_k} \approx 10^5 (10\alpha)^{-1} \left(\frac{R}{100H} \right)^2 \left(\frac{R}{R_A} \right)^{3/2} \text{s} \, . \tag{6.35}$$

This inequality is valid for most of the observed accreting pulsars.

In order to estimate the role of precession, we must compare the time t_r of radial motion of matter in the disk with the time-scale of precession $t_{\mathrm{pr}} = 2\pi/\Omega_r$. It follows from (6.33) that

$$\frac{t_r}{t_{\text{pr}}} \approx \left(\frac{R_{\text{A}}}{R}\right)^4 \left(\frac{R^2}{R_{\text{d}}^2}-1\right)^{-1/2} . \tag{6.36}$$

Clearly, precession becomes significant at a distance on the order of the Alfvén radius $R \approx R_{\text{A}}$.

In the steady state, the equation for a twisted accretion disk has the form (Lyubarskii 1979)

$$-\frac{\dot{M}}{2\pi}\frac{de_k\omega_k R^2}{dR} = \frac{d}{dR}\left(\eta H R^3 \frac{de_k\omega_k}{dR}\right) + [e_\Omega \times e_k] R^3 \Sigma \Omega_r \omega_k , \tag{6.37}$$

where e_Ω is a unit vector along the angular velocity of precession of the ring.

Let us write down the asymptotic form for the angle α_r between the axis of an annular disk of radius R and the axis of a neutron star, and also for the angle φ defining the position of the angular momentum vector for the ring in space (Lipunov, Shakura 1980):

$$\alpha_r \approx c R^{7/8} \exp\left(-R_{\text{pl}}/R\right)^4 , \quad \varphi \approx (R_{\text{pl}}/R)^{9/4} . \tag{6.38}$$

Here, c is a constant and R_{pl} is defined as

$$R_{\text{pl}} \approx \left(\frac{R_{\text{A}}}{R_{\text{d}}}\right)^{7/9} R_{\text{d}} . \tag{6.39}$$

Under the action of the magnetic forces, the inner parts of the disk are arranged along the equator of a rotating neutron star. It has been shown that the nature of the solution thus obtained is independent of the angle between the rotational axis and the magnetic axis of the neutron star (Lipunov et al. 1981).

Let us write down the exact solution of (6.37) under the following conditions

$$e_k\omega_k(R\to 0)\to\sqrt{GM_x/R^3}\,n_\omega , \quad e_k\omega_k(R\to\infty)\to\sqrt{GM_x/R^3}\,n_{\text{d}} \tag{6.40}$$

where n_ω and n_{d} are the unit vectors along the rotational axis of the neutron star and the rotational axis of the disk at large distances from the star. Using the notation

$$Y = \omega_k R^2 \, e^{i\varphi} \sin\alpha_r \tag{6.41}$$

we can present (6.37) and the boundary conditions (6.40) in the form

$$\frac{d^2 Y}{dR^2}+\frac{1}{2R}\frac{dY}{dR}+ia\,YR^{-13/2} = 0 , \tag{6.42}$$

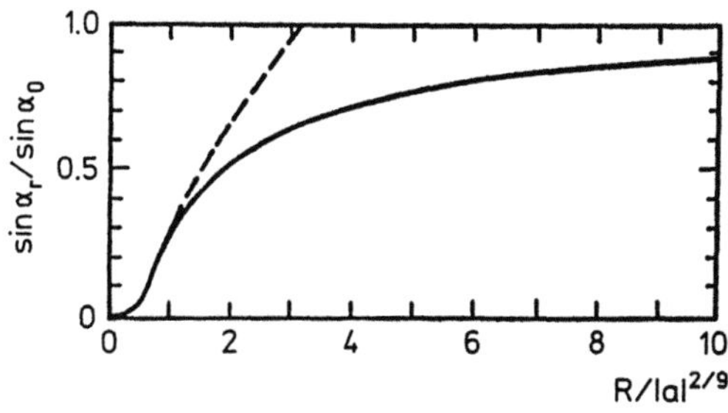

Fig. 6.5. Dependence of the angle between the rotational axis of the star and the rotational axis of the disk ring on the distance from the star

$$Y \xrightarrow[R \to \infty]{} \sqrt{GMR} \sin \alpha_0 \; ,$$

$$Y \xrightarrow[R \to 0]{} 0 \; ,$$

where

$$a = -\frac{24 R_d \mu^2 (3 \cos^2 \beta - 1) \cos \alpha_0}{\pi \sqrt{GM_x} \dot{M}} \; .$$

The solution of the problem (6.42) can be written

$$Y = \sqrt{GM_x} \sin (\alpha_0) \Gamma \left(\frac{8}{9}\right) \left(\frac{A}{2}\right)^{1/9} R^{1/4} [J_{-1/9}(A R^{-9/4})$$

$$- e^{-i\pi/9} J_{1/9}(A R^{-9/4})] \; , \tag{6.43}$$

where

$$A = 4(1+i)\sqrt{a}/(9\sqrt{2}) \; , \quad J_{1/9} \quad \text{and} \quad J_{-1/9}$$

are the Bessel functions of the imaginary argument. The qualitative form of the solution is shown in Fig. 6.5.

6.1.4 Magnetosphere Boundary Stability for Two-Stream Accretion

The penetration of matter from the disk flow in the case of two-stream accretion does not differ significantly from the accretion onto a magnetized neutron star considered above.

Therefore we will analyze the stability of the external part of the magnetosphere (Fig. 5.14) closed by a spherical flow. The Rayleigh-Taylor instability could play an important role in a spherical flow or atmosphere. However, it was mentioned above that in general the magnetic field is of a quasi-quadrupole nature and the magnetosphere of the quadrupole is much more stable than the dipole magnetosphere (Lipunov 1978b). This is because the quadrupole magnetosphere has a larger curvature.

Indeed, the condition of magnetosphere instability relative to the Rayleigh-Taylor instability has the form of (6.9): $\alpha(\theta) > (\gamma - 1)/\gamma$. If the dipole axis is parallel to the disk flow axis, the shape of the magnetosphere can be approximated by the two-dimensional solution (5.105) for which

$$\alpha(\theta) = \frac{\sin^2 2\varphi}{2(1+\sin^2 2\varphi)} \ .$$
(6.44)

The maximum value of this ratio is observed near the zones $|\theta| = \pi/4$. But even in this case its value is $\alpha(\theta) = 0.25$, which means that the magnetosphere is stable even for $\gamma = 4/3$. Thus, the spherical flow through the external part of the magnetosphere is not likely to be stationary. This must facilitate the formation of a shell around the magnetosphere which gradually settles down and mixes with the disk flow. The case of a highly transient burster-type accretion through an external magnetosphere has been considered by Lipunov (1980).

6.2 The Polar Column

We shall now analyze the motion of matter deep inside the magnetosphere where it is completely controlled by the magnetic field. The situation does not depend significantly on the boundary conditions away from the accreting star, i.e., it does not matter much as to whether the accretion regime at a distance $R \gg R_A$ is disk-type, spherically symmetric, or two-stream-type.

Let us estimate the size of the domain over which the matter falls (Lamb et al. 1973). We assume that matter is "frozen" in the magnetic lines of force at a distance $R_e \approx R_A$ near the magnetic equator. We also assume that the polar lines of force have the form of undistorted dipole lines, i.e., are described by (4.8):

$$R = R_e \cos^2 \theta \ .$$

We introduce the angle $\varepsilon_p = \pi/2 - \theta \ll \pi/2$. Putting $R = R_x$ and $R_e = R_A$, we obtain the apex angle of the polar column into which the matter flows:

$$\varepsilon_p \approx \left(\frac{R_x}{R_A}\right)^{1/2} \ .$$
(6.45)

Here, R_A is the radius of the magnetosphere in the vicinity of the magnetic equator:

$$\varepsilon_p \approx \left(\frac{2\dot{M}\sqrt{2GM_x}}{\mu^2} R_x^{7/2}\right)^{1/7} \approx 6° \times R_6^{1/2} \mu^{-2/7} \dot{M}_{18}^{1/7} m_x^{1/14} \ .$$
(6.46)

It should be emphasized that ε_p has a weak dependence on all parameters appearing in this relation. In the entire theory of accretion onto magnetized stars the value of ε_p is estimated with the highest degree of accuracy. The linear radius of the base of the polar column is estimated as $a_p = R_x \varepsilon_p \approx 1$ km.

However, the situation is clouded by the possibility that the accretion channel may be cylindrical (Basko, Syunyaev 1976). Indeed, in the case of disk accretion when the magnetic axis is quite close to the accretion cisk axis, the polar lines of force are not occupied by plasma. Even in the case of spherically symmetric accretion, the column appears to be empty inside since the magnetosphere regions near the poles are the most stable to the Rayleigh-Taylor instability. The thickness of the polar column can be estimated by using the law of conservation of magnetic flux along the hydrodynamic stream lines of the plasma:

$$BS = \text{const} , \tag{6.47}$$

where S is the flow cross section. Writing the condition of magnetic flux conservation in the Alfvén zone and at the pole, we obtain

$$2\pi R_x \varepsilon_p dB_p = 2\pi R_A \delta_m B_A , \tag{6.48}$$

where B_p and B_A are the magnetic field strengths at the magnetic pole and at the magnetopause boundary. Putting these quantities equal to μ/R_x^3 and μ/R_A^3, respectively, we obtain the thickness of the polar channel:

$$d \approx \left(\frac{R_x}{R_A}\right)^{3/2} \delta_m . \tag{6.49}$$

Finally, we write the expression for the ratio of the polar channel thickness to its radius:

$$\frac{d}{a_p} \approx \frac{\delta_m}{R_A} . \tag{6.50}$$

As can be seen δ_m is a few tens percent of the Alfvén radius and the accretion channel in the vicinity of the magnetic poles is filled to the same extent.

The fact that matter falls on the neutron star surface in a highly anisotropic manner leads to the periodic nature of the emitted radiation (effect of rotation of hot spots). In addition, there are other mechanisms in the magnetic field itself that make the radiation directional (Gnedin, Syunyaev 1974; Bisnovaty-Kogan 1973). These mechanisms are associated with a sharp anisotropy in the motion of electrons.

The accretion hydrodynamics in a polar channel were considered by Basko and Syunyaev (1976). We shall present the results of these investigations below. Let us consider the general case of the accretion channel geometry (Fig. 6.6). The analysis and conclusions are based on the solution of a system of hydrodynamic and radiation transport equations. The general pattern is as follows. At a low accretion rate, when the luminosity L_x of the star is less than

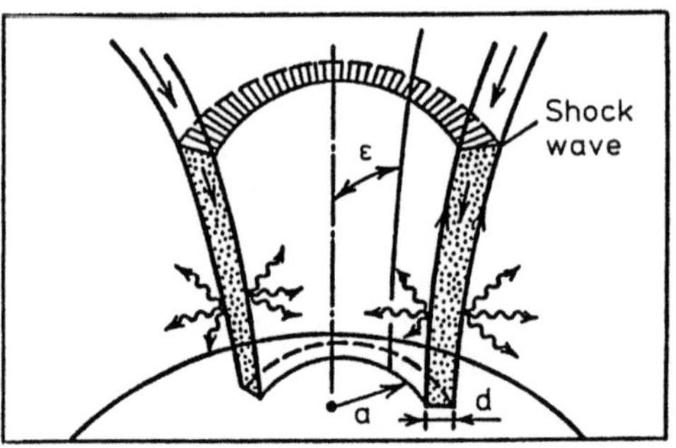

Fig. 6.6. Polar column geometry (Basko, Syunyaev 1976)

a critical value L^*, the gas falls freely on the surface of the neutron star where it gives away its energy as it is decelerated in the atmosphere of the star. For $L_x \simeq L^*$, a radiation shock wave is formed at the surface of the neutron star. The falling gas is stopped by the outgoing radiation and in turn emits radiation of its own. For a given luminosity, the optical thickness across the accretion channel becomes on the order of unity.

Upon a further increase in the accretion rate, the radiation shock wave rises to the Alfvén surface. A slow-settling regime occurs in the channel in this case. The luminosity of the star attains a certain maximum value L_{max} after which it becomes independent of the external accretion rate (the excess energy is carried away in the form of neutrinos). Since there is no spherical symmetry, L_{max} may be several times larger than Eddington's luminosity limit L_{Ed} (3.27).

The solution of the problem on the motion of matter behind the shock wave front was sought under the following assumptions:

1) The accretion channel is stretched along the radius, has a size $d \ll 2\pi a_{\mathrm{p}}$ and height $R - R_x$. It is assumed to be axially symmetric relative to the magnetic axis of the star. The shape of the channel is defined by the equation

$$d(r) = d_{\mathrm{p}} \left(\frac{R}{R_x}\right)^{n/2} \quad ; \quad a = a_{\mathrm{p}} \left(\frac{R}{R_x}\right)^{n/2} \; . \tag{6.51}$$

The case $n = 3$ corresponds to flow along purely dipole field lines.

2) The main contribution to the pressure comes from the radiation: $P = u/3$, where u is the radiant energy density.

3) The gas flow is stationary.

The system of equations of motion, continuity, and energy conservation has the form (Zel'dovich, Raizer 1966)

$$v \frac{\partial v}{\partial R} = -\frac{1}{3\varrho} \frac{\partial u}{\partial R} - \frac{GM_x}{R^2} \; ,$$

$$\varrho v R^n = -s R^n = \mathrm{const} \; , \tag{6.52a}$$

$$\frac{1}{R^n}\frac{\partial}{\partial R}(R^n F_r)+\frac{1}{R}\frac{\partial F_\varepsilon}{\partial \varepsilon}=0 \ . \tag{6.52b}$$

Here, F_r and F_ε denote the energy flux along the directions R and ε (Fig. 6.6):

$$F_r = -\frac{c}{3\kappa\varrho}\frac{\partial u}{\partial R}+\frac{4}{3}uv+\varrho v\frac{v^2}{2}-\varrho v\frac{GM_x}{R} \ ,$$

$$F_\varepsilon = -\frac{c}{3\kappa\varrho}\frac{1}{R}\frac{\partial u}{\partial \varepsilon} \ , \tag{6.53}$$

where $\kappa = \sigma N_e/\varrho = 0.38\,(\sigma/\sigma_T)\,\mathrm{cm}^2/\mathrm{g}$ is the opacity of matter. In the third equation of the system (6.52), we have put $d \ll R\varepsilon$.

Hereafter, all the quantities are referenced to the center of the accretion channel and the transverse diffusion of radiation is averaged over the angle ε:

$$v\frac{dv}{dR}+\frac{1}{3\varrho}\frac{dv}{dR}+\frac{GM}{R^2}=0 \ ,$$

$$\varrho v R^n = -sR^n = \mathrm{const} \ , \tag{6.54}$$

$$\frac{1}{R^n}\frac{d}{dR}(R^n F_r)+\frac{2F_\varepsilon}{d}=0 \ .$$

The corresponding expressions for the radiation flux have the form

$$F_r = -\frac{c}{3\kappa\varrho}\frac{du}{dR}+\frac{4}{3}uv+\varrho v\frac{v^2}{2}-\varrho v\frac{GM}{R} \ ,$$

$$F_\varepsilon = \frac{2c}{3\kappa\varrho}\frac{u}{d} \ . \tag{6.55}$$

The solution of the system of equations (6.54, 55) in different regimes can be obtained in analytical form. Here, we shall confine ourselves to a description of the qualitative flow pattern.

The critical luminosity L^* is defined by

$$L^* = 2\frac{2\pi a_\mathrm{p}GM_x c}{\kappa R}\approx 4\times10^{36}\left(\frac{\sigma_T}{\sigma}\right)\left(\frac{a_\mathrm{p}}{3\times10^4\,\mathrm{cm}}\right)R_6^{-1}m_x\,\mathrm{erg/s} \ . \tag{6.56}$$

For $L_\mathrm{a} = \dot{M}GM/R_x < L^*$, two zones can be distinguished — the free-fall zone and the shock wave zone. In the free-fall zone, matter does not emit radiation and falls freely with a velocity $\sqrt{2GM_x/R}$. The height of the shock wave zone is of the order of the channel thickness d_p, which is much smaller than the size R_x of the star. After the passage of the shock wave, matter loses a part of its kinetic energy. For $L_\mathrm{a} \ll L^*$ the deceleration is caused by Coulomb collisions, as was first observed by Zel'dovich and Shakura (1969). However, for

$L_a \rightarrow L^*$, the contribution of radiation increases the deceleration of electrons which decelerate the ions due to Coulomb forces.

The physical meaning of L^* is that for $L_a = L^*$, the distance over which the radiation stops the incident flux becomes equal to the smallest transverse size of the channel. The optical thickness across the channel then becomes $\tau_T \approx 1$.

For a higher accretion rate $L_a \gg L^*$, the optical thickness becomes larger than unity. The radiation does not manage to carry away energy across the lateral walls and the shock wave rises to such a level that the entire energy liberated in the polar column is emitted through the lateral surface. The most uncertain problem concerns the boundary conditions which must be specified at the bottom of the column. As the accretion rate increases, the shock wave rises until it reaches the Alfvén surface level or attains the maximum height beyond which the accretion channel loses its rigidity. This is possible only for magnetic fields that are not very strong ($\lesssim 10^{13}$ Oe). Upon a further increase in the accretion rate, the luminosity of both polar columns is stabilized at values that are larger than the critical luminosity L^{**} by a factor of 2 or 3:

$$L^{**} = 2\frac{2\pi a_p GM_x c}{d_p \kappa} = \frac{a}{d}L_{\mathrm{Ed}} \approx 10^{39}\left(\frac{a}{10d}\right)\left(\frac{\sigma_T}{\sigma}\right)m_x \, \mathrm{erg/s} \ . \tag{6.57}$$

This expression shows the extent to which the limiting luminosity depends on the channel geometry. For example, $L^{**} = L_{\mathrm{Ed}}/4$ for a continuous axisymmetric channel.

If the magnetic field strength at the poles of a neutron star is $B \gtrsim 2\times 10^{13}$ Oe, the magnetic pressure can confine a plasma of such high density and at such high temperature that the entire energy is carried away by neutrinos. Indeed, equating the radiation pressure to the magnetic field pressure

$$u/3 = B^2/8\pi \ , \tag{6.58}$$

we find ($u = aT^4$) that for $B \gtrsim 2\times 10^{13}$ Oe the field will confine plasma with radiation at a temperature of $T \gtrsim 9\times 10^9$ K. At densities $\sim 10^5 - 10^6$ g/cm^3, which are typical of the conditions considered here, the main contribution to the rate of neutrino losses comes from the process of electron-positron pair annihilation (Beaudet et al. 1967):

$$e^+ + e^- \rightarrow v_e + \bar{v}_e \ . \tag{6.59}$$

The energy loss is given by

$$\varepsilon_v \approx 10^{24} \ \mathrm{erg/cm^3 \, s}$$

and is independent of the density of matter. The formation of radiation spectrum of a polar column is considered in Sect. 6.7.

In conclusion, we would like to mention some of the works in which the flow in a polar column was investigated: Inoue (1975), Shapiro and Salpeter (1975), Wang and Frank (1981), Langer and Rappaport (1982), and Morfill et al. (1984).

6.3 Spin-up, Spin-down and Induced Precession of Accreting Stars

The spin-up and spin-down of an accreting star are determined by the exchange of angular momentum between accreting matter and the star. This exchange is realized through electromagnetic forces in the Alfvén zone.

6.3.1 Spin-up Torque

The spin-up of accreting neutron stars can take place in binary systems in which the trapped matter has an angular momentum due to orbital motion (Shvartsman 1970a, 1971a). In the case of disk accretion, the spinning-up torque can be presented as the flux of angular momentum transferred from the last Keplerian orbit (Pringle, Rees 1972):

$$K_{su}(d) = \dot{M}\sqrt{GM_x R_d} \; . \tag{6.60}$$

The narrower the transition zone in which the accreting matter is decelerated, the better this estimate.

If there is no accretion disk and the matter is captured directly from stellar wind, we can use the estimate (3.84) (Davidson, Ostriker 1973; Illarionov, Syunyaev 1975):

$$K_{su}(w) = \dot{M}\eta_k \Omega R_G^2 \; . \tag{6.61}$$

Moreover, $K_{su}(d) \geqslant K_{su}(w)$. If this were not so, an accretion disk would be formed around the star and the "excess" angular momentum would be carried away by turbulent viscosity.

The last argument indicates that irrespective of the accretion regime there exists an upper limit of torque spinning up the star. It can be easily verified that the maximum torque is given by (Lipunov 1981)

$$K_{su}(\text{max}) = \dot{M}\sqrt{GM_x R_c} \; . \tag{6.62}$$

If matter had a much larger angular momentum, it would not penetrate beyond the corotation radius and could not transfer its torque to the star.

Matter falling on the surface of a neutron star emits its gravitational energy in the form of radiation so that the luminosity of the accreting star is determined by the standard relation $L_x = \dot{M}GM/R_x$. Using this relation as well as the equation $dI\omega/dt = K_{su}(\max)$ for the change of angular momentum, we obtain the maximum value for the spin-up of the accreting star,

$$|\dot{p}_{su}|_{\max} = \frac{R_x p^{7/3} L_x}{(2\pi)^{4/3} (GM_x)^{1/3} I}$$

$$\approx 5.3 \times 10^{-5} m_x^{-1/3} I_{45}^{-1} R_6 p^{7/3} L_{37} \text{ s/year} , \qquad (6.63)$$

where L_x is the accretion luminosity of the star, and R_x and I the radius and the moment of inertia of the star.

Thus we have found the restriction imposed on the relation between the three observed quantities $p, \dot{p}$, and L_x. It should be emphasized that the upper limit (6.63) does not contain quantities depending on the magnetic field structure and is independent of the specific accretion regime.

6.3.2 Spin-down Torque

It is much more difficult to compute the spin-down torque applied to an accreting star having a magnetic field. We recommend the following expresson:

$$K_{sd} = \kappa_t \frac{\mu^2}{R_c^3} , \qquad (6.64)$$

where κ_t is a dimensionless constant of the order of unity.

A similar result is obtained from the consideration of the magnetic field strength in an accreting disk. It was suggested by Lynden-Bell and Pringle (1974) and Lipunov (1981) that the stellar field penetrates the disk. The appearance of a spin-down torque is associated with the emergence of the toroidal component B_φ of the magnetic field:

$$K_{sd} = \int_{R_{\min}}^{\infty} \frac{B_z B_\varphi}{4\pi} R\, dS , \qquad (6.65)$$

where $R_{\min}$ is the minimum distance starting from which the field in the disk is directed in such a way that the torque is against the rotation of the star. Obviously, $R_{\min} = R_c$. Putting

$$B_z B_\varphi = \kappa_t B_d^2 , \quad \text{where} \quad B_d = \mu/R^3$$

is the dipole field, we arrive at (6.64).

Finally, let us consider an entirely different situation. Suppose that the magnetosphere is surrounded by a turbulized plasma shell (Davies, Pringle 1981; Lipunov 1982c). The finite viscosity causes a friction between the magnetosphere and plasma. The spin-down torque can be estimated by using the classical formula for deceleration of a sphere immersed in a viscous fluid

(Landau, Lifshitz 1953). Approximating the magnetosphere by a sphere of radius R_A, we obtain

$$K_{sd} = 8\pi v_t \varrho R_A^3 \omega \; ,$$

where v_t is the turbulent viscosity. Putting this quantity equal to

$$v_t = \tfrac{1}{3} v_t l_t = \kappa_t \omega R_A R_A$$

and assuming that

$$R_A = \left(\frac{\mu^2}{2\dot{M}\sqrt{2GM_x}} \right)^{2/7} \; ,$$

$$\varrho = \frac{\dot{M}}{4\pi R_A^{3/2}\sqrt{2GM_x}} \; ,$$

we again arrive at (6.64).

The spin-down of a neutron star due to the generation of acoustic waves in accreting plasma was considered in detail by Sibgatullin (1984). Acoustic spin-down is less effective than the processes considered above.

6.3.3 Analytical Model of Torques Applied to a Magnetized Accreting Star

The most important aspect in the analysis of many observed features of accreting neutron stars in binary systems is the assumption that spin-up and spin-down torques exist simultaneously at the accretion stage. The idea that a part of the matter may acquire an angular momentum from the star while another part may transfer angular momentum to the star was put forth by Lipunov and Shakura (1976) and later for the case of disk accretion by Ghosh and Lamb (1978).

In the disk accretion regime, matter interacting with the magnetic field inside the corotation radius spins up the star while interaction outside the corotation radius leads to a spin-down. In the case of a two-stream accretion, part of the angular momentum may be carried away by spherical flux or by a turbulized spherical shell.

In the most general form, we can write the equation for the change of angular momentum of an accreting star in the form (Lipunov 1982c)

$$\frac{dI\omega}{dt} = \begin{cases} \dot{M}\eta_k \Omega R_G^2 - \kappa_t \dfrac{\mu^2}{R_c^3} \; , & \text{quasispherical accretion} \\[4mm] \dot{M}\sqrt{GMR_d} - \kappa_t \dfrac{\mu^2}{R_c^3} \; , & \text{disk} \end{cases} \tag{6.66}$$

where $R_d = \varepsilon R_A$. This model of angular momentum contains three dimensionless parameters η_t, κ_t and ε, which are apparently close to unity.

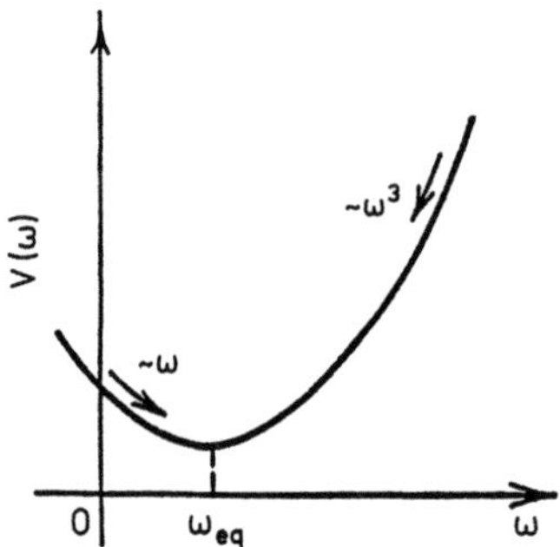

Fig. 6.7. Scalar potential for an accreting star in a binary system

Formula (6.66) is applicable only as long as accretion takes place. In the case of disk accretion, for example, this happens as long as $R_d \leqslant R_c$, i.e., $p \geqslant p_A$ (Chap. 4). If $R_d \gg R_*$ the change in the moment of inertia can be neglected. In this case, using the relations

$$p = 2\pi/\omega, \quad L = \dot{M}GM_x/R_x,$$

$$A \approx 5 \times 10^{-4}(3\kappa_t)\mu_{30}^2 I_{45}^{-1} m_x^{-1} \text{ s/year,}$$

$$B_d \approx 7.7 \times 10^{-5} \varepsilon^{1/2} m_x^{-3/7} R_6^{6/7} I_{45}^{-1} \text{ s/year .}$$

$$B_w \approx 5.2 \times 10^{-6} R_6^2 \varphi(m) I_{45}^{-1} \dot{M}_{-6}\eta \text{ s/year,} \quad \varphi(m) \approx \frac{M_0^{2/3}/M_x^2}{10^{2/3}} M_\odot ,$$

we obtain

$$\dot{p} = \begin{cases} A - B_d p^2 L_{37}^{6/7} , & \text{disk} \\ A - B_w p^2 L_{37} T_{10}^{1/3} , & \text{quasispherical accretion} \end{cases} \tag{6.67}$$

Let us now turn to (6.66) describing the change in the angular momentum. It follows from the form of these equations that we can introduce a scalar potential $V(\omega)$ such that (Lipunov 1987a)

$$\frac{d\omega}{dt} = -\nabla_\omega V(\omega) . \tag{6.68}$$

For the torque (6.66), the scalar potential is

$$V(\omega) = \begin{cases} -A_2\omega + A_3\omega^3 + \text{const} , & \omega \geqslant 0 , \\ -A_2\omega - A_3\omega^3 + \text{const} , & \omega < 0 . \end{cases} \tag{6.69}$$

The advantage of the scalar potential lies not only in the simplicity of the energy interpretation (Fig. 6.7) according to which the neutron star strives to attain a state corresponding to the minimum of $V(\omega)$, but also in that it is convenient to describe the random fluctuations of the period of rotation of an accreting star (Sect. 6.8).

6.3.4 Equilibrium Period

The evolution equation presented above indicates that an accreting neutron star must endeavor to attain an equilibrium state in which the resultant torque vanishes (Davidson, Ostriker 1973; Lipunov, Shakura 1976). This hypothesis is confirmed by the observations of X-ray pulsars (see below).

Equating the right-hand side of (6.66) to zero, we obtain the equilibrium value of the period:

$$p_{eq} \approx 7.8\,\pi \sqrt{\kappa_t/\varepsilon^2}\,(GM_x)^{-5/7} y^{-3/7}\,\text{s}\ ,\quad \text{disk} \tag{6.70}$$

$$p_{eq} = \sqrt{A/B_w}\,L_{37}^{-1/2}\,T_{10}^{-1/6}\,\text{s}\ ,\quad \text{quasispherical accretion} \tag{6.71}$$

where $y = \dot{M}/\mu^2$ is the gravimagnetic parameter.

Alternatively, we can present the period in a form convenient for computations:

$$p_{eq} \approx 1.0\,L_{37}^{-3/7}\,\mu_{30}^{6/7}\,\text{s}\ ,\quad \text{disk} \tag{6.72}$$

$$p_{eq} \approx 10\,\eta_k^{-1/2}\,\dot{M}_{-6}^{-1/2}\,\varphi(m)^{-1/2}\,L_{37}^{-1}\,T_{10}^{-1/6}\,\mu_{30}\,\text{s}\ ,\quad \text{stellar wind} \tag{6.73}$$

where $T_{10} = T/10$ is the period of the binary system, $\dot{M}_{-6} = \dot{M}_0/10^{-6}M_\odot$/year is the stellar wind rate, $\varphi(m) \approx m_{10}^{2/3}/m_x^2, m_{10} = M_0/10M_\odot$ the mass of the optical star, $\kappa_t = 1/3$, and $\varepsilon = 0.45$.

Let us turn to the case of disk accretion. The above model of spin-up and spin-down torques possesses an unexpected property. The equilibrium period obtained by setting the torque to zero is connected with the critical period p_A through a dimensionless factor:

$$p_{eq} = \sqrt{2\sqrt{2}}\,\frac{\kappa_t^{1/2}}{\varepsilon^{7/4}}\,p_A\ . \tag{6.74}$$

The parameters κ_t and ε must be such that $p_{eq} > p_A$.

Since $\kappa_t \approx \varepsilon \approx 1$, the equilibrium period in the case of disk accretion is close to the critical period p_A separating the accretion stage A and the propeller stage p. This means that a small variation in the accretion rate (a few tens or a few hundreds percent) will transform the neutron star from the accretion regime to the propeller state (due to a change in p_A and not the period of the star!). These small variations in $\dot{M}$ lead to catastrophic changes in the luminosity of the accreting star. Hence such a stage was termed "catastrophic equilibrium" (Lipunov 1987b).[1]

This can be illustrated by the following scheme (Fig. 6.8). Upon a random variation of the accretion rate, the neutron star oscillates from one state to another. This is accompanied by an enormous change (by a factor of several hundreds) in luminosity:

---

[1] In the theory of catastrophes, a catastrophe means an abrupt change in an internal parameter of the system (luminosity) upon a slight variation of the external parameter (accretion rate) (Poston, Stewart 1978).

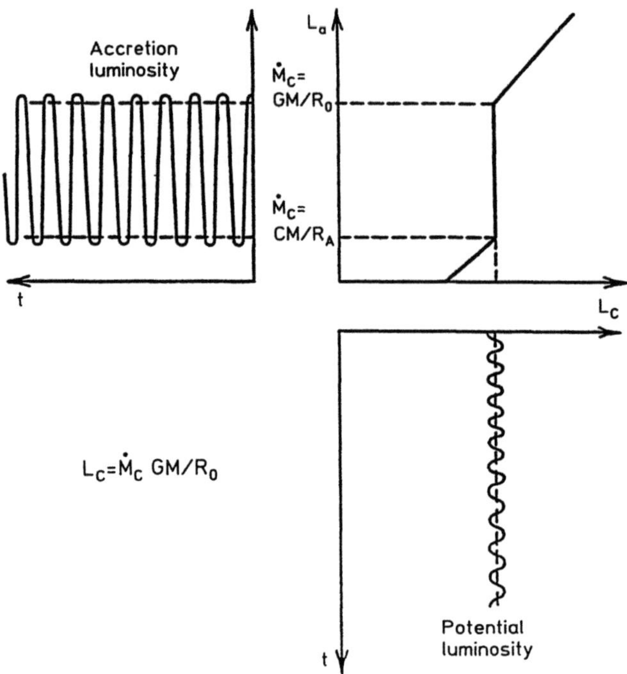

Fig. 6.8. The catastrophic equilibrium: a slight variation in the accretion rate (potential luminosity, *lower right*) leads to a sharp change in the X-ray luminosity (accretion luminosity, *upper left*)

$$A \rightleftarrows P\,,$$

$$L = \frac{\dot{M}GM_x}{R_x} \rightleftarrows L = \frac{\dot{M}GM_x}{R_A}\,.$$

In the case of accretion from stellar wind, p_{eq} is apparently much larger than p_A and the equilibrium is not catastrophic.

6.4 Observed Properties of X-Ray Pulsars

X-ray pulsars were discovered aboard the specialized X-ray satellite Uhuru (Schreier et al. 1972). About 20 X-ray pulsars are known to exist at present (Table 6.1). The main observed quantities are: X-ray luminosity L_x, pulsar period p and variations $\dot{p}$ of this period, pulsar masses M_x, periods of rotation T of binary systems (all pulsars are undoubtedly members of binary systems). This should be supplemented by detailed information about the kind of spectrum (usually thermal, with temperatures of $10-20$ keV) and the pulse shape (Fig. 6.9). Most of the X-ray pulsars have been identified with massive

Table 6.1. X-ray pulsars

Pulsar	Pulsation period [s]	Orbital period [days]	Normal component	Time scale for spin-up [yrs]	Luminosity [10^{37} erg/s]
A 0538−66	0.069	16.66	Be	?	80
SMC X-1	0.71	3.892	B0 I	1400	60
Her X-1	1.24	1.7	HZ Her	$3 \cdot 10^5$	1
H 0850−42	1.8				
4U 0115+63	3.61	24.31	B	30000	3
V 0332+53	4.375	34	Be	200	
Cen X-3	4.84	2.087	06 II−III	3400	5
1E 2259+59	6.98	0.03			1
4U 1627−67	7.68	0.0289	KZ TrA	5000	
2S 1553−54	9.26	30.7 (?)			
LMC X-4	13.5	1.408		$>10^3$	35
2S 1417−62	17.6	>15			<4
OAO 1653−40	38.2			200	>0.04
4U 1700−37	67.4				
A 0535+26	104	111(?)	09.7 IIe	1000	2
GX 1+4	122	15	M6 IIIe	47	4
4U 1230−61	191				
GX 304−1	272	135(?)	B0−B5		0.2
Vela X-1	283	8.965	O−B	3000	0.15
4U 1145−61	292	187(?)		>1000	0.03
1E 1145.1−61	297	>12	HEN 715	>300	0.3
A 1118−61	405		Mira-type		0.5
4U 1907+09	437	8.4	O−B		
4U 1538−52	529	3.73	B0 I	>500	0.4
GX 301−2	696	41.4	A 977	>100	1
X Per	835	580(?)	O9e	1400	0.001

OB-stars (Bradt et al. 1979). However, some examples of pulsars in low-mass close binary systems with red dwarfs (4U 1626−67, see Joss et al. 1978; Middleditch et al. 1981) and red giants of M 6-type (GX 1+4, see Glass, Feast 1973; Davidsen et al. 1976) arc also known. In the next section we briefly consider the main observational data listed in Table 6.1.

6.5 Energy Parameters of Pulsars and Transport of Matter in Binary Systems

The luminosity of X-ray pulsars varies form $\sim 10^{33}$ erg/s (for pulsar X Per) to $\sim 10^{39}$ erg/s (characteristic representatives SMC X-1, A 0538−66, etc.).

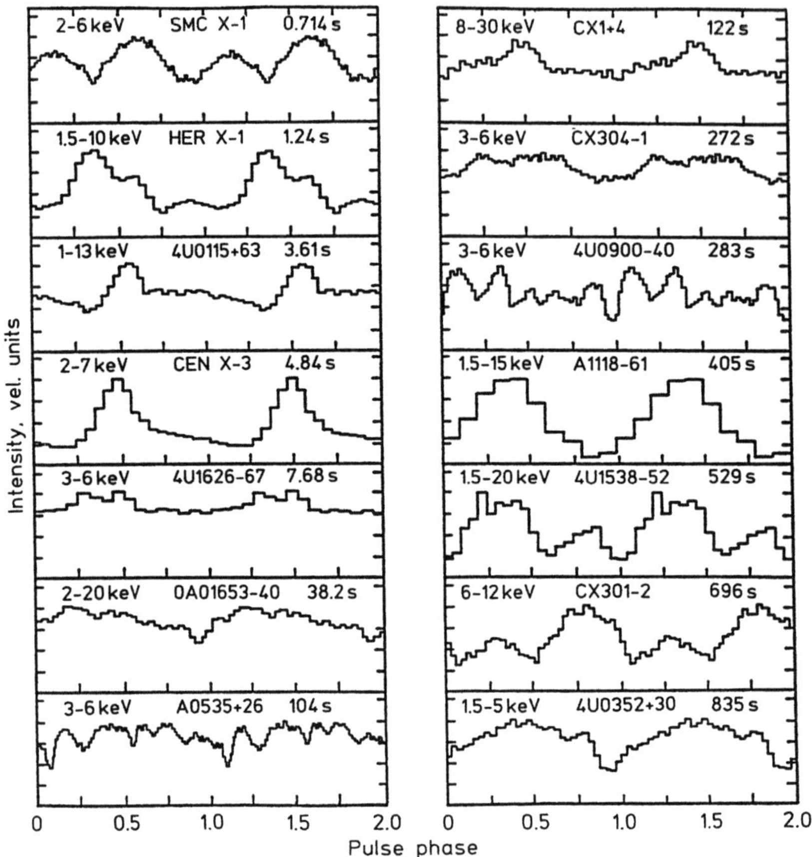

Fig. 6.9. Pulse profiles of several X-ray pulsars (Rappaport, Joss 1983)

In order to ensure such a luminosity, matter must accrete to the surface of the neutron star at a rate of $10^{13} - 10^{19}$ g/s, which corresponds to $10^{-13} - 10^{-7} M_{\odot}$/year. Such mass flows may come about in three different ways depending on the type of the normal star (Sect. 3.8). If the normal component is a supergiant whose Roche lobe has not been filled and which intensively runs out in the form of spherical stellar wind, the trapping of matter is described by the Bondi-Hoyle-Littleton formula (Sect. 3.3). The X-ray luminosity of the pulsar is defined by the expression [see (3.81)]

$$L_x \approx 10^{36} \frac{\xi_1 m_x^3 \dot{M}_{-6}}{(m_0+m_x)^{2/3} T_{10}^{4/3} v_8^4 R_6 (1+\tan^2 \beta)} \text{ erg/s} . \tag{6.75}$$

Here, ξ_1 is a dimensionless factor of the order of unity, m_x and m_0 are the masses of the neutron star and the optical component in units of solar mass,

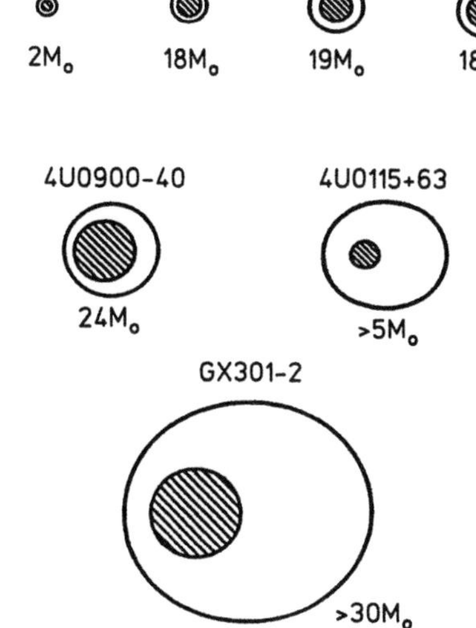

Fig. 6.10. Orbits of X-ray pulsars and size of optical components (shown on the same scale) (Rappaport, Joss 1983)

$\tan \beta = v_{\mathrm{orb}}/v_{\mathrm{w}}$ is the ratio of the orbital velocity to the stellar wind velocity at the site of the accreting star.

Formula (6.75) indicates that quite specific (though not very stringent) conditions must be satisfied for the emergence of a bright source with luminosity $\sim 10^{38} - 10^{39}$ erg/s. It can be seen that for $\dot{M}_{-6} \approx 1$, we can vary only with the velocity of the stellar wind: the condition $v_{\mathrm{w}} \approx 200 - 300$ km/s must be satisfied. This estimate is much lower than the observed velocities in the stellar winds of blue supergiants at infinity: $v_{\infty} \approx 1000 - 3000$ km/s. Hence a neutron star must be as close as possible to a supergiant where the stellar wind velocity is lower. In general, these concepts are confirmed by observations. In systems of the type Cen X-3, SMC X-1, etc., the neutron star is so close to the supergiant that the latter almost completely fills its Roche lobe (see the binary systems in Fig. 6.10).

A considerable part of X-ray pulsars constitutes binary systems with Be-stars (Λ 0535 + 26, A 0538 − 66, etc.). All these pulsars are characterized by very sharp variations in luminosity (of a transient nature) and a very high luminosity in the active state: $\sim 10^{37} - 10^{38}$ erg/s. These values are quite astonishing since the rate of outflow of Be-stars is very low: $\dot{M} \approx 10^{-8} - 10^{-9}\, M_\odot$/year. In order to ensure a luminosity of $10^{37} - 10^{38}$ erg/s, it must be assumed that a neutron star in the active state captures a considerable part of the matter released by the normal star. For this purpose, the stellar wind velocity must be very low. In systems with supergiants, this condition is met because the components are close. However, systems with Be-stars are quite different: their periods may be as large as tens or even hundreds of days! Apparently, this is due to a very extraordinary regime of outflow of Be-stars. These stars rotate rapidly and dissipate matter quite anisotropically in the equatorial rotational plane. Most probably, the velocity of expansion of the shell is not very high (it cannot be much higher than the Keplerian velocity, otherwise there is no reason for the formation of a disk shell) and is of the order of $100-300$ km/s. Moreover, outflow is a highly transient process and occurs on time scales varying from a few months to several years.

Finally, the third regime occurs in low-mass binary systems (Her X-1, GX1+5, etc.), where the mass of the optical component does not exceed a few solar masses. Such stars have an extremely weak stellar wind ($\sim 10^{-12}-10^{-14} M_{\odot}$/year) and may be the source of matter for a bright pulsar only if they fill their Roche lobe. In this case, the rate of outflow may be determined by one of the three characteristic times (Yungel'son, Masevich 1983), viz., the nuclear time t_{nucl}, the thermal time t_{KH}, and the time associated with the dissipation of angular momentum of a binary system due to the emission of gravitational waves or magnetic stellar wind.

Following Yungel'son and Masevich (1983), we can estimate the thermal and nuclear times as $t_{\text{nucl}} \approx 10^{10} m_0^{-2}$ years, and $t_{\text{KH}} \approx 3 \times 10^7 m_0^{-2}$ years. It follows from these estimates that for $m_0 \approx 1$, the outflow rate may attain the value $10^{-10}-10^{-8} M_{\odot}$/year corresponding to an X-ray luminosity of $10^{36}-10^{38}$ erg/s.

6.6 Spectrum and Magnetic Fields

The spectra of X-ray pulsars (Fig. 6.11) are of a thermal nature with a characteristic exponential cutoff at high energies (in some cases, however, a "hard tail" is observed). The characteristic photon energy in the vicinity of the maximum of the spectrum is ~ 10 keV. This is quite naturally in agreement with the observed luminosity of pulsars and the accretion pattern described above, in which matter falls on a small region near the pole on the surface of the neutron star.

We assume that the radiation emitted by hot spots at the poles is thermalized. Let us estimate the characteristic energy of photons by proceeding from the Stefan-Boltzmann formula

$$\varepsilon_\gamma \approx 3 k_{\text{B}} T_{\text{bb}} \approx 3 k_{\text{B}} \left(\frac{L_x}{\sigma_{\text{SB}} S} \right)^{1/4} \approx 30 L_{38}^{1/4} S_{10}^{-1/4} \text{ keV} ,$$

where $S_{10} = S/10^{10}$ cm^2 is the area element from which energy is emitted. A strong magnetic field leads to a radiation anisotropy (Bisnovaty-Kogan 1973; Lamb et al. 1973; Gnedin, Syunyaev 1974; Basko, Syunyaev 1975; Wang 1975). The problem concerning the formation of beamed radiation is still at the stage of initial investigation (see, for example, Yahel 1980). Moreover, it is clear that for a high luminosity ($10^{38}-10^{39}$ erg/s), the mechanism of formation of

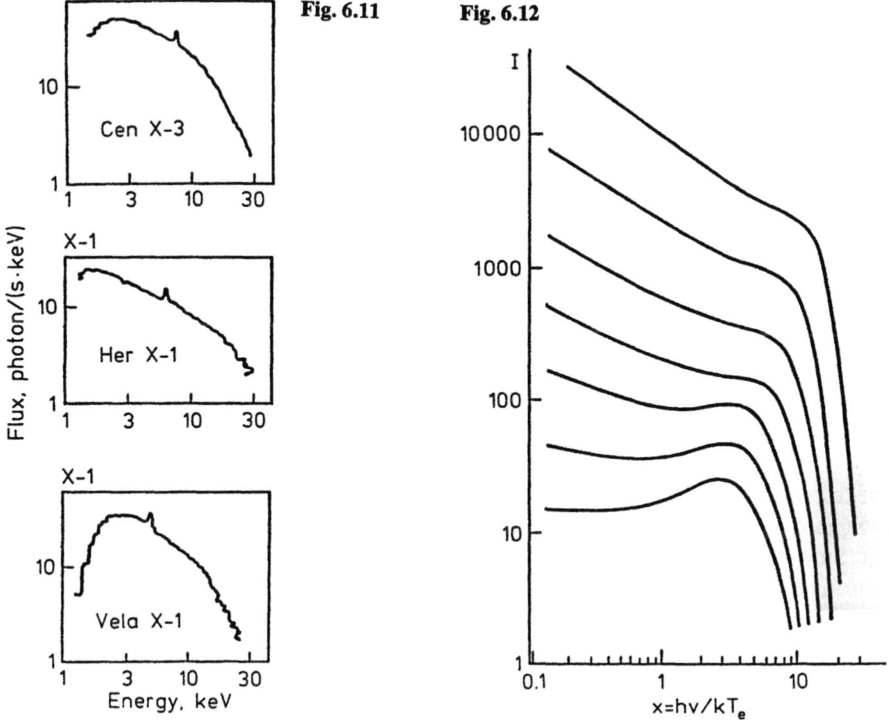

Fig. 6.11. Spectra of three X-ray pulsars obtained on board the specially equipped X-ray satellite TENMA (Japan) (Nagase 1985). The emission feature is the iron line

Fig. 6.12. Spectra formed in radiation-dominated shock waves (Lyubarskii, Syunyaev 1982)

beamed radiation by a magnetic field is no longer valid (Basko, Syunyaev 1976) since the radiation is strongly thermalized. Here, the geometrical factors (while considering the rotation of a neutron star, we observe the polar column at different angles) or a screening of poles by matter flowing in the magnetosphere may play a significant role.

The existence of gyrolines in the spectra of X-ray pulsars at energies corresponding to the cyclotron frequency

$$E = \hbar\omega_c \approx 12\,B_{12}\,\text{keV}$$

was remarkably predicted by Gnedin and Syunyaev (1974) and discovered by Trümper et al. (1978) (Fig. 1.9). The spectral feature observed in the radiation of Her X-1 lies in the interval $30-50$ keV corresponds to a magnetic field $B \approx (3-5) \times 10^{12}$ Oe (the gravitational red shift cannot change this estimate significantly). Features of the same type (but manifested less strongly) were also observed in two pulsars: 4U 0115+63 (Wheaton et al. 1978) and

4U 1626−67 (Pravdo et al. 1977). The magnetic field strengths corresponding to the positions of these lines are $\sim 2 \times 10^{12}$ Oe.

The analysis of radiation in such strong fields has been carried out in a number of papers and reviews. We shall not go into the details of this problem but refer the reader to the review by Pavlov and Gnedin (1983) and the monograph by Dolginov et al. (1979). The effects of variation of the scattering cross-section in a magnetic field, vacuum polarization, and the formation of gyrolines in the emission spectrum are the most significant.

The shape of the spectra of most of the X-ray pulsars indicates that comptonization of the radiation-dominated shockwave plays an important role in the formation of these spectra. It should be recalled that the importance of comptonization in the formation of spectra of accreting neutron stars was emphasized by Zel'dovič and Shakura (1969) (Chap. 3). Comptonization of radiation was investigated firstly by Kompaneets (1956) and, in detail by Syunyaev and Titarchuk (see the review by Pozdnyakov et al. 1982). The formation of the emission spectrum in a radiation-dominated shock wave which is transparent to true absorption was studied by Lyubarskii and Syunyaev (1982) who investigated in detail the fate of energy in the flow pattern proposed by Basko and Syunyaev (1976). Plasma is slowed down as a result of scattering of photons by electrons moving together with the plasma. Photons acquire energy on account of the Doppler effect. When the energy acquired by photons becomes sufficiently high, most of it is transferred to the electrons (Compton effect). Practically all of the energy of incident plasma is concentrated in protons which "drag" the electrons through the photon gas due to Coulomb forces, heating the gas and the electrons, though the protons themselves are the last to be heated as a result of collisions with electrons (Lyubarskii, Syunyaev 1982).

Figure 6.12 shows the typical spectra formed in the radiation-dominated shock waves. The spectra are characterized by small values of the spectral indices in $I \propto \nu^{-\alpha}$ ($0 < \alpha < 1$) and by an exponential cutoff at $h\nu \gtrsim 100$ keV. The solution obtained by Lyubarskii and Syunyaev (1982) is applicable either for weak fields ($\sim 10^{11}$ Oe) or strong fields for which the gyrofrequencies are higher than the emission frequencies (this corresponds to fields $\gtrsim 10^{13}$ Oe).

Several X-ray sources exhibit FeXXIV emission lines. The formation of such lines on the Alfvén surface was considered by Basko (1980).

6.7 Periods of X-Ray Pulsars and Their Variation

Among the characteristics of X-ray pulsars, the ones that are measured most accurately are their periods and the rate of variation of these periods. The periods of X-ray pulsars vary over a wide range from tens of milliseconds to thousands of seconds. The distribution over periods (in spite of its sketchiness) leads to the conclusion that most of the pulsars have large periods (larger than ~ 100 s, Table 6.1). Such periods are more characteristic of white dwarfs than

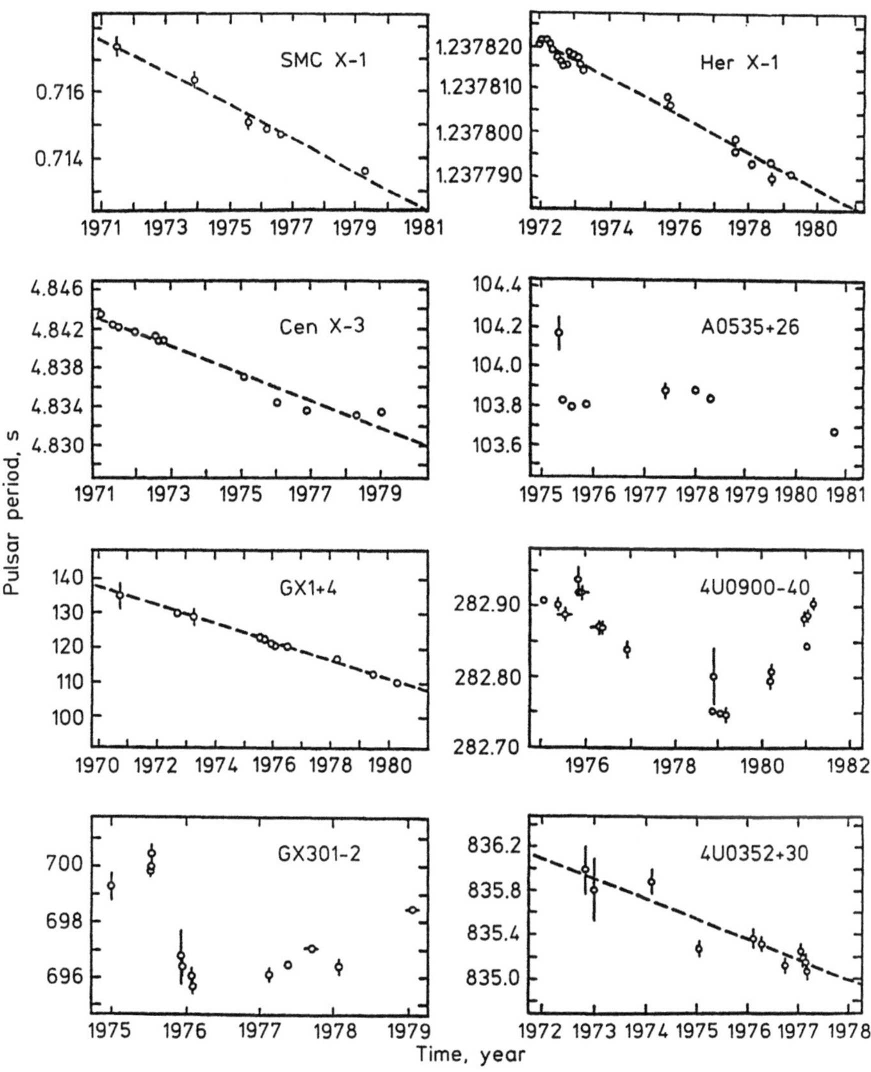

Fig. 6.13. Change in the periods of X-ray pulsars over time (Rappaport, Joss 1983)

of neutron stars. Hence it can be naturally asked if these X-ray pulsars are indeed neutron stars. The answer to this question is given by the following properties of X-ray pulsars.

The pulsar periods (after subtracting the orbital motion) do not remain transient (Fig. 6.13) and vary on time scales ranging from tens of years (for long-period pulsars) to hundreds of thousands of years. The maximum spin-up of an accreting neutron star was estimated in Sect. 6.4. Analyzing the structure of (6.63) for limiting spin-up, we observe that the spin-up $|\dot{p}_{\mathrm{su}}|_{\max}$ for neutron stars is about $100-1000$ times larger than for white dwarfs. Figure 1.10 (Lipunov 1981) shows the plot of the dependence of $|\dot{p}_{su}|_{max}$ on the parameter $p^{7/3} L$. A comparison of the theoretical upper limit with the observed values of pulsar spin-up clearly indicates that these X-ray pulsars are accreting neutron stars and not white dwarfs.

The variation in the pulsar period is a random phenomenon. However, on the average pulsars spin-up (this is true for most pulsars for which measurements of $\dot{p}$ were carried out). There are three important questions to which we seek answers in the theory of X-ray pulsars:

1) How can the observed distribution of pulsars over periods be explained?

2) The long-period pulsars ($p \gtrsim 100$ s) are components of massive stars whose lifetime is $\sim 10^7$ years. Assuming that neutron stars are created with small periods (Sect. 11.2), the spin-down time to a period of ~ 100 s calculated according to the magnetic dipole law (4.12) in a standard field $\mu_{30} \approx 1$ is found to be much larger than the lifetime of a normal component: $\gg 10^7$ years. How can we then explain the paradox that neutron stars do manage to spin-down to such large periods during the lifetime of the normal component?

3) Why is a spinning-up observed in most of the pulsars?

6.7.1 Equilibrium of X-Ray Pulsars

The change in the angular momentum of an accreting neutron star is associated with spinning-up and spinning-down torques:

$$\frac{dI\omega}{dt} = K_{su} - K_{sd} \ . \tag{6.76}$$

Let $\dot{p}_{su}$ be the spin-up in the absence of spin-down torques ($K_{sd} = 0$) and, conversely, let $\dot{p}_{sd}$ be the spin-down for $K_{su} = 0$. In this case, we can write

$$\dot{p} = \dot{p}_{su} + \dot{p}_{sd} \ . \tag{6.77}$$

It was shown in Sect. 6.4 that under the action of these torques, the accreting star attains an equilibrium state in which $\dot{p} = 0$. Such a rigorous equality could be ensured only when the external parameters (e.g., the accretion rate) remain unchanged.

Indeed, the transfer of matter in a binary system is subjected to a number of random variations which could be associated with the properties of a normal star. The change in the accretion rate also changes the inflow of the angular momentum. Hence the equilibrium described in Sect. 6.4 for a real binary system can be maintained only on the average. Hence the change in the period must approach zero only when averaged over a sufficiently long interval of time:

$$\langle \dot{p} \rangle = 0 \ . \tag{6.78}$$

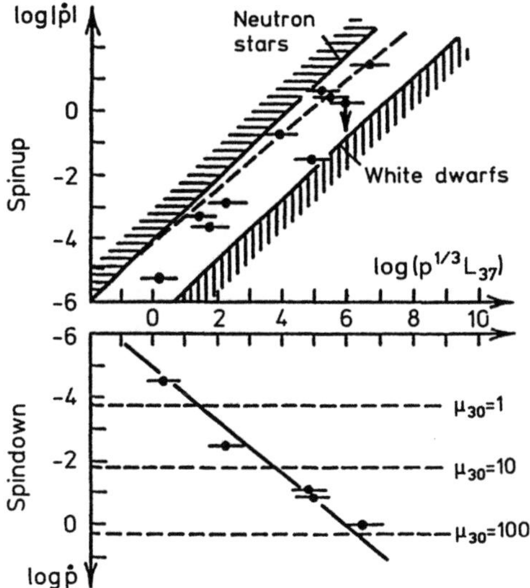

Fig. 6.14. Dependence of the spin-up and spin-down of X-ray pulsars on the observed characteristics of pulsars. The upper (*hatched*) region corresponds to the maximum spin-up for a neutron star and the lower one to a white dwarf (Lipunov 1981 a)

Pulsars must exhibit a reversal in the sign of $\dot{p}$, i.e., a transition from spin-up to spin-down, and vice versa (Lipunov, Shakura 1976). Such jumps in the values of $\dot{p}$ have been observed so far in five X-ray pulsars. This in itself is a strong argument supporting the fact that they are in an equilibrium state. It follows from (6.77, 78) that the rate of variation of the period during spin-down must correlate with the spin-up rate (Lipunov 1981 a):

$$|\dot{p}_{\rm sd}| \sim |\dot{p}_{\rm su}| \; . \tag{6.79}$$

The correlation (6.79) is indeed confirmed by observations (Fig. 6.14). This proves that the pulsar periods are close to their equilibrium values. The value of the equilibrium period $p_{\rm eq}$ depends strongly on the accretion regime [see (6.72, 73)].

It is qualitatively clear that the equilibrium period is longer, the larger the torque spinning down the neutron star and the smaller the spinning-up torque. Accordingly, there are two ways of explaining the large values of pulsar periods: either we can assume that for some reason the spin-down torques are large or that the spin-up torques are small in the case of long-period pulsars. Shakura (1975) proposed that long-period pulsars have superstrong magnetic fields of $\sim 10^{14}$ Oe. An alternative explanation was offered by Basko and Syunyaev (1976), who assumed that the spin-up torque in the case of direct accretion from stellar wind is small and the pulsar will spin-down to periods of hundreds and thousands of seconds.

The first interpretation seems to be preferable in spite of its nontrivial nature. If we accept this interpretation, it becomes clear that neutron stars manage to spin-down to such large periods (the spin-down torque is large, although the spin-down time will be small).

Let us write the approximate expression for the equilibrium period corresponding to the analytical model for spin-up and spin-down torques (6.6) $\varepsilon = 0.45$; $\kappa_t = 1/3$; $\eta_k = 1/4$:

$$p_{\text{eq}} \approx 1\, \mu_{30}^{6/7} L_{37}^{-3/7}\ \text{s} \qquad\qquad \text{disk}$$

$$p_{\text{eq}} \approx 20 \dot{M}_{-6}^{-1/2} \varphi(m)^{-1/2} L_{37}^{-1} T_{10}^{-1/6} \mu_{30}\ \text{s} \qquad \text{quasispherical accretion .}$$

$$\tag{6.80}$$

It should be emphasized that in the case of disk accretion, the gravimagnetic parameter is uniquely linked with the equilibrium period through the relation

$$y \approx 10^{-43} p_{\text{eq}}^{-7/3}\ \text{s/cm}^5\ . \tag{6.81}$$

It follows from (6.80) that a long-period pulsar with luminosity $L_x = 10^{36} - 10^{37}$ erg/s may be formed in the disk accretion regime only if the neutron star has a sufficiently high field: $\mu_{30} \approx 100$.

A strong argument in favor of the hypothesis on the equilibrium of X-ray pulsars was put forth by Gosh and Lamb (1979a). Between 1970 and 1980, the pulsar Her X-1 had a characteristic spin-up time $t_{\text{su}} = |p/\dot{p}| \approx 300\,000$ years. However, if the pulsar was only spun-up (from the accretion disk) and if the

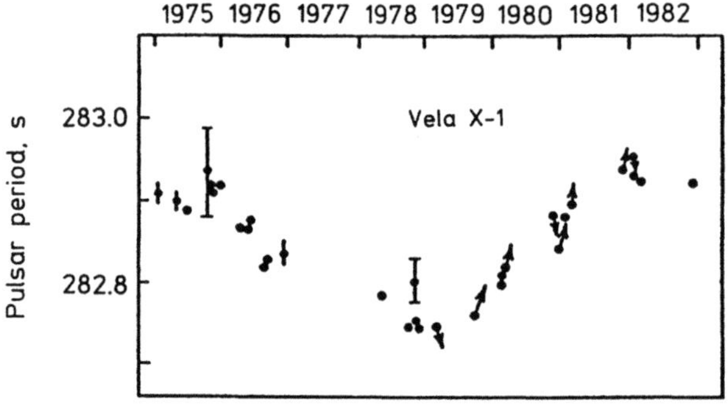

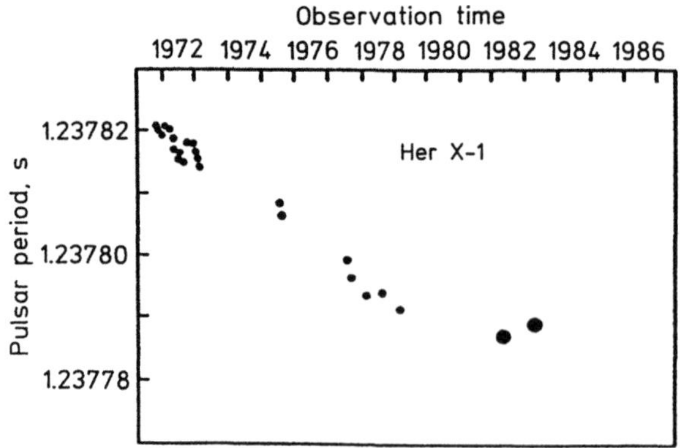

Fig. 6.15. Observations made in the 1980s reveal that, as expected, the periods of X-ray pulsars do not change on the average. The latest points correspond to observations from the Japanese satellites HAKUCHO and TENMA

spin-down torques were equal to zero, the spin-up time would be given by $|\dot{p}/\dot{p}_{su}| \approx 8000$ years.

The spin-up of the pulsar Her X-1 is about 40 times lower than that expected from "pure" spin-up from a disk in a standard magnetic field. Hence it can be stated that the spin-up and spin-down torques in the case of Her X-1 are balanced to within $\sim 1/40$.

Finally, the equilibrium of pulsars is a natural consequence of their evolution. This can be verified directly from model calculations (Chap. 11). In short, evolution arguments lead to the following. Most of the observed X-ray pulsars have a characteristic spin-up time of ~ 100 or ~ 1000 years. If the spin-up were monotonic and not occasional, the probability of observing a pulsar in this stage would be $10^2 - 10^3$ times lower than in the equilibrium state (in which it spends $\gtrsim 10^5$ years). Obviously, the observations do not match these conclusions. Hence the spinning-up process is only brief episode in the life of a pulsar.

The observation of X-rays from pulsars Vela X-1 and Her X-1 made on board the Japanese satellites TENMA and HAKUCHO and the European satellite EXOSAT (X-ray Observatory Satellite) showed that these pulsars, which were spun up almost monotonically during the first ten years, suddenly began to spin-down and reverted within $2-3$ years to their initial period observed ten years earlier (Fig. 6.15). These observations are completely in line with the expected results.

6.7.2 Magnetic Fields of X-Ray Pulsars

The position of cyclotron lines in the spectrum of a pulsar can be used to determine the magnetic field strength in the vicinity of the magnetic pole of a neutron star. At present, cyclotron lines have been observed in only three pulsars (reliably, in only one pulsar, Her X-1). An analysis of the spin-up and spin-down processes in neutron stars shows that the rate of variation of an X-ray pulsar period and the value of the period itself contain information about the magnetic field of a neutron star (to be more precise, about its dipole magnetic moment).

A method of measurement of the magnetic field (with a chronometer) based on the hypothesis of the equilibrium state and using an equation of the type (6.80) was proposed by Lipunov and Shakura (1976). Ghosh and Lamb (1979) adopted a different approach in which the magnetic field was estimated from the variation $\dot{p}$ of the period. This method was based on the numerical models of neutron star spin-up and spin-down under the action of a torque. However, Ghosh and Lamb used only the data on the spin-up of X-ray pulsars. Firstly, this approach does not provide unique values of the magnetic dipole moment. Secondly, the spin-up of a neutron star depends extremely weakly on its magnetic field. In the case of disk accretion [see (6.67)], $\dot{p} \propto \sqrt{R_d} \propto \mu^{2/7}$. For quasispherical accretion, $\dot{p}_{su}$ is independent of the magnetic dipole moment of the star [see (6.67)]. If we also take experimental errors into consideration, it becomes clear that such estimates are extremely rough.

Table 6.2. Magnetic fields of X-ray pulsars

Pulsar	Magnetic dipole moment [10^{30} Oe cm^3]			
	Ghosh and Lamb (1979)		Lipunov (1982c)	
	Slow rotator	Fast rotator	Disk accretion	Stellar wind without disk
SMC X-1	0.50	0.50	1.0	4.7*
Her X-1	–	0.47	0.6	–
4U 0115 + 63	3×10^{-3}(?)	1.4	3.5	1–2
Cen X-3	10^{-2}(?)	4.5	5.7	2
A 0535 + 26	3.3	148	150	30
GX 1 + 4	0.93	170	180	?
GX 304 – 1	–	–	140	?
Vela X-1	–	86	120	3*
2S 1145 – 62	–	–	260(?)	?
1E 1145 – 61	–	–	260(?)	?
A 1540 – 53	–	–	430	20
GX 301 – 2	0.3	394	1000*	110
X Per	4.8	4.8	35	–

* less reliable value

Table 6.2 presents the values of the magnetic dipole moment obtained from (6.72, 73) on the basis of the equilibrium hypothesis. Data on $\dot{p}$ can be used for an independent verification. Using the values of spin-down, we can choose the correct value of the magnetic dipole moment (Lipunov 1982c) (Fig. 6.16).

Two points are worth noting here. Firstly, the magnetic dipole moment of the X-ray pulsar Her X-1, $\mu \approx 6 \times 10^{29}$ Oe cm^3, is in agreement with the independent estimate of the magnetic field strength $B_0 \approx (3-5) \times 10^{12}$ Oe for a neutron star of radius $R_x = 6-7$ km. In view of the approximate nature of estimates, the agreement seems to be satisfactory. Secondly, the magnetic field of long-period pulsars was indeed found to be much higher than the standard value of $\mu \approx 10^{32}$ Oe cm^3. Assuming the radius of the neutron star to be $R_x = 10$ km, we obtain the estimate $\mu \approx 10^{13.5} - 10^{14.5}$ Oe for the magnetic field at the pole. This is much higher than the critical value of $\sim 4.3 \times 10^{13}$ Oe and the magnetic fields of radiopulsars. However, this is not unexpected (Sect. 8.6).

It was mentioned in Sect. 6.4 that a catastrophic equilibrium is realized in the case of spin-up from a disk. Since the equilibrium period p_{eq} is close to the critical period p_A, i.e., since the disk radius is close to the corotation radius, a slight change in the accretion rate leads to a catastrophic variation of the pulsar luminosity (by several orders of magnitude). The model of the magnetosphere of an X-ray pulsar whose period is close to the equilibrium period is shown in Fig. 6.17. It is interesting to note that over an enormous dynamic

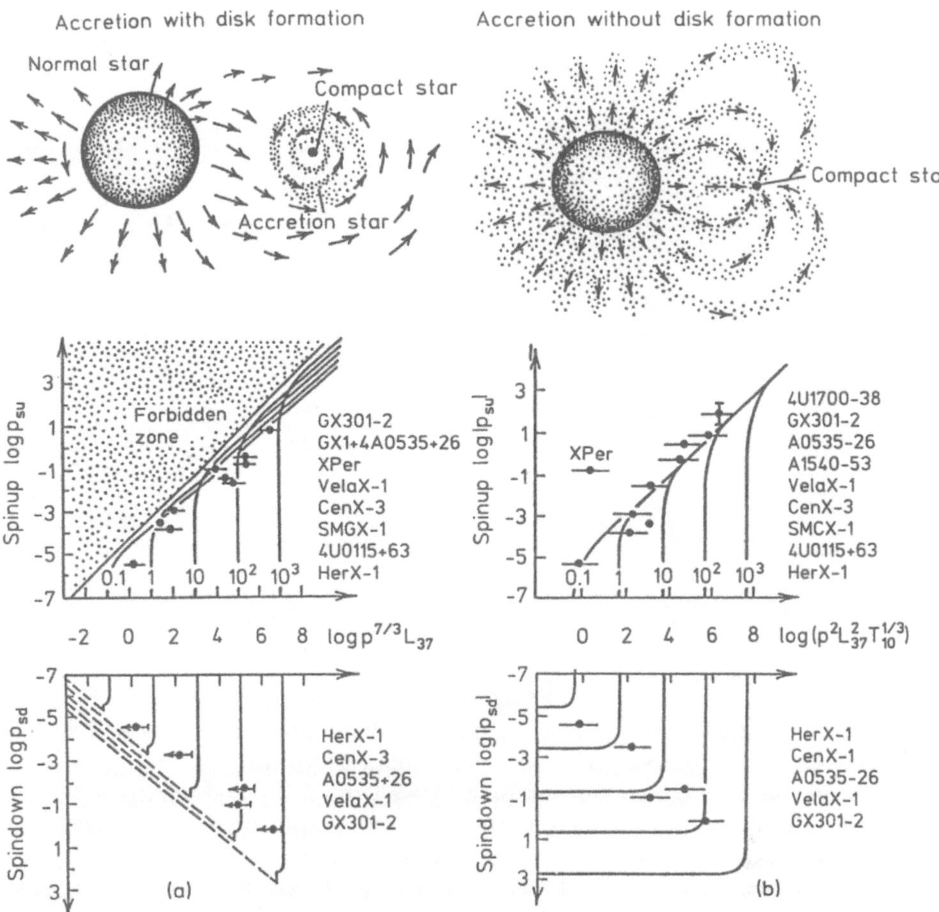

Fig. 6.16 a, b. Dependence of spin-up and spin-down of pulsars for two accretion regimes on the observed parameters of X-ray pulsars. The experimental data are shown by *points*. The *solid lines* correspond to the analytical model of torque (6.66) for different values of the magnetic dipole moment (shown in units of 10^{30} Oe cm^3) (**a**) Disk accretion, (**b**) Accretion without disk

range of variation (decrease) of X-ray luminosity of the pulsar, the value of $\dot{p}$ remains constant and positive (Lipunov 1987 b).

6.7.3 Reasons Behind the Average Spin-up of X-Ray Pulsars

It is borne out by observations that the pulse period varies over an extremely wide time scale, from a few hours to $10-15$ years (maximum observation period). Apparently, variations over longer periods of time also take place. Thus, there exist strong arguments suggesting that pulsars are on the average in an equilibrium state. However, a complete averaging over the times at our

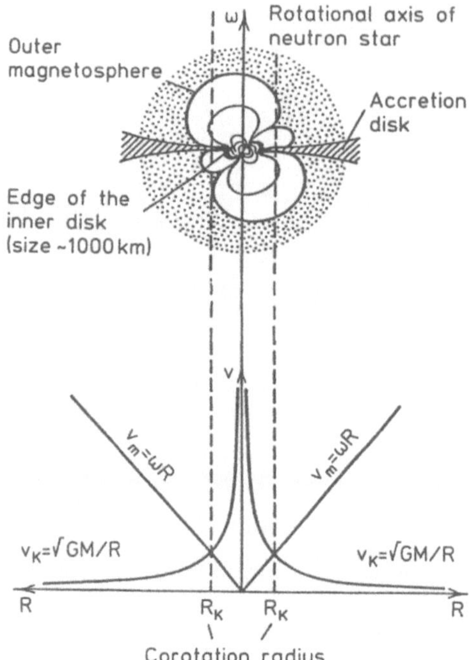

Fig. 6.17. Structure of the magneto-sphere of an X-ray pulsar

disposal revealed a spin-up in most pulsars which, however, is slower than that expected in the complete absence of spinning-down torques.

Three different explanations were offered for this phenomenon. The first one is purely a selection effect. The neutron star is spun up as long as accretion takes place, i.e., as long as the pulsar operates, while it is spun down in the pro-peller regime, i.e., when the luminosity falls by a factor of several hundreds. Obviously, it is easier to observe a spinning-up pulsar than a spinning-down neutron star. The second explanation was offered by Ghosh and Lamb (1979). According to the model of spin-up and spin-down by torque, pulsars may be spun up or spun down. However, if the processes are not symmetric, i.e., if a pulsar is spun down more rapidly than it is spun up, the probability of its being observed in the spin-up stage will be higher. Finally, Syunyaev and Shakura (1977) remarked that the period of an equilibrium pulsar may vary due to the evolution of the optical companion. It is clear, however, that such evolutionary changes (which are inevitable) must have an extremely large time scale, $10^4 - 10^6$ years. The observed spin-up time, however, is $10^2 - 10^3$ years. Ap-parently, the first two mechanisms are operative.

Let us consider the fluctuations of the period of a neutron star (Lipunov 1987a). Suppose that the change in the angular velocity of an accreting star occurs under the action of a random torque:

$$\frac{d\omega}{dt} = F(\omega) + \Phi \; , \tag{6.82}$$

where $F(\omega)$ is a certain constant torque, and Φ is the fluctuating torque such that its mean value is equal to zero, $\langle \Phi \rangle = 0$. We shall also assume that the "force" is potential,

$$F(\omega) = -\nabla_\omega V ,$$

where V is the scalar potential (Sect. 6.4). Equation (6.82) is essentially the Langevin equation for a random motion in the frequency space (Haken 1980). Let $f(\omega)$ denote the probability that a rotator will have a frequency ω. In the case under consideration, the function $f(\omega)$ satisfies the Fokker-Planck equation (Zel'dovich, Myshkis 1973; Haken 1980):

$$\frac{\partial f}{\partial t} = \frac{df F(\omega)}{d\omega} + D\frac{\partial^2}{\partial \omega^2} f , \tag{6.83}$$

where D is the "diffusion" coefficient determined by the correlation of a random force Φ:

$$\langle \Phi(t)\Phi(t') \rangle = 2D\delta(t-t') . \tag{6.84}$$

The stationary solution of (6.83) has the form

$$f(\omega) = Ne^{-V(\omega)/D} , \tag{6.85}$$

where N is determined from the normalization condition

$$\int\limits_{-\infty}^{+\infty} f(\omega)d\omega = 1 . \tag{6.86}$$

The scalar potential for an accreting star can be written [see (6.69)]

$$V(\omega) = \begin{cases} A_1(\omega^3 - \omega_{eq}^3) - B_1(\omega - \omega_{eq}) & , \quad \omega \geqslant 0 \\ A_1(-\omega^3 - \omega_{eq}^3) - B_1(\omega - \omega_{eq}) & , \quad \omega < 0 , \end{cases} \tag{6.87}$$

where

$$A_1 = \frac{\kappa_t \mu^2}{3GM_x I} ; \quad B_1 = \frac{\langle \dot{M} k_{su} \rangle}{I_1} ; \quad \omega_{eq} = \sqrt{\frac{B_1}{3A_1}} .$$

Here $\omega_{eq} = 2\pi/p_{eq}$ is the equilibrium frequency, and k_{su} is the specific spin-up torque in the accreting matter. The probability distribution over frequencies is shown in Fig. 6.18. It can be seen that this distribution is not symmetric about the equilibrium frequency (nonsymmetric potential).

We introduce the dimensionless frequency

$$x \equiv \frac{\omega - \omega_{eq}}{\omega_{eq}} .$$

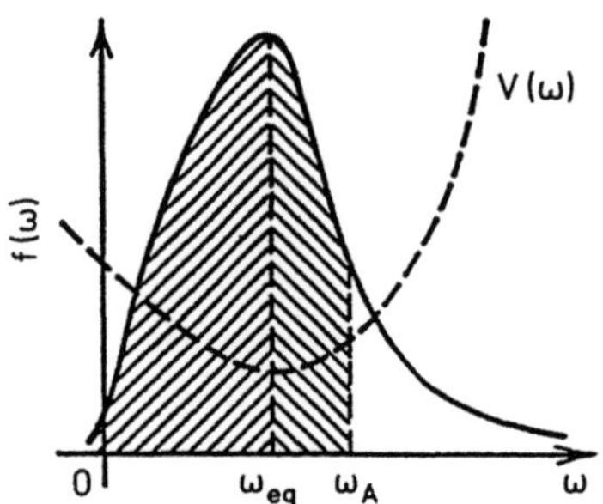

Fig. 6.18. The probability distribution function is non-symmetric about the equilibrium frequency

In this case, the distribution (6.85) becomes

$$f(x) = \begin{cases} Ne^{-\gamma(x^3+3x^2)} , & x \geqslant -1 , \\ Ne^{\gamma(x^3+3x^2+6x+2)} , & x < -1 , \end{cases}$$

(6.88)

where

$$\gamma \equiv \frac{2A_1\omega_{eq}^3}{D} = \frac{2B_1\omega_{eq}}{3D} .$$

The quantity γ can be estimated as

$$\gamma \approx \frac{t_{su}}{\Delta t} ,$$

where t_{su} is the characteristic spin-up time for the pulsar in the absence of spinning-down torques, and Δt is the characteristic time in which the value of $\dot{p}$ changes considerably. For large γ ($\gamma \gg 1$), the distribution has a practically symmetric form:

$$f(x) = \sqrt{\frac{3\gamma}{\pi}}\, e^{-\gamma(x^3+3x^2)} , \quad x \geqslant -1 .$$

(6.89)

Asymmetry appears only for small values of γ.

The ratio of the number of accreting pulsars with $\omega < \omega_{eq}$ to the number of pulsars with $\omega > \omega_{eq}$ is given by

$$\frac{N_+}{N_-} = \frac{\displaystyle\int_0^{\langle x_A \rangle} e^{-\gamma(x^3+3x^2)}\,dx}{\displaystyle\int_{-\infty}^{-1} e^{\gamma(x^3+3x^2+6x+2)}\,dx + \int_{-1}^{0} e^{-\gamma(x^3+3x^2)}\,dx} ,$$

where

$$x_A = (\omega_A - \omega_{eq})/\omega_{eq}, \quad \omega_A = 2\pi/p_A .$$

Thus, the observed dominance of spinning-up (on the average) pulsars is a consequence of (a) the closeness of critical frequencies ω_A and ω_{eq}; and (b) the asymmetry of the scalar potential $V(\omega)$.

The value of γ can be estimated for each X-ray pulsar from the observed fluctuations of the period (this was pointed out by M.E. Prokhorov in a private communication). As $x \to 0$, we obtain the approximate equality

$$\gamma \approx \frac{1}{3x^2} \; .$$

In general, the introduction of the scalar potential $V(\omega)$ allows us to look at an accreting neutron star from the position of synergetics (Lipunov 1987 a).

6.7.4 Rapid Fluctuation of Periods and Internal Structure of Neutron Stars

While considering the change in the angular momentum of a neutron star, we have so far assumed that it rotates as a rigid body. This assumption is valid only for processes taking place over a time considerably longer than the time scale of interaction of the crust with the superfluid core of the neutron star (τ_c).

Baym et al. (1969) proposed a simple two-component model in which I_s is the moment of inertia of the superfluid component and I_c the moment of inertia of the core:

$$I_c \dot{\omega}_c = K(t) - I_c(\omega_c - \omega_s)/\tau_c \; ,$$
$$I_s \dot{\omega}_s = I_c(\omega_c - \omega_s)/\tau_c \; . \tag{6.90}$$

It was shown by Lamb et al. (1978) that if there are no internal sources of angular momentum (such sources could be formed as a result of the rupture of superfluid vortices), the power spectrum of the rotational frequency variations in a pulsar contains important information about the structure of the neutron star. If external disturbances occur very rapidly (as compared to τ_c), the neutron star will respond to them as a rigid body with moment of inertia I_c. On the other hand, if the disturbances are very slow, the neutron star will react to them as a rigid body with moment of inertia $I_s + I_c = I$. The spectral analysis of fluctuations of the angular velocities of X-ray pulsars allows probing the core of neutron stars (Lamb 1979 and Pines 1980).

6.8 Variability of X-Ray Sources. Transients

The time variation of X-ray sources is extremely diverse and covers practically all possible situations. Fluctuations of the brightness (periodic, quasiperiodic, or random) occur over quite different periods varying from a few milliseconds to tens of years. This interval is more or less uniformly filled. This is not surprising: the magnetohydrodynamic equations in a gravitational field are exceptionally rich in transient solutions. This is supplemented by the natural nonstationary nature of boundary conditions which are specified for a normal star supplying plasma to an accreting system.

We have endeavored to present in Table 6.3 all the known mechanisms leading to a variation of the X-ray flow, with a brief account of the nature of this variation. The mechanism can be divided into two classes, extrinsic and intrinsic. The division is quite arbitrary since these mechanisms often work together. The extrinsic mechanism involves fluctuations caused by the variations outside the gravitational capture radius $R > R_G$. Practically all X-ray sources exist alternately in "high" and "low" states. Sources in which the high states are much shorter than the low states are generally called transient sources. Unfortunately, this is not the most suitable criterion for isolating objects of the same type. It is quite clear that the box labelled "transients" will also contain sources with a different nature of variation.

For a number of sources, the reason behind the high and low states is known. For example, the fastest X-ray pulsar A 0538−66 is activated after time intervals that are multiples of 16.6 days. There is no doubt that we are dealing with a binary system in which the neutron star moves in an orbit with a large eccentricity.

However, the reason behind the sharp increase (by a factor of 100) in the flow is not clearly known so far. According to some authors, sudden activations are associated with the fact that at the instant of the passage of a periastron, the normal star fills its Roche lobe (to be more precise, its analog) and "ejects" matter (Brown, Boyle 1984). The pulsar luminosity attains values of $L_x \approx 8 \times 10^{38}$ erg/s at the peak. In another model, however, the triggering mechanism associated with a transition from the propeller to accretion regime may be realized (Lipunov, Shakura 1976). In this case, a sharp increase in the loss of the normal star mass is not required, and it may even remain constant (Maraschi et al. 1983; Gnusareva, Lipunov 1985). It was suggested by Maraschi et al. (1983) that in such a binary system the pulsar may go over from the accretion stage A to the ejection stage E and back.

It was shown by Gnusareva and Lipunov (1985) that the existence of such neutron stars with mixed states is the inevitable result of evolution of a neutron star in a binary system with a very large eccentricity. In such a mixed state, the change $\dot{p}$ in the pulsar period over a period of the binary system is equal to zero. For a random or periodic variation of the accretion rate, the neutron star performs horizontal oscillations on the $p-L$ diagram (see Fig. 3.11).

Thus, three different situations are possible depending on the amplitude of the accretion rate oscillations. For a very small amplitude (or away from the catastrophic equilibrium), oscillations are not accompanied by a transition to another regime. In this case, the change in the X-ray luminosity exactly repeats (albeit with a delay) the variations of the accretion rate. Near the catastrophic equilibrium or for large oscillations of $\dot{M}_c$ (by several orders of magnitude), a transition to the intermediate stage is possible. Oscillations with an even larger amplitude may lead to a transition to the ejector state E. Another

Table 6.3. Mechanisms of variation of X-ray sources

Mechanism	Nature of variation	Time scale	Variation of X-ray flux, stellar magn.	Author
Fall of drops of matter onto the surface	random	$\dfrac{R_x}{v_{ff}} \approx 10^{-5}$ s	$\lesssim 1$	Shvartsman (1971 b)
Rotation of neutron star	periodic	$p \approx (10^{-3} - 10^{3})$ s		Amnuel' and Guseinov (1968)
Alfvén zone instability	random	$\dfrac{R_A}{v_{ff}} \approx (10^{-1} - 10^{-2})$ s	up to $(1-5)$	Lamb and Lamb (1977)
Thermonuclear bursts	quasi-periodic	hours – days	~ 5	Rosenbluth et al. (1973)
Polar column instability	quasi-periodic	seconds	$\sim 10^{-3}$	Livio (1984)
Centrifugal barrier ($A \leftrightarrow P$ transition)	depends on type of ext. variation		~ 5	Lipunov and Shakura (1976)
Stellar wind heating at $R \approx R_G$	quasi-periodic	$\dfrac{R_G}{v_{ff}} \approx 10^{4} \left(v_w/300 \, \dfrac{\mathrm{km}}{\mathrm{s}} \right)^{-3}$ s	~ 5	Syunyaev (1978)
Nonstationary disk accretion	quasi-periodic	$\dfrac{1}{2\pi\alpha} \left(\dfrac{R}{H} \right)^{2} T$	~ 5	Syunyaev and Shakura (1977 a, b)
Binary orbit eccentricity	periodic	T		Guseinov (1970)
Optical star precession	periodic	$(10-20)\,T$	~ 5	Shakura (1972)
Nonstationary ejection of matter by a normal star	random, quasi-periodic	days – years	$0 - \infty$	Fabian et al. (1975)
Tilting of orbital plane of accreting star towards the equatorial plane of normal companion	periodic	$T/2$	$0 - 5$	

qualitatively different situation may be realized in the case when the accretion rate is very high or when the magnetic field of the neutron star is weak. It was mentioned in Sect. 4.8 that the propeller stage may not occur at all.

Thus, the following transitions are possible in the mean equilibrium state upon a variation of the accretion rate:

$$A \leftrightarrow P \; , \quad A \rightleftarrows P \rightleftarrows E \; , \quad A \leftrightarrow E \; .$$

The transition to or from the ejection stage is accompanied by a hysteresis. It should be recalled that the pressure of an accreting plasma inside the capture radius increases as $R^{-5/2}$, i.e., more rapidly than the pressure of the relativistic wind $(\propto R^{-2})$. It was mentioned by Shvartsman (1970c) that a transition from the ejection stage occurs when the plasma pressure becomes equal to the stellar wind pressure at the capture radius. If, on the contrary, a transition takes place to the E-state (from P or A), the change in the regime occurs when these pressures become equal deep inside, at distances of the order of the light cylinder radius.

Another flare-type variation mechanism is the explosive burning of the hydrogen- and helium-rich accretion matter at the surface of a neutron star (Rosenbluth et al. 1973). These processes are currently being studied in connection with the theory of X-ray bursters (Sect. 6.11). Van Horn und Hansen (1974) suggested the same mechanism to explain the X-ray transients.

6.9 Generation of Relativistic Particles

In recent years, a number of communications have reported the discovery of ultrahigh-energy radiation ($10^{11} - 10^{17}$ eV) emitted by the binary X-ray systems Vela X-1 (Protheroe et al. 1984), Cyg X-3 (Samorski, Stamm 1983; Lloyd-Evans et al. 1983), and Her X-1 (Dowthwaite et al. 1984). It was shown by Eichler and Vestrand (1984) that high-energy γ-radiation is generated in the vicinity of the binary systems as a result of collisions of relativistic particles with the flux of matter or with a neighboring star. These particles must be protons or nuclei with energies of $\gtrsim 10^{16}$ eV. In this case, such X-ray systems may make a significant contribution to the overall flux of the galactic cosmic rays (Wdowczyk, Wolfendale 1983). Using the published values of fluxes, we find that the luminosities are of the order of 10^{37} erg/s.

The reported discovery of pulsations in the ultrahigh-energy range in the X-ray pulsar Her X-1 is quite astonishing.

The possibility of acceleration of relativistic particles by accreting neutron stars was indicated by Lipunov (1980a, b). This hypothesis is based on the fact that the magnetosphere of a neutron star in the purely disk accretion regime is open (Lipunov 1978a, b) just like for single neutron stars. Some magnetic field lines freely reach the light cylinder and relativistic particles may freely move to infinity along these lines (see Fig. 5.9). This should result in the emergence of cosmic rays and synchrotron radiation.

However, it is clear that the acceleration of relativistic particles differs considerably from the accretion of particles in the polar gap generally accepted for radiopulsars. Apparently, we must take into consideration the initial ideas concerning the acceleration of relativistic particles on the light cylinder (Pacini 1967; Ginzburg 1971), or the Blandford-Znajek mechanism.

It is quite striking that the pulsars Her X-1 and Vela X-1 have quite different angular velocities: their rotational energies are in the ratio 280^2:1. Nevertheless, both of them are powerful sources of relativistic particles. The energy to which relativistic particles are accelerated is proportional to the electric field strength ε (Sect. 8.3):

$$\varepsilon \approx \frac{v}{c} BeR \ ,$$

where R is the characteristic size of the acceleration zone.

Let us estimate the maximum power of relativistic wind under various assumptions concerning the acceleration mechanism. We shall consider two situations.

1) The particles derive their energy from the rotational energy of the neutron star, as in the case of single neutron stars. The power of magnetorotational radiation is restricted by the magnetic dipole losses (Sect. 8.3):

$$L_\mathrm{m} \approx \frac{\mu^2}{R_\mathrm{l}^3} \omega \approx 4 \times 10^{31} \mu_{30}^2 p^{-4} \ \mathrm{erg/s} \ , \tag{6.91}$$

where, as before, ω is the rotational frequency of the star, R_l the radius of the light cylinder, and μ the dipole moment.

2) Suppose that the relativistic particles are accelerated in the Alfvén zone. Their energy cannot exceed the gravitational energy liberated by the accreting matter at the magnetosphere:

$$L_\mathrm{a} = \dot{M} \frac{GM}{R_\mathrm{A}} \approx \frac{\mu^2}{R_\mathrm{c}^3} \omega \approx 10^{37} p^{-3} \ \mathrm{erg/s} \ . \tag{6.92}$$

Here we have used the assumption that the neutron star is in an equilibrium state and $R_\mathrm{A} \approx R_\mathrm{c}$.

The acceleration of relativistic particles emitted by accreting stars was considered by Kundt (1984) and Tsygan (1981).

6.10 X-Ray Bursters

An entirely new type of X-ray sources was discovered by Grindlay et al. (1976) who registered a burst of X-rays towards the globular cluster NGC 6624 (see Fig. 1.12). So far, dozens of burster-type soft X-ray (~ 1 keV) sources have been

Table 6.4. X-ray bursters (Lewin, Joss 1983)

Burster	Surrounded by	Remarks
MXB 0512 − 40	NGC 1851	
MXB 14?? − 6?		one type I burst
MXB 1455 − 31	blue star	transient
MXB 1535 − 29		only one burst, probably not of type I
XB 1608 − 52	star	transient
MXB 1636 − 53	blue star	forty optical and many synchronous X-ray bursts
MXB 1659 − 29	blue star	transient, star is observed only in high X-ray state
XB 1702 − 42	4U 1702 − 42	
MXB 1715 − 32		
XB 1724 − 30	Tarzan 2	
MXB 1728 − 34	Grindlay-Hertz 1	some bursts have two peaks
MXB 1730 − 335	Liller 1	rapid burster (type I and II bursts)
XB 1732 − 30	Tarzan 1?	only bursts are observed
MXB 1735 − 44	blue star	two optical bursts (1 synchronous)
MXB 1742 − 29		associated with a transient?
MXB 1743 − 28		three rapid-sequence bursts (4 and 17 min)
MXB 1743 − 29		transient, two-peak bursts
XB 1744 − 26	GX 3 + 1	
XB 1745 − 24	Tarzan 5	only bursts
MXB 1746 − 37	NGC 6441	associated with a constant source?
XB 180? − 2?	NGC 6553?	no constant X-ray sources in the error box
XB 1813 − 14	GX 17 + 2	radiosource most likely
MXB 1820 − 30	NGC 6624	bursts observed only in low state
MXB 1837 + 05	blue star	synchronous optical and X-ray bursts observed
MXB 1850 − 08	NGC 6712	
MXB 1906 + 00		
XB 1908 + 00	blue star	transient Aquila X-1
MXB 1916 − 05		4U 1915 − 05

discovered. Most of these have either been identified with known globular clusters or the globular clusters corresponding to them were later detected in the direction of the sources (Table 6.4).

The main peculiarity of the X-ray bursters is their time behavior: repeated outbursts are observed against the background of a quasistationary (background) flow. The spectra of X-ray bursters are much softer than those of X-ray pulsars (Fig. 1.13). All bursts are divided into two unequal groups: type I bursts which repeat in intervals of several hours, and type II bursts which are much more rapid (a few minutes) and are observed only in one rapid burster MXB 1730 − 335.

Let us briefly consider the main observed properties of bursters (see also reviews by Grindlay 1981; Lewin, Clark 1980; Ergma 1982).

6.10.1 Localization and Spatial Distribution

Identification of the first burster 3U 1820−30 with the globular cluster NGC 6624 raised hopes that all bursters belong to globular clusters. Indeed, ten bursters were found to belong to known globular clusters. However, 15 bursters have not been identified as yet with any of the known globular clusters (Lewin et al. 1977).

The general distribution of X-ray bursters follows the pattern of distribution of the "bulge" sources which also have a soft spectrum but which do not exhibit flare activity (Fig. 6.19). The most important investigations in the localization of bursters in globular clusters were carried out aboard the X-ray observatory "Einstein" (Grindlay 1981).

Soon after the discovery of X-ray bursters, two groups of models were proposed: the model of the accreting neutron star and the model of a massive $(100-1000)M_\odot$ black hole (Bahcall, Ostriker 1977). If we assume the second hypothesis, it is obvious that such a heavy object must settle towards the center of the globular cluster, its position being closer to the center, the more massive the object. It follows from general thermodynamic considerations that the kinetic energy of globular cluster "particles" must be the same (temperature equalization). Hence we find the mean velocity v_x of the heavy particle (Landau, Lifshitz 1976):

$$v_x = \sqrt{\frac{\langle M \rangle}{M_x}} \langle v \rangle , \qquad (6.93)$$

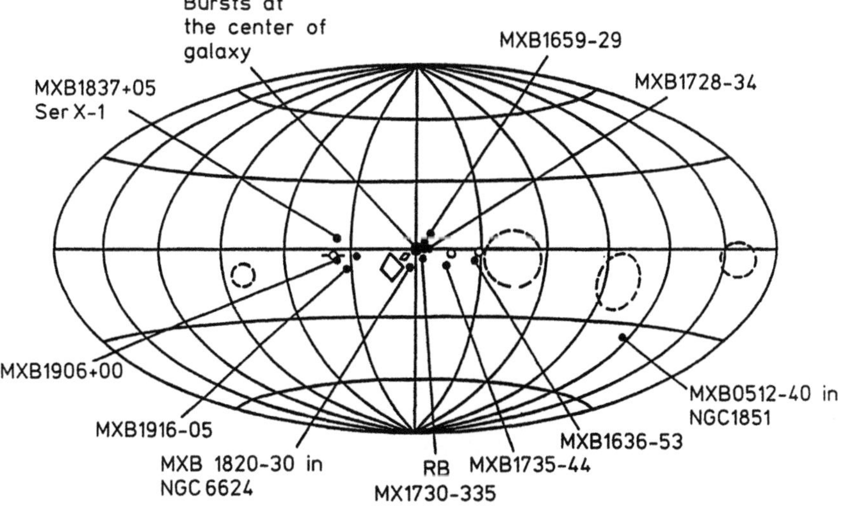

Fig. 6.19. Distribution of bulge X-ray sources (Lewin et al. 1977)

where $\langle M \rangle$ and $\langle v \rangle$ are the average mass and velocity of a globular cluster star and M_x is the mass of a heavy particle. Using this relation, Bahcall and Wolf (1976) showed that the mass of a test particle (X-ray source) is inversely proportional to the square of its mean distance R from the center of the cluster:

$$M_x \approx 0.9 \left(\frac{R_{\mathrm{core}}}{R} \right)^2 \langle M \rangle \, , \tag{6.94}$$

where R_{core} is the radius of the core of globular cluster. In actual practice, we deal with angular separations. The radius R for clusters containing X-ray sources is less than $\sim 6''$. The position of the X-ray source inside a cluster must be determined with an accuracy of $\sim 1''$. Such an accuracy was attained by using a high resolution imager (HRI) on board the observatory Einstein.

Observations revealed that X-ray sources do not coincide with the centers of globular clusters. The average separation R was found to be $\sim 0.5\, R_{\mathrm{core}}$, and hence the masses of X-ray sources cannot exceed the average stellar mass by a factor of more than four. Putting $\langle M \rangle \approx 0.5\, M_\odot$, we find that the mass $M_x \approx 2 M_\odot$. This sealed the fate of the model of massive black hole, at least for most of the sources (Grindlay 1981).

6.10.2 Periodic Variations of X-Ray Flux. X-Ray Eclipses

A striking feature of the X-ray bursters is that the radiation emitted by them is not subjected to any periodic variations. None of the observed bursters revealed any X-ray pulsations. Even traces of X-ray eclipses could not be observed for a long time.

In recent years, eclipses have been observed in several X-ray bulge sources (including X-ray bursters). Traces of eclipse-like events have been observed in four bulge sources and/or bursters: MXB 1659−29 (Lewin et al. 1978; Cominsky et al. 1983), 4U 1915−0.5 (Walter et al. 1982; White, Swank 1982), 4U 1755−33 (White et al. 1983), and MXB 1820−30 (Cominsky et al. 1985). A period of 4.4 hours is revealed for 4U 1755−33. For MXB 1659−29, the duration of the eclipse was found to be 15 minutes, while the period was found to be equal to $7.114104 + 0.000168$ hrs (Cominsky, Wood 1984). The source 4U 1915−05 had a much smaller eclipse period (50 minutes), while the period of 4U 1820−30 was even smaller (just 11 minutes).

In order to present a compact binary system, we recall, for example, that the Keplerian rotational period at the surface of the Sun is 2 hrs 40 min. The binary system must be very compact: $a \approx 10^{10}$ cm. The absence of X-ray pulsations (to be more precise, the lack of their observation) and the rarity of X-ray eclipses can be explained more or less naturally (see below).

6.10.3 Luminosity and Spectra of Bursters

The luminosity of a burster during a flare is estimated as $\sim 10^{38}$ erg/s. This value is close to the Eddington limit (3.27) for a star of solar mass. The key

Table 6.5. Characteristic burst repetition time t_b, burst energy E_b and the ratio of the background luminosity to the burst luminosity averaged over t_b

Source	t_b	$E_b, 10^{39}$ erg	$\gamma_b \equiv \dfrac{L_0}{\langle L_b \rangle} = \dfrac{E_b}{t_b}$
MXB 1728−34	$3.0-7.8^h$	∼6.0	50−100
MXB 1730−335	6^s-450^s	0.08−12	<2
MXB 1735−44	$50^m-7.5^h$	1−3	≤100
MXB 1742−29	∼13^h	0.5−3	≤100
MXB 1743−29	∼35^h	∼5	≤100
3U 1820−30	$2.2-4.4^h$	1.3−1.8	∼35
MXB 1837+05	∼6.3^h	∼1	∼150
MXB 1906+00	∼8.9^h	∼1.4	∼80

to the problem on the nature of a burster is a quantity equal to the ratio of the background luminosity L_0 to the luminosity of a burst $\langle L_b \rangle$ averaged over the period between the bursts.

Table 6.5, borrowed from Lamb and Lamb (1977) shows the characteristic time t_b of burst recurrence, burst energy, and the ratio $\gamma_b = L_0/\langle L_b \rangle$ for 8 bursters. The luminosity has been calculated under the assumption that the distance from all sources is equal to 10 kpc [with the exception of 3 U 1820−30 (NGC 6624) for which the separation is 6 kpc, and MXB 1730−335, for which the distance is assumed to be 11 kpc].

Bursters have active and inactive states (when the background luminosity is reduced). The burst recurrence time is maintained only on the average. It was observed (Oda 1981) that the energy of type II bursts (rapid bursters) is proportional to the interval between bursts (Fig. 6.20). It seems that there exists a certain energy reservoir whose contents are poured out following the attainment of a certain (sufficiently high) level. It follows from Table 6.5 that for all bursters except the rapid burster MXB 1730−335, the ratio $\gamma_b \approx 10^2$. The origin of this "magic" number becomes quite clear if we consider that the background radiation is the result of energy liberation upon gas accretion at the surface of a neutron star, while a burst results from the nuclear explosion of matter accumulated between bursts. Indeed, in the case of accretion to a star with a radius of 8−10 km and mass of $1.5-2M_\odot$, an energy of 150−250 MeV/nucleon is released, while the thermonuclear explosion of hydrogen-rich matter gives ∼6 MeV/nucleon. The ratio of these energies is just equal to ∼γ_b, as was first mentioned by Maraschi and Cavaliere (1977) and Woosley and Taam (1976).

Figure 1.13 shows a typical spectrum of X-ray bursters. Roughly, the spectrum is similar to black-body radiation with temperature of ∼2 keV (Van Paradijs 1978). The temperature is slightly increased during bursts. In the emitting region of bright X-ray bursters, the Compton scattering thickness exceeds the optical thickness with respect to the true absorption. Hence the photons

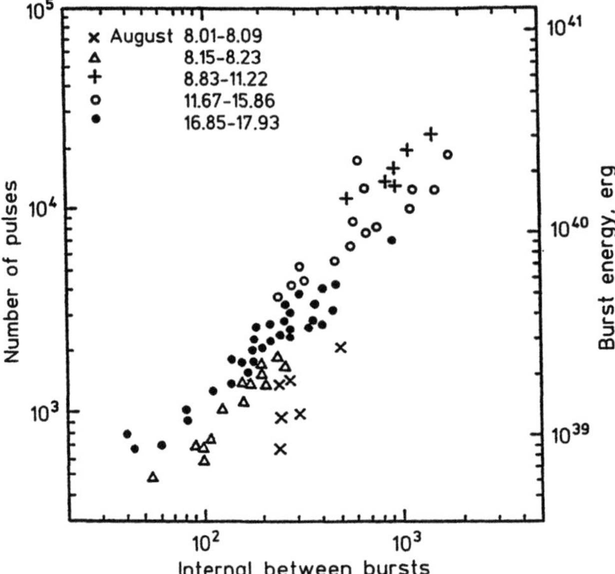

Fig. 6.20. The energy of burster flashes is proportional to the time elapsing up to the next flash (Oda 1981)

experience a number of collisions before thermalization. Consequently, the temperatures determined from the observed spectrum correspond to deeper layers ($\tau \approx 4$) of the emitting region. The spectra of several bursters revealed iron lines with an energy of ~ 6.5 keV.

While analyzing the spectra of bursters, we must take into account the effects associated with the general theory of relativity (Van Paradijs 1978).

6.11 Nuclear Burning at the Surface of Neutron Stars. Spherically Symmetric Model

Bursts of type I, which are characterized by a high value of $\gamma_b \approx 10^{1.5}$, are associated with the burning of helium at the surface of an accreting neutron star. All the calculations made so far are based on the assumption of spherical symmetry (we shall later analyze the closeness of this assumption to reality). Moreover, it is assumed that the thickness of the shell in which the accreting matter is accumulated is small in comparison with the radius of the neutron star. In other words, we assume the plane-parallel layer approximation. Rosenbluth et al. (1973) were the first to consider nuclear burning as an additional source of energy liberation in the case of accretion onto a neutron star.

Let the thickness of the shell of accreting matter be $\Delta R_{sh} \ll R_x$. Obviously, the following relations hold in this case:

$$M_{sh} = 4\pi R_x^2 \Delta R_{sh} \varrho_{sh} \; ,$$

$$P_{sh} = \varrho_{sh} g \Delta R = \frac{G M_x \varrho_{sh} \Delta R_{sh}}{R_x^2} \; , \tag{6.95}$$

where M_{sh} is the mass of the shell, and ϱ_{sh}, P_{sh} are the density and pressure at the bottom of the shell. Putting $M_{sh} = \dot{M}t$, we find that the density and pressure increase monotonically at the bottom of the shell:

$$\varrho_{sh} = \frac{\dot{M}t}{4\pi R_x^2 \Delta R_{sh}} \; , \tag{6.96}$$

$$P_{sh} = \frac{G M_x \dot{M}t}{4\pi R_x^2} \; . \tag{6.97}$$

For a shell thickness $\Delta R_{sh} \approx 50$ m, the pressure at the bottom attains values up to $P_{sh} \approx 10^{24}$ dyne/cm^2, and the density $\varrho_{sh} \approx 10^7$ g/cm^3. The limiting Fermi energy for electrons exceeds 1.24 MeV (including the rest energy). This is sufficient for the inverse β-decay reaction

$$e + p \rightarrow n + \nu_e \; . \tag{6.98}$$

The neutrons are then captured by protons and form deuterium. If the temperature exceeds 10^6 K, a rapid thermonuclear reaction accompanied by the formation of helium takes place:

$$n + {}^3\mathrm{He} \rightarrow \alpha + \gamma \; ,$$

$$p + D \rightarrow {}^3\mathrm{He} + \gamma \; .$$

The burning of hydrogen in these reactions is accompanied by the liberation of about 7 MeV per nucleon. Fusion in a degenerate gas is called pycnonuclear burning and takes place at low temperatures.

The nature of nuclear synthesis depends critically on the relation between the energy liberation rate Q_+ (in nuclear burning) and the rate Q_- of cooling by radiation. The quantities Q_+ and Q_- are calculated per gram of matter. For a shell thickness of 50 m and density $\varrho_{sh} \approx 10^7$ g/cm^3, the surface density on the shell is $\Sigma_{sh} = \varrho_{sh} \Delta R_{sh} \approx 5 \times 10^{10}$ g/cm^2. The optical thickness corresponding to, say, Thomson scattering is found to be enormous:

$$\tau_T = \kappa_T \varrho_{sh} \Delta R_{sh} \approx 2 \times 10^{10} \; .$$

The value of thickness obtained from true absorption is also quite large. Consequently, heat is removed by diffusion, and hence the loss rate is given by [see (3.52)]

$$Q_- = \frac{c}{3\kappa\varrho}\frac{d\varepsilon_r}{dR}\frac{4\pi R_x^2}{M_{sh}} \, ,$$

$$(6.99)$$

where ε_r is, as before, the radiation energy density. Substituting $d\varepsilon_r/dR \approx a T_{sh}^4/\Delta R_{sh}$ and $\varrho = \varrho_{sh}$, we obtain from (6.96)

$$Q_- = \frac{16\pi^2 a c R_x^4 T_{sh}^4}{3\kappa M_{sh}^2} \, .$$

$$(6.100)$$

In this formula, we have taken into account the fact that the layer emits on both sides.

The condition for the onset of thermonuclear instability is determined by the equality

$$Q_+ = Q_- \, .$$

$$(6.101)$$

The shell does not get time to cool down and the nuclear reaction occurs in the detonation regime.

Detailed analytical and numerical calculations were carried out by Joss (1977) and Lamb and Lamb (1978). The temperature of the shell between bursts is determined from the balance of the energy liberation rate and cooling rate due to radiation. The shell may be heated due to energy liberation (a) as a result of accretion, (b) on account of radioactivity of the upper hydrogen-helium shell left after the outburst, and (c) in principle from the star itself. For the energy inflow, we can use

$$L_+ = \eta \dot{M} c^2 \, .$$

$$(6.102)$$

For example, $\eta \approx 0.1$ in the case of accretion. We equate this expression with the cooling rate due to radiation $L = Q_- M_{sh}$ (Ergma, Tutukov 1980),

$$T_{sh} = \begin{cases} 10^{9.1}\left(\dfrac{\eta\dot{M}\kappa}{M_x}\right)^{1/4}\left(\dfrac{\varrho_{sh}}{\mu_e}\right)^{5/12} \text{K} \, , & \text{degenerate nonrelativistic gas} \, , \\[4mm] 10^{9.6}\left(\dfrac{\eta\dot{M}\kappa}{M_x}\right)^{1/4}\left(\dfrac{\varrho_{sh}}{\mu_e}\right)^{1/3} \text{K} \, , & \begin{array}{l}\text{degenerate relativistic} \\ \text{electron gas} \, .\end{array} \end{cases}$$

$$(6.103)$$

In deriving the above equation, we have used the equation of state of the degenerate electron gas:

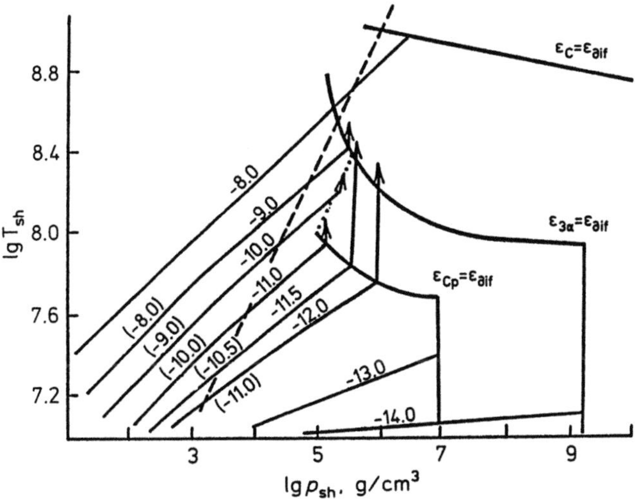

Fig. 6.21. Change in log T_{sh} and log ϱ_{sh} in the shell of an accreting neutron star for different $\dot{M}$ for $\alpha_1 = 0.01$ and $\alpha_1 = 10^{-3}$. The numbers near the lines stand for $\dot{M}(M_\odot/\text{year})$. The *solid lines* show the conditions for the ignition of hydrogen, helium and carbon, while the *dashed line* corresponds to the steady burning of hydrogen (Ergma 1982)

$$P \approx \begin{cases} 10^{13} \left(\dfrac{\varrho_{sh}}{\mu_e} \right)^{5/3} \text{dyne/cm}^2 & \text{for a degenerate non-relativistic electron gas ,} \\[4mm] 1.5 \times 10^{15} \left(\dfrac{\varrho_{sh}}{\mu_e} \right)^{4/3} \text{dyne/cm}^2 & \text{for a degenerate relativistic electron gas .} \end{cases} \tag{6.104}$$

Figure 6.21 shows the results of temperature and density computations made for a shell by using the above formulas. The critical lines of ignition of hydrogen, helium and carbon are indicated (Ergma, Tutukov 1980). It is assumed that the hydrogen ignition line (dashed) intersects the quasi-stationary burning line of hydrogen and the helium ignition line. It follows from Fig. 6.21 that various burning regimes are possible. We shall consider these in detail.

1) **Hydrogen-helium burster.** The burning of hydrogen is accompanied by the liberation of a sufficiently large amount of energy to ignite helium. In turn, the burning of helium raises the temperature to 10^9 K. Helium burning takes place at a high temperature and differs considerably from the normal CNO burning cycle in stars. For simultaneous burning of hydrogen and helium, the parameter $\gamma_b = L_0/\langle L_b \rangle$ increases:

$$\gamma_b = \frac{GM_x}{R_x} \bigg/ (q_H X + q_{He} Y) , \tag{6.105}$$

where $q_H = 6 \times 10^{18}$ erg/g, $q_{He} = 5.7 \times 10^{17}$ erg/g, X, Y are the hydrogen and helium concentration respectively. The value of γ_b depends strongly on the chemical composition. (The hydrogen-helium burster regime is characterized by $\gamma_b \lesssim 100$ and a very long burst attenuation time: $\sim 10^5$ s).

For $Y = 1$, $\gamma_b = 191\, m/R_6$, while for $Y = 0.28$, $\gamma_b \approx 30\, m/R_6$.

The energy liberation during the burning of helium in a 3α-reaction is

$$Q_+ \approx 4.55 \times 10^8 \varrho^2 T_9^{-3} Y^3 \exp\left(-4.41/T_9\right) , \tag{6.106}$$

where

$$T_9 = T_{sh}/10^9\,\text{K} .$$

2) During burning of hydrogen, the amount of helium increases and so does the energy liberated in a $3\,\alpha$-reaction: $Q_+ \propto Y^3$. Self-ignition of helium may occur over a narrow interval of density values. In this regime, the burst duration is large as before, but $\gamma_b \gtrsim 100$.

3) Helium burster. Judging by its parameters, a helium burster is similar to the observed bursts of type I. The burst duration is $\Delta t_b \approx 10$ s and $\gamma_b \gtrsim 100$. The necessary condition for the emergence of a helium burster is that all hydrogen must be consumed in a quasi-stationary regime. For a solar-type chemical composition, steady burning (hot CNO-cycle) takes place over a time of $\sim 10^5$ s. However, several bursters have a much smaller time interval between bursts: $t_b \approx 10^3 - 10^4$ s. Burning-out of hydrogen in such a short time is possible only if the accreting matter has a high carbon and oxygen content. According to the calculations made by Ergma (1982), a hydrogen-helium burster (type I) with a fixed helium concentration emerges for $\dot{M} \lesssim 10^{-11.5}\, M_\odot/\text{year}$, a hydrogen-helium burster with self-ignition of helium is formed for $\dot{M} \approx 10^{-11.5}\, M_\odot/\text{year}$, and a helium burster is formed in the interval $10^{-11.5}\, M_\odot/\text{year} \lesssim \dot{M} \lesssim 10^{-9.5}\, M_\odot/\text{year}$. For $\dot{M} \gtrsim 10^{-9.5}\, M_\odot/\text{year}$, the helium and hydrogen burning zones overlap. Quasi-stationary burning of hydrogen takes place over a wide range of accretion rate.

Numerical computations carried out by various authors (Joss 1978; Taam, Piklum 1979; Taam 1980; Ergma 1982) confirm the helium burster model.

Another important result obtained by the authors of some earlier works is the conclusion that the peak luminosity of a burster is very close to the Eddington limit. Indeed, this is quite obvious. As soon as the luminosity of a burster exceeds the Eddington limit, the role of radiation pressure which effectively reduces the gravitational force increases sharply. The shell expands, the pressure, density and temperature decrease, and the energy liberation in nuclear reactions drops.

The peak luminosity of bursters can be used as a standard candle for measuring distances in the Galaxy.

It should be observed that all the calculations were carried out under the assumption of spherical symmetry. However, there are at least three factors which violate spherical symmetry: (1) nonsymmetric deposition of matter;

(2) rapid rotation of a neutron star; and (3) the magnetic field of the neutron star. Hence these calculations can at best be treated as model computations.

6.12 Accretion to X-Ray Bursters

The absence of bright optical companions near X-ray bursters and bulge sources indicates that normal stars in these binary systems are stars with an exceptionally low luminosity and hence a low mass. In the most popular model of a system with an X-ray burster, the normal component is a low-mass red dwarf with $M_0 \approx (0.1-0.3)M_\odot$ which fills the Roche lobe. In contrast to massive binary systems in which the ratio of component masses $q = M_x/M_0 \ll 1$, the situation is quite different in systems with X-ray bursters: $q \gg 1$ (Joss, Rappaport 1979). Incidentally, the large value of the mass ratio is in agreement with the absence of eclipses in most of the bursters: because of its small size, an optical star covers only a small part of the celestial sphere of a neutron star and the probability of eclipse is low (Milgrom 1978). Apparently, an additional factor explaining the absence of eclipses is associated with the fact that the main part of the radiation emitted by an accreting disk propagates along its axis on account of Compton scattering. Hence the system can be seen only from the pole. To be more precise, we do not observe the system from the equator.

The possibility of accretion of stellar wind of a red giant in some cases cannot be ruled out. The velocity of the stellar wind of red giants is not high and in this case also we can expect the formation of an accretion disk.

The radiation emitted by X-ray bursters and many other bulge sources does not experience strictly periodic pulsations. Comparing bursters and X-ray pulsars, we observe that their luminosities do not differ significantly (naturally, we compare the luminosity of a pulsar with the luminosity of a bursters between bursts). The absence of pulsations can be explained only by the smallness of the magnetic field of the X-ray bursters. It was shown above that the models created by the supporters of nuclear burning did not take into account the field at all.

However, we must not go too far. It would seem that the absence of pulsations indicates the absence of a magnetosphere, i.e., $R_A < R_x$, and the magnetic fields of bursters must satisfy the following conditions [see (4.31)]:

$$\mu < \mu_{\min} \approx 10^{26} L_{37}^{1/2} \text{ Oe cm}^3 \ ,$$

$$B < B_{\min} \approx 10^7 L_{37}^{1/2} \text{ Oe} \ .$$

Actually, this is not the case. The above estimates are only extremal. Even in a field $B \approx 10^{10}$ Oe, it is practically impossible to detect pulsations from the available data. Bursters are components of close binary systems in which the normal star fills the Roche lobe so that the accretion takes place in the disk regime. In the disk regime, however, the equilibrium period of the neutron star [see (6.80)] is $p_{eq} \approx 10^{-3} \mu_{26}^{6/7} L_{37}^{-3/7}$ s. For $\mu_{26} \approx 1-10$, the rotational period of the neutron star is found to be $1-10$ ms. The spin-up time of neutron star to such a period is

$$t_{su} \approx \frac{I \omega_{eq}}{\dot{M}\sqrt{GM_x R_A}} \approx \frac{M_x}{\dot{M}} \alpha_m^{-3} \; , \tag{6.107}$$

where $\alpha_m \equiv R_A/R_x$ is the ratio of the Alfvén radius to the neutron star radius. For $\alpha_m \approx 10$, the spin-up time is found to be $\leqslant 10^7$ years, which is much smaller than the lifetime of low-mass binary systems ($\sim 10^9 - 10^{10}$ years). X-ray bursters must be rapidly rotating neutron stars (Camenzind 1982). It is extremely difficult to detect such rapid pulsations in X-ray sources. For example, pulsations of the superfast X-ray pulsar A 0538−66 ($p \approx 68$ ms) were detected only after about ten years following its discovery and were observed only once (Skinner et al. 1982).

Thus, the only constraints imposed by the observations are

$$B \ll 10^{12} \, \text{Oe} \; ,$$

$$\mu \ll 10^{30} \, \text{Oe cm}^3 \; .$$

In this case, it may happen that $B \gg B_{min}$ and $\mu \gg \mu_{min}$. Accordingly, we must consider two situations: (a) $\mu \lesssim \mu_{min}$ and (b) $\mu > \mu_{min}$. In the former case, the magnetic field really does not exert any influence on the accretion pattern.

6.12.1 Accretion for $\mu < \mu_{min}$

We shall consider disk accretion. It would seem that the accretion pattern is extremely simple in the absence of a magnetic field: the accretion disk comes to the surface of the star. A boundary layer emerges in which matter is slowed down from Keplerian velocity to the rotational velocity of the star.

However, the situation may turn out to be more interesting and more complicated (Lipunov, Postnov 1984). In the first place, if the neutron star rotates slowly, the last stable orbit in the disk is situated at a distance of $3R_g$, as in the case of a Schwarzschild black hole. For $M_x = (1.5-2)M_\odot$, we obtain $R_g = 13.5 - 18$ km, which is much larger than the radius of the neutron star itself for almost every equation of state known to us (Chap. 2). The following regime is then observed (Fig. 6.22). Up to distances corresponding to the last stable orbit, standard disk accretion is observed. After this, viscosity vanishes and matter tends to move in a helical trajectory towards the surface in accordance with the general theory of relativity. The thickness of this zone is not larger than $3-8$ km and matter almost does not radiate in it. After passing through the free-fall region, the plasma impinges on the surface of the neutron star at a small but finite angle, loses its kinetic energy through radiation and gives away its angular momentum to the neutron star.

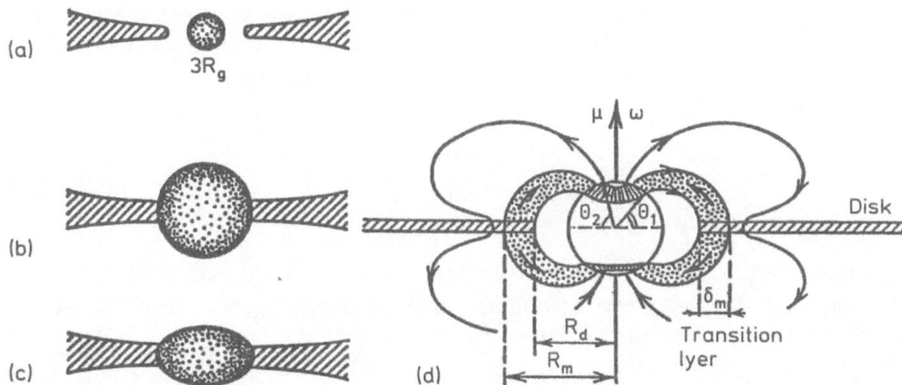

Fig. 6.22a–d. Three possible regimes of accretion to a neutron star without a magnetic field. (a) the radius of the neutron star is smaller than $3R_g$, (b) the radius of the neutron star is larger than $3R_g$, in which case a boundary layer exists, (c) the neutron star is flattened as a result of rapid rotation, while the disk approaches the surface (Lipunov, Postnov 1984). (d) Disk accretion to a star with a weak magnetic field

In the second situation, the disk reaches the surface of the neutron star where matter is decelerated in the boundary layer. It should be emphasized that such a situation is either not realized at all, or observed only for very light neutron stars: $M_x < 1\,M_\odot$.

Finally, the third regime is associated with a rapidly rotating neutron star. In view of the entrainment of the reference frame, the inner edge of the disk approaches the neutron star (it is assumed that the disk and the star rotate in the same direction). Two situations can arise in this case, and it is not possible to choose between them without a knowledge of the exact equations. If the spinning-up of the neutron star first leads to a Roche-type instability accompanied by outflow of matter from the equator, a symbiosis of the disk and the star is obtained, the disk is transformed smoothly into the star without any boundary layer. If bifurcation (or a higher-order instability) sets in earlier, the rotational period is "frozen" due to strong gravitational radiation (Sect. 6.13). A thin transition layer between the disk and the surface of the neutron star may be formed in this case.

6.12.2 Accretion in a Weak Magnetic Field ($\mu \gtrsim \mu_{\min}$)

For $\mu \gtrsim \mu_{\min}$, the Alfvén surface may lie between the surface of the star and the last stable orbit in the disk. In this case, the penetration of matter is facilitated due to the Rayleigh-Taylor instability. In any case, the picture will look like this.

The plane of the accretion disk coincides with the equator of rotation and, apparently, with the magnetic equator (Sect. 6.13) of the neutron star. Matter will fall on the surface of the neutron star along two spherical zones:

$\theta_1 \leqslant \theta \leqslant \theta_2$. The boundaries of these zones are determined under the assumption that matter flows along the dipole lines of force (Fig. 6.22 d):

$$\cos \theta_1 = \left(\frac{R_x}{R_d}\right)^{1/2} \; ,$$

$$\cos \theta_2 = \left(\frac{R_x}{R_m}\right)^{1/2} \; . \tag{6.108}$$

We also use the condition of conservation of magnetic flux: over the star

$$\int (n\boldsymbol{B})\,dS = \int_{R_d}^{R_m = R_d + \delta_m} B\, 2\pi R\, dR \; . \tag{6.109}$$

If the transient layer is thin, i.e., $\delta_m \ll R_m$, matter will fall along a narrow ring at a magnetic latitude

$$\theta_a = \arccos \left(\frac{R_x}{R_d}\right)^{1/2} \; .$$

The thickness, or the width, of the accretion ring is determined from (6.109) by substituting into it formulas (4.5, 7):

$$|\boldsymbol{B}_d|\, S_a \cos \chi = \frac{\mu}{R_d^3}\, 2\pi R_d \delta_m \; , \tag{6.110}$$

$$S_a = 2\pi R_x^2 \frac{R_x \delta_m}{R_d^2} \left(\frac{1 + 4\cot^2 \theta_a}{1 + 3\cos^2 \theta_a}\right)^{1/2} \; . \tag{6.111}$$

The ratio of the surface on which matter falls to the total surface of the star is

$$\frac{S_a}{4\pi R_x^2} = \frac{1}{2} \left(\frac{R_x \delta_m}{R_d^2}\right) \left(\frac{1 + 4\cot^2 \theta_a}{1 + 3\cos^2 \theta_a}\right)^{1/2} \; . \tag{6.112}$$

Let us estimate the optical thickness of the accretion flow in the Alfvén zone. In a weak magnetic field, the Alfvén surface is close to the star and hence the density of matter is large. Thus it can be stated that the Alfvén surface of weakly magnetized neutron stars is opaque to X-rays emanating from the polar rings. Indeed, let v be the velocity of accreting flow (directed along the magnetic field), $\delta(R)$ the flow thickness, and $\delta(R_d) = \delta_m$. Since $\delta_m \ll R_d$, we can write the following equations:

continuity equation

$$2\pi R \varrho v \delta(R) \cos \theta = \dot{M} \; ,$$

equation for line of force

$$R = R_d \cos^2 \theta \; ,$$

law of energy conservation

$$\frac{v^2}{2} - \frac{GM}{R} = -\frac{GM}{R_d} \; .$$

In this case, we get

$$\tau \approx \frac{\kappa L}{2\pi GM_x v \cos^2 \theta} \left(\frac{R_x}{R_d} \right)$$

$$\approx 4 \frac{L}{L_{Ed}} \left(\frac{\kappa}{\kappa_T} \right) \left(\frac{R_x}{R_d} \right) \left(\frac{c}{v_{ff}(R_d)} \right) \cdot \frac{1}{\sin 2\theta \cos \theta} \; . \tag{6.113}$$

Putting $M_x = 1 M_\odot, \kappa = \kappa_T, R_x = 10 \text{ km}$ and denoting $R_{10} = R_d/10 \text{ km}$ we obtain

$$\tau \approx 8 R_{10}^{-1/2} L_{38}/\sin (2\theta) \cos \theta \; . \tag{6.114}$$

The minimum optical thickness is attained for $\theta = \pi/6 = 30°$. The polar diagram ($I \propto e^{-\tau}$) formed only on the basis of the absorption properties of the magnetosphere (the radiation emitted by the star is assumed to be isotropic) is found to be a four-sheet surface.

The approximate calculations presented here clearly demonstrate that for $L \gtrsim 10^{37}$ erg/s, $R_d \lesssim 100 \text{ km}$ and $\theta \lesssim 10 - 20°$ the radiation from the neutron star is emitted from the Alfvén surface only. Since only soft X-rays are emitted by weakly magnetized neutron stars, we must take true absorption into consideration. The minimum thickness at $\theta = 30°$ is

$$\tau \approx 10 R_{10}^{-1/2} L_{38} \; . \tag{6.115}$$

Thus, the X-rays from a burster are emitted in the direction of the polar axis which, apparently, coincides with the rotational axis of the binary system.

6.12.3 Nonstationary Spherically Symmetric Accretion

The model of nonstationary spherically symmetric accretion was proposed by Lamb and Lamb (1977) to explain type II bursts.

According to this model, a magnetosphere, whose structure was discussed in Sect. 5.5, is formed around a magnetized neutron star in the case of spherically symmetric accretion. It was shown in Sect. 6.1 that the magnetosphere boundary is stable relative to the RT-instability if the temperature is of the order of, or higher than, a certain critical value (6.6):

$$T_{\text{out}} > T_{\text{cr}} \approx (0.1 - 0.3)\, T_{\text{ff}}(R_{\text{m}}) \ . \tag{6.116}$$

If the temperature of a plasma heated by a shock wave is higher than the critical temperature, the boundary is closed and there is no accretion at the surface of a neutron star. However, if the plasma is cooled below the critical temperature as a result of some processes (see below), the RT-instability comes into play and accretion begins. The growth time for X-ray luminosity must be of the order of the free-fall time in the magnetosphere:

$$\delta t_{\text{b}} \approx t_{\text{r}} \approx \left(\frac{R_{\text{m}}^3}{2GM_x} \right)^{1/2} \approx 0.06\, R_8^{3/2} m_x^{-1/2}\ \text{s} \ .$$

The luminosity increases to a maximum value (which is below the Eddington limit). When the temperature of X-ray radiation exceeds the plasma temperature at the Alfvén surface, the radiation will heat the accretion flow in such a way that the inequality (6.116) is restored and the boundary is closed. Even if the plasma temperature is higher than the spectral temperature of radiation in the Alfvén zone itself, there always exists a zone at a distance $R \gg R_{\text{m}}$ where the radiation heats the plasma, thus "locking" the accretion. It is difficult to know at present the extent to which this is applicable to type II bursts from a rapid burster. On the one hand, it seems that the probability of spherically symmetric accretion is extremely low for a low-mass binary system. On the other hand, type II bursts are almost certainly a consequence of nonstationary accretion. The reason behind such an accretion may well be the magnetic field of the neutron star. Observations reveal that type II bursts in a rapid burster are equivalent to the standard background radiation from other bursters.

In the model of spherically symmetric accretion, the GTR effects may turn out to be quite significant. Indeed, for a weakly magnetized neutron star, $R_{\text{m}} \approx R_x \approx (3-5)R_g$ and the GTR effects may be as large as $\sim 10\%$. Qualitatively, these effects may boil down to the following. In the post-Newtonian approximation, the relativistic correction makes the effective gravitational force increase more rapidly than $1/R^2$. The difference from the Newtonian force is larger, the closer the magnetosphere boundary approaches the surface of the neutron star. The effective acceleration due to gravity is defined as (Landau, Lifshitz 1973)

$$g = \frac{GM_x}{R^2 \sqrt{1 - R_g/R}} \ .$$

Moreover, the magnetic field also has a weight in the field of gravity. This results in an effective decrease in the magnetic dipole moment. We believe that as a result of both these effects, the boundary will have a larger curvature than in the Newtonian case. Paradoxical though it may seem, the GTR effects will hamper accretion since the magnetosphere will be even more stable to the RT-instability.

6.13 Spinning-up of Weakly Magnetized Neutron Stars

Following the discovery of millisecond radiopulsars (Backer et al. 1982), the idea of accretion spin-up of neutron stars in binary systems gained popularity (Alpar et al. 1982; Joss, Rappaport 1983). This raised several important questions: What is the maximum frequency to which a neutron star can be spun up? How fast can this be done?

The answer to the first question depends most significantly on the structure of the neutron star and hence on the equation of state of matter at nuclear densities. It is well known that an incompressible fluid rotating like a rigid body at a certain (bifurcation) frequency is transformed from a Maclaurin ellipsoid into a Jacobian ellipsoid (Chandrasekhar 1969). It was mentioned by Chandrasekhar (1970) that since a Jacobian ellipsoid has a quadrupole moment, a rapidly rotating neutron star emits gravitational waves until it is transformed into a Maclaurin ellipsoid. Obviously, in the case of accretion spin-up, equilibrium must be established between the inflow of angular momentum from the accreting matter and its outflow in the form of gravitational waves.

Figure 6.23 shows the qualitative form of the scalar potential describing the rotational evolution of a neutron star. At a frequency equal to bifurcation frequency, the potential sharply rises. In fact, the emission of gravitational waves increases so sharply with frequency that the angular velocity of a neutron star cannot increase significantly (say, by $\sim 1\%$) beyond this critical frequency. The bifurcation frequency for an incompressible fluid is defined by

$$\frac{\omega_b^2}{\pi G}\bar{\varrho} \approx 0.374 \ . \tag{6.117}$$

Putting $\bar{\varrho} = M_x \left/ \left(\frac{4}{3}\pi R_x^3\right)\right.$, we obtain the bifurcation frequency

$$\nu_b = \frac{\omega_b}{2\pi} \approx 973\, m_x^{1/2} R_6^{-3/2}\,\mathrm{Hz}\ . \tag{6.118}$$

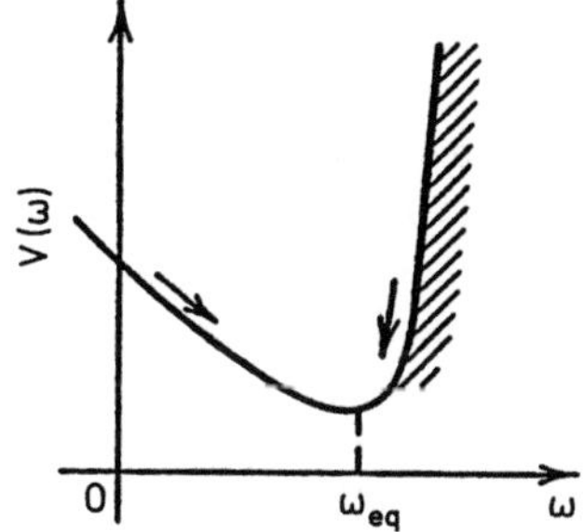

Fig. 6.23. The qualitative shape of the scalar potential for an accreting neutron star without a magnetic field but taking into account the instability leading to the emission of gravitational waves

Qualitatively, it is clear that the compressibility of stellar matter affects the bifurcation frequency. If the compressibility is low, the star is practically homogeneous and it is energetically more advantageous for the star to stretch itself to the form of a cucumber. Conversely, if the compressibility is high, the star has a high concentration of matter, the outer layers do not play a significant role and the bifurcation does not take place. However, Roche instability does set in and this also restricts the maximum rotational frequency of a neutron star.

Recent calculations involving a realistic equation of state and taking into account the rotational and GTR effects have led to unexpected conclusions (Sect. 2.6). It turns out that before the quadrupole disturbances, higher-order disturbances are excited first ("trifurcation"). According to another result, the frequency corresponding to Roche instability is close to the trifurcation frequency.

Let us now turn to the second question concerning the speed at which we can spin-up a neutron star to its highest frequency before it collapses into a black hole. (The star may collapse into a black hole earlier than it attains this frequency.) The results presented below are based on the computations made by Lipunov and Postnov (1984).

While considering the spinning-up of weakly magnetized stars, we cannot neglect the change in the moment of inertia of the neutron star. We can write (6.66) in the form

$$\frac{dI\omega}{dt} = \dot{M}k_{\mathrm{su}} - \kappa_t \frac{\mu^2}{R_{\mathrm{c}}^3} \ . \tag{6.119}$$

The most interesting case is when the magnetic field is so weak that the size of the magnetosphere satisfies the inequality

$$R_{\mathrm{m}} < \max\{R_x, R_{\min}\} \ ,$$

where $R_{\min}$ is the radius of the last stable circular orbit.

Calculations were carried out under the following assumptions: (a) the structure of the neutron star is determined for nine model equations of state without taking into account the rotational effects; (b) GTR effects, including the contribution from the "entrainment of the reference frame" are taken into consideration.

Figure 6.24 shows the plot of spinning-up of a neutron star (with a zero initial frequency). The notation of models corresponds to different equations of state (Arnett, Bowers 1977). Calculations show that the neutron star attains its critical frequency after accumulating just 10% of the mass. Such a rapid spinning-up can be understood with the help of simple estimates.

Let us consider the equation for spin-up without taking into account the relativistic effects

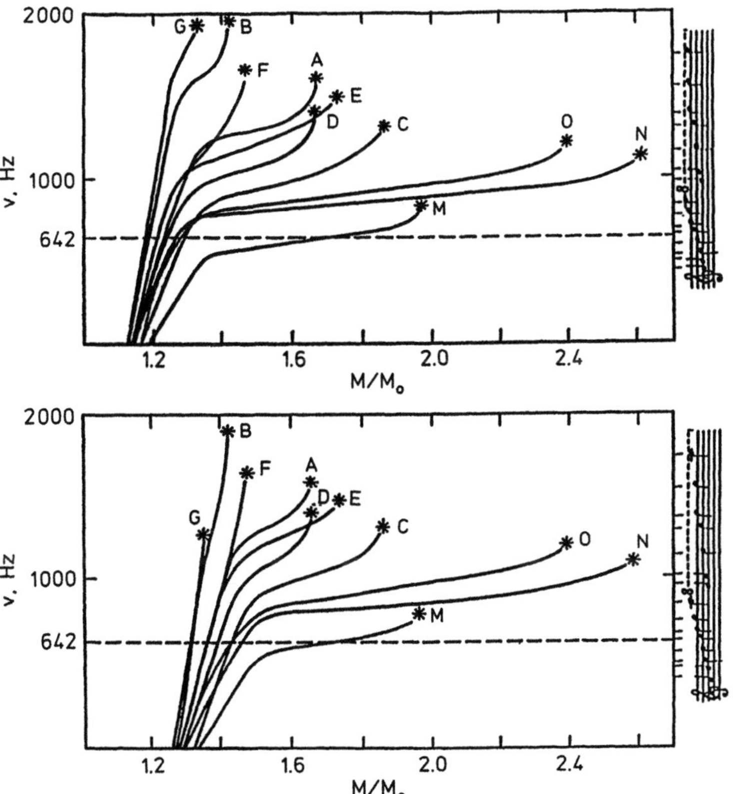

Fig. 6.24. Dependence of the rotational frequency of a neutron star without a magnetic field on $M/M_\odot$ in the case of spin-up from an accretion disk for different equations of state of the substance (Lipunov, Postnov 1984). Results of computations for two values of the initial masses of neutron stars are presented

$$\frac{dI\omega}{dt} = \dot{M}\sqrt{GMR_\mathrm{d}} \ .$$

Obviously, we can rewrite this equation as

$$\frac{dI\omega}{dM} = \sqrt{GMR_\mathrm{g}} \ .$$

We put $R_\mathrm{d} = 3R_\mathrm{g}$ (radius of the last stable orbit):

$$\frac{dI\omega}{dM} = \sqrt{6}\,\frac{GM}{c} \ .$$

Integrating, we obtain

$$I\omega = \frac{\sqrt{6}\,G(M^2 - M_0^2)}{2c} = \frac{\sqrt{6}\,G}{c}\,M\Delta M \ .$$

Substituting $\omega = \omega_{\mathrm{cr}} = \sqrt{0.374\,\pi\,G\hat{\varrho}}$, we find that in order to be spun up to the cricital frequency, a star must accumulate the mass

$$\frac{\Delta M}{M_0} \approx \frac{cI\omega}{\sqrt{6}\,GM_0^2} \approx 0.3\,I_{45}\,m^{-3/2}R_6^{-3/2} \ .$$

For $m = 1.5$ and $R = 1$, we obtain $\Delta M/M_0 \approx 0.15$. The neutron star acquires the critical frequency in a time which is an order of magnitude shorter than the time in which its mass is doubled:

$$t_{\mathrm{su}} \approx 0.1\,\frac{M}{\dot{M}} \approx 10^8 L_{37}^{-1} \text{ years} \ ,$$

where $L_{37} = L_x/10^{37}$ is the X-ray luminosity of the source.

Calculations for models with a soft equation of state show that the star collapses early. However, for more realistic equations of state with the Oppenheimer-Volkoff limit $M_{\mathrm{OV}} \approx 1.5 - 2.5\,M_\odot$ the critical frequency is attained earlier. Hence bursters or other X-ray sources with $\mu < \mu_{\min}$ must have close rotational periods. Each frequency can be juxtaposed to a musical note. Speaking in terms of these "high tones", we can state that the discovery of pulsars in the third octave would be decisive for the choice of the equation of state.

This assumption received unexpected support in a recent work (Lipunov, Postnov 1988). The newly discovered millisecond pulsar PSR 1957+20 has a period $p = 1.607$ ms, which differs only by 3% from the period of the first millisecond pulsar PSR 1937+21 (Fruchter et al. 1988). It is interesting to note that for PSR 1957+20, $\dot{p} < 10^{-20}$, which means that the pulsar has an anomalously low magnetic field, $\mu < \mu_{\min}$! It was revealed as a result of a more detailed analysis that according to these data, a rigorous equation of state of the type L is most probable [see (2.5)] (Lipunov, Postnov 1988).

Another important question concerns the apprehension that a rapid rotation may obstruct the collapse of a neutron star into a black hole. It is well-known that the angular momentum of a black hole is characterized by the Kerr parameter

$$a = \frac{J}{GM^2/c} \ .$$

The parameter $a \leqslant 1$. For all equations of state, $a \lesssim 0.64$. The possible errors associated with simplifying assumptions cannot change this value by more than 10%. This is confirmed by recent calculations (Friedman et al. 1985). Thus, as the Oppenheimer-Volkoff limit is reached, a neutron star rotating like a rigid body inevitably collapses into a black hole.

Before concluding this section, let us consider the mutual orientation of the rotational axis, the dipole axis, and the axis of the disk. First of all, it is obvious that in the process of prolonged evolution, the rotational axes of the neutron star and the binary system (and hence of the disk) are aligned in the same direction. The inflow of the rotational angular momentum takes place in the direction of rotation of the binary system, while the loss of angular momentum takes place in the direction of rotation of the star. The accreting star forgets the initial direction of rotation. There are arguments supporting the assumption that the magnetic axis and the rotational axis coincide. It was shown (see p. 152) that for a very slow initial rotation, the inner parts of the disk may be aligned along the magnetic equator: the inflow of the angular momentum makes the star rotate about the magnetic axis. It is clear that the effect of the X-ray pulsar will not be observed if the magnetic axis and the rotational axis coincide.

The change in the direction of the magnetic axis and the rotational axis under the action of magnetic forces was considered by Wang and Robnik (1982), Goldreich (1970), and Lamb et al. (1975). The results of these investigations, however, are contradictory and further investigations are necessary in this direction.

6.14 Low-Mass X-Ray Sources. "Noisars"

Although galactic bulge sources have much in common with X-ray bursters, they do not exhibit burst activity. Above all, this concerns sources like Sco X-1, Cyg X-2, etc. Like bursters, these sources have a relatively soft spectrum, bright optical companions, and no strictly periodic pulsations of X-ray radiation. The same is true of X-ray eclipses. The binary nature of these sources is generally established with great difficulty. Suffice it to say that the binary nature of the brightest X-ray source Sco X-1 was established only ten years after its discovery. As expected, Sco X-1 turned out to be a close binary system with a period of 0.818 days.

An interesting effect was observed in Sco X-1 in the radiofrequency range. At such frequencies, this source was closer to a quasar than to a normal binary system. Apart from the fact that it is a radiosource on its own, there also exist two "radio ears" arranged symmetrically near it at a distance of $\sim 1'$ (Hjellming, Wade 1971; Hjellming 1978; Velusamy et al. 1985). The flux of lateral components is equal to 20 mJy and has a power spectrum $I_\nu \sim \nu^{-\alpha}(\alpha \approx 1)$. This unique phenomenon has been known for the last 15 years, but no satisfactory interpretation has been found for it. Recently, the radiotelescope VLA was used in an attempt to measure the intrinsic motion of the lateral components (Fomalont et al. 1983). Their velocity was found to be less than $20d$ km/s, where d (kpc) is the distance from Sco X$-$1. This phenomenon is generally not propounded by theorists (recall the "boom" surrounding SS 433), although it reveals an equally strong connection with the active galactic cores.

A recently discovered phenomenon is apparently typical of bulge sources. Van der Klis et al. (1985) detected quasi-periodic oscillations of the X-ray flux in the source GX 5-1 (4U 1758-25) belonging to the class of low-mass X-ray systems. The quest for these oscillations was started in view of the conviction that weakly magnetized stars must rotate rapidly. It should be noted that attempts to find a periodicity in the variations have proved futile so far (Lewin 1979; Sadeh et al. 1982; Leahy et al. 1983; Langmeier et al. 1985). Observations were made on board the specially equipped X-ray satellite EXOSAT in the $1-18$ keV interval. The satellite has a high sensitivity ($\sim 9.2 \times 10^{-12}$ erg cm^{-2} s^{-1}) in this range (or 0.3 mJy in the interval $2-11$ keV). The X-ray flux oscillates between 700 and 1100 mJy.

Investigations of the variations in GX 5-1 for time periods varying between 0.5 ms and 2 s show that the power spectrum consists of three components:

1) a low-frequency component for $\nu < 15$ Hz (the power decreases with increasing frequency),
2) a broad peak with its center at a frequency of ~ 30 Hz, and
3) a plane spectrum above 100 Hz corresponding to Poisson's distribution.

The position ν_c of the center of the peak is 20 Hz $< \nu < 40$ Hz. However, no periodic pulsations were detected in the frequency range 0.5 Hz $- 2$ kHz.

A reliable model for "noisars" has not been proposed so far. Van der Klis et al. (1985) considered a number of possible mechanisms leading to variations in the $\sim 20-40$ Hz range.

A glance at Table 6.4 presenting the mechanisms leading to pulsations of the accreting stars shows that the best agreement with the characteristic frequency is attained for the time of free-fall from the Alfvén surface. The corresponding frequency is given by

$$\nu_{\mathrm{ff}} \approx \frac{v_{\mathrm{ff}}}{R_{\mathrm{A}}} \approx 4 \mu_{30}^{-6/7} L_{37}^{3/7} \; \mathrm{Hz} \; ,$$

while the frequency of Keplerian rotation in the Alfvén zone is

$$\nu_{\mathrm{K}} = \frac{\nu_{\mathrm{ff}}}{2\pi\sqrt{2}} \approx 0.11 \, \nu_{\mathrm{ff}} \; .$$

For the sake of simplicity, we assume that $M_x = 1 \, M_\odot$, and $R_x = 10$ km. For a frequency $\nu_c = 30$ Hz and luminosity $L_{37} \approx 10$ (observational estimate for GX 5-1), a field $\mu = 10^{29}$ ($B \approx 10^{11}$ Oe) is required at $\nu_{\mathrm{ff}} \approx 30$ Hz, while $\mu = 10^{28}$ ($B \approx 10^{10}$ Oe) for $\nu_{\mathrm{K}} = 30$ Hz.

The absence of a pulsating component can be explained by the fact that the rotational and magnetic axes either coincide or are inclined to each other at a small angle ($\leqslant 10°$), but the radiation emitted from the surface of the neutron star cannot be observed because of a large optical depth of matter flowing over the magnetosphere of the neutron star [see (6.114)].

Alpar and Shaham (1985) suggested that the random pulsations of GX 5-1 appear at the beat frequency of the magnetosphere rotation and the Keplerian motion of matter in the disk. In this case,

$$\nu_c = 0.37 \mu_{30}^{-6/7} L_{37}^{3/7} \, \text{Hz} - \nu$$

and the rotational frequency of the neutron star is $\nu \approx 100\,\text{Hz}$.

Analogous effects were recently discovered in Sco X-1 (Middleditch, Priedhorsky 1985) and Cyg X-2, which had characteristic frequencies of $\sim 6\,\text{Hz}$ and $\sim 20\,\text{Hz}$ respectively.

6.15 $p-y$ Diagram for Accreting Neutron Stars

The evolutionary status of binary X-ray systems will be considered in Chap. 11. For the time being, we shall confine ourselves to a description of the general picture in which each of the objects discussed in this chapter occupies a definite position.

The $p-y$ diagram is the most convenient object for investigation. It should be recalled that the use of the gravimagnetic parameter y saves us from taking into account the differences in the magnetic fields of neutron stars. (However, this statement is true only for subcritical accretion regimes.) Nevertheless, it can be mentioned even at this stage that the difference in the magnetic fields of neutron stars has not been explained unambiguously so far. A generally prevalent point of view on the dissipation of magnetic fields (in a period of $\sim 10^6 - 10^7$ years) of neutron stars (Sect. 8.6) seems to be confirmed by the radiopulsar observations. However, there are a number of arguments against this point of view and the differences in magnetic fields can just as well be due to differences *not* in age, but in the initial conditions.

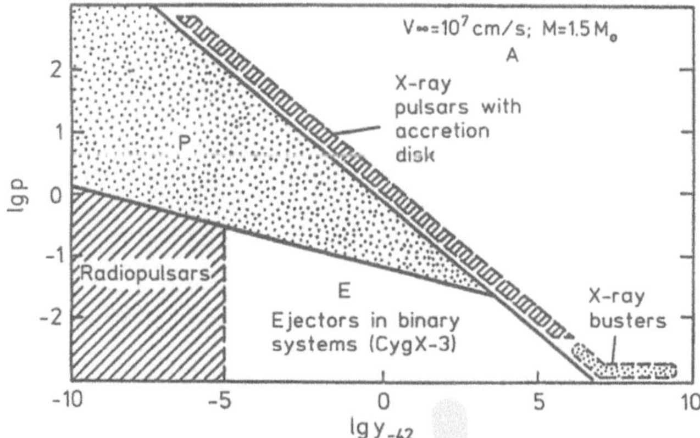

Fig. 6.25. The $p-y$ diagram for neutron stars

Figure 6.25 shows the $p-y$ diagram in which X-ray pulsars, X-ray bursters and other sources of the galactic bulge (noisars) are shown together. It can be assumed that non-bursting, and non-pulsating sources (Sco X-1, etc.) are neutron stars having intermediate magnetic fields between X-ray pulsars and X-ray bursters. In other words, their fields are sufficiently strong to prevent an explosive burning of matter, but too weak to result in the formation of X-ray pulsars. In this case, noisars on the $p-y$ diagram must lie between X-ray pulsars and X-ray bursters (the two domains may overlap). Neutron stars on which disk accretion takes place lie along the p_A-line. The objects accreting from stellar wind may lie much higher than this critical line.

7. The "Propeller" Regime

Shvartsman (1970a) was the first to point towards the necessity of an intermediate regime separating the ejection and accretion stages, in which a rapidly rotating magnetic field obstructs stationary accretion of the plasma. According to the ideas put forth by Shvartsman, after the ejection stage a neutron star still rotates so rapidly that the particles falling under the light cylinder are entrained by the magnetic field and accelerated to relativistic energies. The synchrotron radiation emitted in this case must be pulsating as in the case of radiopulsars (second generation pulsars). Later, Illarionov and Syunyaev (1975) considered a slower rotation and suggested that matter is ejected by the magnetic field with a parabolic velocity. They called this phenomenon the "propeller" effect. The question of such an ejection of matter still remains disputable, but this is not of basic importance. What is more important is that there exists a regime in which stationary accretion is not possible, and the neutron star slows down even more effectively than as a result of magnetic dipole losses. Following the notation presented in Chap. 4, we shall call this regime the "propeller" (P) regime irrespective of whether matter is ejected to infinity or is accumulated near the magnetosphere of the neutron star. The existence of the propeller regime is also confirmed by observations: for a neutron star to be spun down during the lifetime of the optical component to a period characteristic of X-ray pulsars, a mechanism ensuring a more effective spinning down than the standard magnetic dipole mechanism (4.11) is required. This is also confirmed by the numerical experiment on the simulation of neutron star evolution in binary systems (Chap. 11). According to these calculations, a considerable part of neutron stars in massive binary systems is in the "propeller" regime. Hence it becomes obvious that a detailed investigation of the propeller regime must be carried out.

The propeller regime was discussed in a number of earlier publications (Pringle, Rees 1972; Davidson, Ostriker 1973; Shakura 1975; Fabian 1975; Kundt 1976; Lipunov, Shakura 1976; and Savonije, Van den Heuvel 1977). Each of these publications presents a different qualitative picture and formula for spinning-down torque. More detailed investigations were carried out by Holloway et al. (1978), Davies et al. (1979), Lipunov (1980a), Davies and Pringle (1981), Lipunov (1982d), Sibgatullin (1984), and Wang and Robertson (1985). The problem is made more complicated because the inverse effect of the boundary on the motion of accreting flux is of considerable importance in the propeller stage and hence cannot be neglected. This means that the value of the critical period p_A determined in Chap. 4 must be treated as preliminary and a more precise value must be determined by solving the self-consistent problem.

Several versions of the hydrodynamic flow of matter in the propeller regime were considered. These include (a) a steady flow of matter to some sectors and depletion from some other sectors (Shvartsman 1970a; Illarionov, Syunyaev 1975; Shakura 1975), (b) transient regime in which the inflow of the matter is replaced by its outflow from the boundary (Shakura 1975), (c) the formation of a quasistatic shell (Davies et al. 1979; Lipunov 1980a; Davies, Pringle 1981; and Sibgatullin 1984), and (d) the formation of a heavy shell of increasing mass.

Let us assume that matter shed by the magnetic field accumulates in a very thin layer around the magnetosphere. In the spherically symmetric case, the condition of equilibrium of the magnetosphere boundary and the shell can be written in the form (Lamb et al. 1973)

$$\frac{\mu^2}{8\pi R^6} \approx \frac{GM_{sh}M_x}{4\pi R^4} \, , \tag{7.1}$$

where $M_{sh} = \dot{M}t$ is the mass of the shell. The magnetosphere boundary has a radius

$$R_m \approx \left(\frac{\mu^2}{2GM_{sh}M_x}\right)^{1/2} . \tag{7.2}$$

For $R_m > R_c$, a centrifugal barrier obstructing accretion is formed. If the friction between the shell and magnetic field is small and the accreting matter is cooled effectively, the magnetosphere will gradually contract: $R_m \propto t^{-1/2}$. When the corotation radius is reached, matter is deposited on the surface of the neutron star. After this, the process is repeated. This mode of nonstationary spherically symmetric accretion was studied by Tsygan (1976), Baan (1977), and Lipunov et al. (1982).

The energy liberated during the shedding of matter by the shell can be easily calculated (Lipunov et al. 1982):

$$E_b = \frac{M_{sh}GM_x}{R_x} = \frac{\mu^2 \omega^{4/3}}{2(GM_x)^{2/3}R_x} \approx 2 \times 10^{37} \mu_{30}^2 \, m^{-2/3} R_6^{-1} p^{-4/3} \text{ erg} . \tag{7.3}$$

Similarly, we can also calculate the characteristic duration Δt_b of a burst and the burst recurrence period t_b:

$$\Delta t_b \approx \frac{R_c}{v_{ff}} \approx \frac{P}{2\sqrt{2\pi}} . \tag{7.4}$$

$$t_b \approx \frac{ER_x}{GM\dot{M}} \approx 0.1 \mu_{30}^2 \, m_x^{-11/3} v_6^3 p^{-4/3} \varrho_{-24}^{-1} \text{ years} .$$

Here, we have used (3.39) for the accretion rate in the case of a star moving with a velocity v in a medium with density $\varrho_{-24} = \varrho/10^{-24}\,\mathrm{g/cm^3}$. However, such a pattern can be realized only if the magnetic viscosity at the magnetosphere boundary is quite low and there exists an effective way of cooling the matter of the shell. If cooling is not effective, a quasistatic atmosphere of constant mass is formed around the magnetosphere. If magnetic viscosity and cooling are taken into account, a light shell may be formed which effectively takes away the angular momentum from the magnetosphere. In this case, the shell will most probably be transformed into a ring or a disk. In the following sections, we describe the results of investigations of such processes.

7.1 Quasistatic Shells

In this section, we shall mainly follow the work of Davies and Pringle (1981). Suppose that the angular momentum of the matter captured in a neutron star is quite small so that a spherically symmetric accretion regime exists. Matter is captured from the stellar wind flowing from the optical component at a velocity v_{w} and an accretion rate $\dot{M}_0$. The dissipation of rotational energy in the boundary layer adjoining the magnetosphere leads to a heating of the spherically symmetric flow and the formation of an atmosphere (envelope). Moreover, the rotation of the magnetosphere gives rise to circular motion in the envelope which is converted into turbulence (Fig. 7.1). Consequently, the angular momentum is transferred from the bottom of the atmosphere to its outer boundary where it is carried away by the stellar wind from the optical star. The emergence of turbulence seems to be quite real since the characteristic Reynolds number is large:

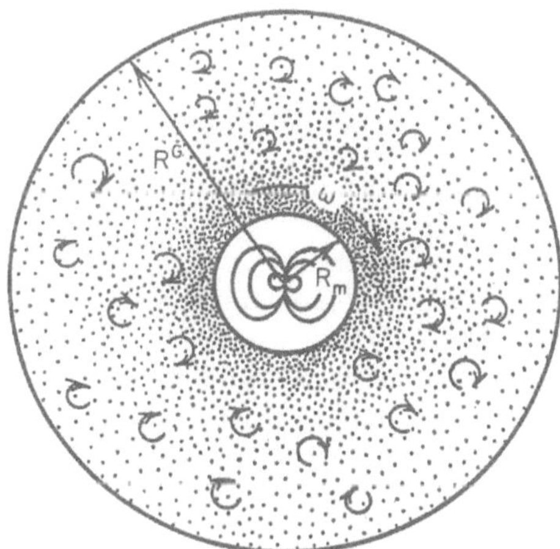

Fig. 7.1. Turbulent envelope around a rapidly rotating star in the propeller regime

$$\mathrm{Re} = \frac{\omega R_{\mathrm{m}} R_{\mathrm{m}}}{\nu} \approx 10^5 - 10^7 \ , \tag{7.5}$$

where ωR_{m} is the characteristic velocity at the magnetosphere boundary, R_{m} is the radius of the magnetosphere and its characteristic size, and ν is the kinematic viscosity. It is assumed that $R_{\mathrm{m}} \lesssim R_{\mathrm{G}}$. Using the concept of a mixing length (Schwarzschild 1958), we can estimate the flux of energy transported by convective or turbulent motions:

$$L(R) \approx 4\pi R^2 \tfrac{1}{2} v_{\mathrm{t}}^3(R)\varrho(R) \ , \tag{7.6}$$

where v_t is the characteristic velocity of turbulent flow and ϱ is the density of matter. Obviously, $v_{\mathrm{t}} < a_{\mathrm{s}}$, the velocity of sound, because otherwise the energy of turbulent motion will attenuate due to energy dissipation in shock waves.

Suppose that the density of matter in the shell is quite low and that the losses due to radiation can be neglected. In this case, we can write

$$L(R) = \mathrm{const} \ . \tag{7.7}$$

Since the atmosphere is quasi-stationary (the characteristic time of rearrangement is much larger than the free-fall time), we can use the hydrostatic equilibrium equation:

$$\frac{dP}{\varrho dR} = -\frac{GM_x}{R^2} \ . \tag{7.8}$$

Writing the equation of state in the polytropic form

$$P = A\varrho^{\gamma} \ , \tag{7.9}$$

we can write the equilibrium equation (7.8) as

$$\frac{d \ln P}{d \ln R} = -\frac{\gamma}{2}\frac{v_{\mathrm{ff}}^2}{a_{\mathrm{s}}^2} \ ; \tag{7.10}$$

where v_{ff} is the free-fall velocity. It can be easily seen from this equation that for cold matter, i.e., for $a_{\mathrm{s}} \ll v_{\mathrm{ff}}$, large pressure gradients are generated, i.e., pressure changes significantly for $\Delta R \ll R$. Conversely, for $a_{\mathrm{s}} \gg v_{\mathrm{ff}}$ (hot matter), pressure remains practically constant. In the intermediate case $a_{\mathrm{s}} \approx v_{\mathrm{ff}}$, the characteristic radii are of the same order: $\Delta R \approx R$.

The solution depends significantly on boundary conditions which can be prescribed in the following form at the outer boundary of the atmosphere:

$$\begin{aligned} P &= P_{\infty} = \tfrac{3}{4}\varrho_{\mathrm{w}} v_{\mathrm{w}}^2 \ , \\ \varrho &= \varrho_{\infty} = 4\varrho_{\mathrm{w}} \ , \quad R \to \infty \ . \end{aligned} \tag{7.11}$$

Here, ϱ_w is the density of stellar wind before the passage of a high-intensity shock wave which is assumed to be adiabatic (Zel'dovich, Raizer 1966). For a polytropic law $\gamma = 1 + 1/n$, we obtain the solution of the hydrostatic equilibrium equation in the form

$$P = P_\infty \left[1 + \left(\frac{1}{1+n} \right) \frac{R_1}{R} \right]^{n+1} ,$$

$$\varrho = \varrho_\infty \left[1 + \left(\frac{1}{1+n} \right) \frac{R_1}{R} \right]^{n} , \tag{7.12}$$

where R_1 is determined from the equality $GM_x/R_1 = P_\infty/\varrho_\infty$. Hence $R_1 = (8/3)R_G$. Putting $n = 3/2 \, (\gamma = 5/3)$, we obtain

$$P = P_\infty \left(1 + \frac{16}{15} \frac{R_G}{R} \right)^{5/2} ,$$

$$\varrho = \varrho_\infty \left(1 + \frac{16}{15} \frac{R_G}{R} \right)^{3/2} . \tag{7.13}$$

For $R \ll R_G$, this solution coincides with the one obtained in Sect. 5.5 [see (5.64)].

In the general case, the structure of the atmosphere depends on the energy dissipation rate at the inner boundary $R_{in} = R_m$ of the atmosphere. Depending on the rotational velocity of the neutron star, various regimes are possible.

7.1.1 Supersonic Propeller

Suppose that the rotational velocity of the neutron star is so high that the linear velocity of rotation of the magnetosphere is much higher than the velocity of sound in matter:

$$\omega R_m \gg a_s(R_m) . \tag{7.14}$$

It is logical to assume that the velocity v_t of turbulent motion at the magnetosphere boundary is close to the velocity of sound a_s. In this case, the rate of dissipation of the rotational energy at the lower boundary of the atmosphere is given by (7.6):

$$L(R_m) \approx 4\pi R_m^2 \frac{1}{2} \varrho(R_m) a_s^3(R_m) . \tag{7.15}$$

Further, we assume that $a_s(R) \approx v_{ff}(R)$ for $R_m \leqslant R \leqslant R_{out}$ (the case when $a_s \gg v_{ff}(R)$ will be considered below). In this case, (7.7) is satisfied everywhere:

$$4\pi R^2 \varrho(R) v_t^3 = \text{const} . \tag{7.16}$$

Assuming that the turbulence is acoustic everywhere on the shell, i.e., $v_t \approx a_s$, we directly obtain from (7.16) the relation $\varrho \propto R^{-1/2}$ according to which the atmosphere has an effective polytropic exponent $n = 1/2$. Suppose that this is not true and that the velocity of turbulent flow falls more rapidly than the velocity of sound with increasing R, say, as $v_t/a_s \propto R^{-\alpha}$. In this case, we obtain from (7.16) $\varrho \propto R^{3\alpha - 1/2}$, i.e., $n = 1/2 - 3\alpha$. However, if the turbulent flow velocity decreases, so does its effect on the structure of the atmosphere, and the atmosphere must be adiabatic with $n = 3/2$. This leads to a contradiction. Hence we are left with the assumption that the turbulence is subsonic everywhere: $v_t \approx a_s$, i.e., $n = 1/2$. Consequently, the atmospheric pressure is given by

$$P = P_\infty \left(\frac{R_G}{R}\right)^{3/2} . \tag{7.17}$$

The outer boundary of the atmosphere is at a distance $R_{out} \approx R_G$. It should be recalled that if the reciprocal effect of the magnetosphere is not taken into account, we obtain $P \propto R^{-5/2}$.

Let us now estimate the inner radius of the atmosphere from the condition of equality of magnetic and gas pressures:

$$\frac{\mu^2}{8\pi R_m^6} = P(R_m) . \tag{7.18}$$

Substituting (7.17) into this equation, we find the radius of the magnetosphere in the case of a supersonic propeller:

$$R_m \approx \left(\frac{R_G}{R_A}\right)^{2/9} R_A > R_A . \tag{7.19}$$

As before, R_A is the Alfvén radius (4.23) for $R_A < R_G$. Qualitatively, it is quite clear that heating of the atmosphere leads to a decrease in its density and pressure, as a result of which the magnetosphere swells. It follows from (7.16) that the dissipation rate of rotational energy is independent of the rotational frequency and is equal to

$$L \approx \dot{M}_c v_w^2 \approx 10^{33} \dot{M}_{17} v_8^2 \, \text{erg/s} . \tag{7.20}$$

The corresponding spinning-down torque is given by

$$K_{sd} \approx \frac{\dot{M}_c v_w^2}{\omega} \approx \left(\frac{R_c}{R_m}\right)^{3/2} \frac{\mu^2}{R_m^3} , \tag{7.21}$$

where $\dot{M}_c = \pi R_G^2 \varrho_w v_w$ is, as before, the accretion rate of the captured matter. The characteristic spinning-down time $t_{sd} = I\omega/K_{sd}$ is then defined as

$$t_{sd}^{(1)} \approx 1.6 \times 10^7 \, \dot{M}_{15}^{-1} \, v_8^{-2} \, I_{45} p^{-2} \text{ years} \; . \tag{7.22}$$

Here, p is the initial period of rotation of the star in seconds.

The above regime is sustained until the rotational velocity of the magnetosphere boundary exceeds the free-fall velocity. This is equivalent to the requirement that $R_m > R_c$ (R_c is the corotation radius). Hence a supersonic propeller exists as long as

$$p \lesssim p_1 \approx 23 \, \mu_{30}^{2/3} \, \dot{M}_{15}^{-1/3} \, v_8^{-2/3} \text{ s} \; . \tag{7.23}$$

Otherwise, a subsonic propeller regime occurs.

7.1.2 Subsonic Propeller

Let us now assume that the stopping radius (magnetosphere radius) is smaller than the corotation radius. In this case, the neutron star rotates slowly and the shape of the magnetosphere can be approximated quite well by the solution obtained in Sects. 5.5 and 5.9. Investigation of stability relative to the Rayleigh-Taylor instability shows that if there are no effective cooling mechanisms, the magnetosphere is stable. We shall now consider a case of this type in which the cooling is not effective and, moreover, the dissipation of rotational energy produces additional heating. Hence, we can assume that in the case of spherically symmetric accretion matter does not fall on the surface of the neutron star even if the stopping radius is smaller than the corotation radius. An extended atmosphere is formed around the magnetosphere and the velocity of sound at its boundary is much higher than the rotational velocity of the magnetosphere: $v_s \gg \omega R_m$. The structure of such an adiabatic atmosphere was considered by us in Sect. 5.5 and is described by (7.13). The radius of its inner boundary is equal to the ordinary Alfvén radius.

The spinning-down relation has the following form (Davies, Pringle 1981; Lipunov 1982):

$$K_{sd} = \kappa_t \frac{\mu^2}{R_c^3} \; , \quad \kappa_t = \frac{v_t l_t}{3 \omega R_A^3} \; , \tag{7.24}$$

where R_c is the corotation radius. The characteristic spin-down time in the subsonic propeller regime is $t_{sd}^{(2)} \approx 10^3 \mu_{30}^{-2} m_x I_{45} p$ years. It follows from the law of energy conservation (7.16) that in the atmosphere surrounding the subsonic propeller, the Mach number for turbulent motion increases with distance from the star:

$$M_t = \frac{v_t}{a_s} \propto R^{1/3} \; .$$

Obviously, the applicability of the subsonic propeller regime is determined by the condition

$$M_t(R_G) < 1 \ .$$

When does accretion set in? The magnetosphere becomes very unstable if the plasma cooling is more effective than its heating. The characteristic cooling time due to free-free transitions is

$$t_{br} \approx 2 \times 10^{11} \, T^{1/2} n^{-1} \text{ s} \ .$$

The heating time is defined as t_{sd}. It can be seen easily that

$$\frac{t_{sd}}{t_{br}} \propto R^{1/2} \ .$$

Hence the cooling will be more effective first at the "bottom" of the atmosphere. Putting $t_{sd}/t_{br} = 1$, we obtain for $R = R_A$ the critical period beyond which accretion becomes possible:

$$p_{br} \approx 60 \mu_{30}^{16/21} \dot{M}_{15}^{-5/7} m_x^{-4/21} \text{ s} \ . \tag{7.25}$$

Of course, the result obtained here depends very strongly on the stability criterion for the magnetosphere. For $R_m \approx R_c$, the structure of the magnetosphere has not been calculated and the stability criterion has not been established. Besides, the Kelvin-Helmholtz instability also comes into play in this case.

7.1.3 Very Rapid Propeller

Suppose that a neutron star rotates so rapidly and the heating of the magnetosphere is so effective that the velocity of sound is much higher than the free-fall velocity: $a_s \gg v_{ff}$. The thermal energy of a substance is much higher than its gravitational energy (the matter is unaffected by gravity). In such an atmosphere, there is no pressure gradient: $P = \text{const}$ [see (7.10)]. Putting $a_s = \omega R_m$, we find from the condition of conservation of energy flux for $R = R_m$ that $a_s \propto R^{-2}$ and $\varrho \propto R^4$. The density increases in the outward direction. This in fact means that a hot cavern is formed outside the magnetosphere. The radius R_{out} of the outer boundary of the cavern is determined by the boundary condition

$$R_{out} = R_m \left(\frac{\omega R_m}{v_\infty} \right)^2 \approx 6 \times 10^9 v_8^{-7/4} m_x^{1/2} p^{-1/2} \dot{M}_{15}^{-1/4} \mu_{30}^{1/2} \text{ cm} \ . \tag{7.26}$$

Since the pressure inside the cavern is constant, the magnetosphere radius (radius of the inner boundary of the cavern) is determined by (4.23) for $R_A > R_G$:

$$R_m \approx 4 \times 10^9 \mu_{30}^{1/3} v_8^{-5/6} m_x^{1/3} \dot{M}_{15}^{-1/6} \text{ cm} \ . \tag{7.27}$$

The rate of dissipation of the rotational energy is

$$L_{\mathrm{m}} \approx \frac{\mu^2}{R_{\mathrm{m}}^3}\,\omega \approx 2\times10^{31}\mu_{30}p^{-1}m_x^{-1}\dot{M}_{15}^{1/2}v_8^{5/2}\ \mathrm{erg/s}\ . \tag{7.28}$$

Among other things, this leads to an expression for the spinning-down torque:

$$K_{\mathrm{sd}} \approx \frac{\mu^2}{R_{\mathrm{m}}^3}\ . \tag{7.29}$$

The characteristic spinning-down time corresponding to this torque is

$$t_{\mathrm{sd}}^{(3)} \approx 6\times10^7 I_{45}\mu_{30}^{-1}m_x v_8^{-5/2}\dot{M}_{15}^{-1/2}p^{-1}\ \mathrm{years}\ . \tag{7.30}$$

The radius of the outer boundary of the atmosphere decreases in the course of spinning down, and for $R_{\mathrm{out}}\approx R_{\mathrm{G}}$ gravitation becomes important: a transition to the supersonic "propeller" regime takes place (p. 212). The period corresponding to this transition is

$$p_{31} \approx 2.2\,m_x^{-1}v_8^{1/2}\dot{M}_{15}^{-1/2}\mu_{30}\ \mathrm{s}\ . \tag{7.31}$$

This regime, which is accompanied by the formation of a cavern, was first studied by Kulsrud (1971) in the course of his investigations on the spinning down of magnetic stars.

Formulas (7.21, 24, 29) for the spinning-down torque have been presented in another form which differ from that of Davies and Pringle (1980). In the most general case, there exists a universal form of notation, e.g., $\kappa_t\mu^2/R_t^3$, for the spinning-down torque (Sect. 11.2). It should also be noted that all formulas for critical periods contain the gravimagnetic parameter $y = \dot{M}/\mu^2$.

Following Davies and Pringle (1981), we have so far assumed that the magnetosphere radius is smaller than the gravitational capture radius: $R_{\mathrm{m}} < R_{\mathrm{G}}$. In actual practice, however, the inverse inequality may also hold.

7.1.4 Nongravitating Propeller

Suppose that $R_{\mathrm{m}} > R_{\mathrm{G}}$. In this case, the graviational force is not significant and the pressure in the envelope around the rotating neutron star is constant. Consequently, as in the previous case, the magnetosphere radius is equal to the Alfvén radius (7.27). The structure of the atmosphere is also conserved as well as the formulas (7.28) for the dissipation of rotational energy and (7.29) for the spinning-down torque. However, the difference is that after spinning down such a neutron star goes to the georotator regime (Chap. 4) rather than the accretion regime. The condition for the transition is $R_{\mathrm{m}} \leqslant R_{\mathrm{c}}$ [see (4.53): $R_{\mathrm{A}} > R_{\mathrm{G}}$].

7.2 Spinning-Down in the Boundary Layer

In Sect. 7.1, we were able to determine the spinning-down torque without considering the detailed pattern of interaction between the plasma and the magnetic field of the neutron star. For example, the structure of the boundary layer between the shell and the magnetosphere was found to be entirely insignificant in view of the fact that the energy flux transported by the shell was assumed to be constant (7.7): energy must not be emitted or reflected (for example, by a shock wave) en route. This approximation is valid for a rather low accretion rate when the energy loss due to radiation can be neglected (as a rule, free-free radiation is meant). Otherwise, we must take into account the detailed structure of the boundary layer resulting from the penetration of plasma into rapidly rotating magnetosphere of the neutron star.

The spinning down of a neutron star depends significantly on the stability of the rotating magnetosphere, the penetration depth of plasma into the magnetosphere and the speed at which it is entrained by the magnetic field. In the limit of an ideally conducting plasma and an absolutely stable magnetosphere, the spinning-down torque is equal to zero. It would seem that the rapidly rotating nonsymmetric magnetosphere can shed matter on its own. This is possible only for a finite magnetic viscosity. Otherwise, no forces emerge (cf. D'Alembert's paradox in fluid mechanics).

The first attempts at numerical and qualitative investigations of the processes of plasma penetration into the rapidly rotating magnetic field of a neutron star and at the computation of boundary layer parameters were made by Wang and Robertson (1985). This approach consists of two steps. First the nonstationary problem of the penetration of plasma and its mixing with magnetic field is solved numerically in the cylindrical symmetry approximation. After this the solution of the idealized problem is applied to the real three-dimensional case. The following value of the spinning-down torque was obtained

$$K_{\rm m}^- \approx \frac{2\zeta}{\pi\beta} \dot{M}\omega R_{\rm m}^2 \ , \tag{7.32}$$

where $\zeta = 0.5$ and $\beta \lesssim 1$. The characteristic spinning-down time is

$$t_{\rm sd} \approx \frac{\pi}{5} \frac{\beta}{\zeta} \left(\frac{R_x}{R_{\rm m}}\right)^2 \frac{M}{\dot{M}} \approx 3\times 10^6 \frac{1}{\eta^{1/2}} \left(\frac{\beta}{\zeta}\right)^{3/4} \omega^{3/4} \dot{M}_{15}^{3/4} m_x^{3/4} \mu_{30}^{1/2} \ .$$

Formula (7.32) can be written

$$K_{\rm sd} \approx \frac{1}{4\pi} \left(\frac{\delta}{R_{\rm m}}\right) \frac{\mu^2}{R_{\rm m}^3} \approx \frac{\eta^2}{2\pi} \left(\frac{R_{\rm c}}{R_{\rm m}}\right)^3 \frac{\mu^2}{R_{\rm m}^3} \ . \tag{7.33}$$

The torque (7.32) has the same expression as if the rotational energy were carried away by the flow of noninteracting particles (Shakura 1975). However, it is difficult to imagine a stationary regime in a continuous medium with an outflow of matter that initially fell on the neutron star in a spherically symmetric manner. We believe that if the radiation plays a dominating role in the shell (envelope), a situation leading to a two-stream accretion regime must be realized in the shell.

7.3 Two-Stream Flow Formation due to the Propeller Effect

If matter is effectively cooled as a result of radiation, the shell is found to be thin (Sect. 7.2). In this case, the following course of events may occur (Lipunov 1980a; 1982). The spherically symmetric flow of accreting matter collides with the magnetosphere. Matter accelerated by the magnetic field in the transient layer acquires an angular momentum and forms a thin shell which gradually settles down on the equator plane of rotation. A disk flow is formed in this plane.

7.3.1 Stationary Flow from Disks

The problem described above has a time-independent solution. The entire matter falling spherically symmetrically onto a neutron star is condensed to the disk from which it then flows out beyond the capture radius, carrying away the angular momentum of the neutron star. The radius of the inner edge of the disk is constant and equal to the radius at which the specific angular momentum k_{sd} acquired by matter on the magnetosphere is equal to the Keplerian angular momentum $\sqrt{GM_xR}$, i.e.,

$$R_d = \frac{k_{sd}^2}{GM_x} \ . \tag{7.34}$$

Integrating the equation describing the change in the angular momentum, we obtain

$$\dot{M}_c \omega_k R^2 + 2\pi W_{r\varphi} R^2 = C \ ; \tag{7.35}$$

where ω_k is the Keplerian rotational frequency of matter. The boundary condition is specified at the outer edge of the disk under the assumption that the surplus angular momentum is carried away by the external forces (stellar wind or tidal forces):

$$W_{r\varphi}(R_{out}) = 0 \ . \tag{7.36}$$

The stationary outflow from a Keplerian disk is described by the formula

$$W_{r\varphi} = -\frac{\dot{M}_c \omega_k}{2\pi}\left[\left(\frac{R_{\text{out}}}{R}\right)^{1/2} - 1\right] . \tag{7.37}$$

The rate of energy liberation in an accretion disk is

$$L = 2\pi W_{r\varphi}(R_d)\omega_k(R_d)R_d^2 . \tag{7.38}$$

For viscous stresses defined by (7.37), the energy flux incident on the disk is given by

$$L = \dot{M}_c \frac{GM_x}{R_d}\left[\left(\frac{R_{\text{out}}}{R_d}\right)^{1/2} - 1\right] \approx \dot{M}_c \frac{GM}{R_d}\left(\frac{R_{\text{out}}}{R_d}\right)^{1/2} . \tag{7.39}$$

The thermal energy flux emitted by a unit surface area of the disk is

$$Q_- = \frac{1}{2}W_{r\varphi}R\frac{\partial\omega}{\partial R} = \frac{3}{8}\frac{\dot{M}_c}{\pi}\omega_k^2\left[\left(\frac{R_{\text{out}}}{R}\right)^{1/2} - 1\right] . \tag{7.40}$$

We shall calculate the spectrum of such a disk, assuming that its surface emits radiation like a black body. In the case of standard disk accretion, we had $Q \propto R^{-3}$ and $F_\nu \propto \nu^{1/3}$ for $h\nu < kT(R_d)$. In the present case, $Q \propto R^{-5/2}$ and we obtain for the spectral flux

$$F_\nu = \frac{4\pi^2 h\nu^3}{c^2}\int_0^\infty \frac{R\,dR}{e^{h\nu/kT}-1} \tag{7.41}$$

for $h\nu < kT(R_d)$. Since $Q \propto aT^4$, we get $T \propto R^{-5/8}$

$$F_\nu \propto \nu^{-1/5} \tag{7.42}$$

for $h\nu < kT(R_d)$. The spectral flux grows towards the low-frequency region. This is natural since for the thermal energy flux $Q \propto R^{-5/2}$ the main contribution to the radiation comes from the outer (colder) parts of the accretion disk.

Using the standard disk accretion model equation (Table 3.1), we can easily calculate the physical characteristics of the outgoing flux. The radius of the outer edge apparently does not considerably exceed the gravitational capture radius:

$$R_{\text{out}} \approx R_G .$$

7.3.2 Time-Dependent Solution

If the inflow of the angular momentum to the inner edge of the disk is not too large, the inner boundary of the disk flux begins to creep towards the neutron star. The disk slowly spreads out until its inner edge finally reaches corotation radius and accretion to the neutron star surface begins. Accretion continues for a period of time approximately equal to the duration of radial motion in the disk, after which the initial situation is realized again. Apparently, this leads to a burst-type X-ray source (more precisely, a transient source). This process can be described by the time-dependent equations (3.91, 92) of two-stream accretion.

The structure of the disk flux changes slowly and is described in the region $R < R_1$ (as before, R_1 is the radius at which matter intrudes into the disk) by the so-called dead disk or accumulator disk model described below.

7.4 Dead Disks and Accumulator Disks

Let us consider the disk accretion regime and see what happens to a disk when the radius of its inner edge is larger than the corotation radius.

This situation was first analyzed by Syunyaev and Shakura (1977) who showed that a dead disk or an accumulator disk is formed in the propeller regime. The accretion rate in such a disk is negligibly small. The angular momentum extracted from the neutron star flows outwards over the disk from the inner edge. This is accompanied by energy liberation in the disk. However, the radiation power of such a disk is much lower than the energy liberated by an accreting neutron star (by a factor of R_d/R_x).

The structure of a dead disk is calculated as follows. In the equation for the conservation of the angular momentum (p. 218), we put $\dot{M}_c = 0$. This gives

$$W_{r\varphi}(R) = W_{r\varphi}(R_d) \left(\frac{R_d}{R}\right)^2 . \tag{7.43}$$

The energy flux from a unit surface area of the disk is given by

$$Q_- = Q_-(R_d) \left(\frac{R_d}{R}\right)^{7/2} . \tag{7.44}$$

If such a disk emitted radiation as a black body, its spectrum would be described by a power law:

$$F_\nu \propto \nu^{5/7} \quad \text{for} \quad h\nu < kT(R_d) . \tag{7.45}$$

The following results have been obtained in calculations of the structure of a dead disk in the standard disk accretion model (α-model) in which the free-free absorption plays a dominant role (see Table 3.1):

$$H(\text{cm}) \qquad \approx 2.3 \times 10^6 \alpha^{1/4} \omega_k^{-6/7} \Sigma^{3/14} ,$$

$$a_s(\text{cm/s}) \qquad \approx 2.3 \alpha^{1/14} \omega^{1/7} \Sigma^{3/14} ,$$

$$T(\mathrm{K}) \approx 2\times 10^{4}\alpha^{1/7}\omega^{2/7}\Sigma^{3/7} ,$$

$$P\,(\mathrm{dyne/cm^{2}}) \approx 0.7\times 10^{6}\alpha^{1/14}\omega^{8/7}\Sigma^{17/14} ,$$

$$W_{r\varphi}\,(\mathrm{dyne/cm}) \approx 3.3\times 10^{12}\alpha^{8/7}\omega^{2/7}\Sigma^{10/7} ,$$

$$Q_{-}\,(\mathrm{erg/cm^{2}\,s}) \approx 2.5\times 10^{12}\alpha^{8/7}\omega^{9/7}\Sigma^{10/7} . \tag{7.46}$$

Using the transport equation for the angular momentum, we can find the connection between Σ and R for a given $\dot{M}$. The parameter in the transport equation may be the total mass of the disk:

$$M_{\mathrm{d}} = 2\pi \int_{R_{\mathrm{d}}}^{R_{\mathrm{out}}} \Sigma(R)R\,dR . \tag{7.47}$$

From (7.44), we obtain the dependence of disk parameters on the radius:

$$\Sigma(R) = \Sigma(R_{\mathrm{out}})\left(\frac{R_{\mathrm{out}}}{R}\right)^{11/10} ,$$

$$H(R) = H(R_{\mathrm{out}})\left(\frac{R}{R_{\mathrm{out}}}\right)^{21/20} ,$$

$$a_{\mathrm{s}}(R) = a_{\mathrm{s}}(R_{\mathrm{out}})\left(\frac{R_{\mathrm{out}}}{R}\right)^{9/20} , \tag{7.48}$$

$$P(R) = P(R_{\mathrm{out}})\left(\frac{R_{\mathrm{out}}}{R}\right)^{61/20} ,$$

$$T(R) = T(R_{\mathrm{out}})\left(\frac{R_{\mathrm{out}}}{R}\right)^{9/10} .$$

The total luminosity (7.38) of the dead disk is given by

$$L \approx 5.7\times 10^{34}\alpha^{8/7}m_{x}^{1/7}M_{23}^{10/7}R_{11}^{-9/7}\omega\left(\frac{R_{\mathrm{d}}}{R_{\mathrm{c}}}\right)^{-3/2} \mathrm{erg/s} .$$

Here, $R_{11} = R_{\mathrm{out}}/10^{11}$ cm, $M_{23} = M_{\mathrm{d}}/10^{23}$ g, and ω is the rotational frequency of the neutron star.

The time in which a neutron star is slowed down by a dead disk is

$$t_{\mathrm{sd}} \approx \frac{I\omega}{2\pi W_{r\varphi}(R_{\mathrm{d}})R_{\mathrm{d}}^{2}} \approx 600\,\omega\alpha^{-8/7}m_{x}^{-1/7}R_{11}^{9/7}M_{23}^{-10/7} \text{ years } .$$

The position of the internal edge of the disk is determined from the balance of the magnetic and gas pressures at the inner edge

$$\frac{B^2}{8\pi} = \frac{B_{\mathrm{d}}^2}{8\pi}\left(\frac{R_{\mathrm{d}}}{2H}\right) = P = \varrho\, v_{\mathrm{s}}^2 = \varrho_{\mathrm{ff}} v_{\mathrm{ff}}^2\left(\frac{R}{H}\right) ,$$

where ϱ_{ff} and v_{ff} are the density and velocity in the case of spherically symmetric accretion. Here we have taken into consideration the enhancement of the magnetic field strength at the inner edge of the disk (Sect. 5.6). The radius of the disk is found to be of the order of the Alfvén radius [see (4.23)]: $R_{\mathrm{d}} \approx R_{\mathrm{A}}$, i.e., the same as in the case of spherical accretion. It would seem that after accumulating a thin layer of matter, the disk may push the magnetic field deeper. But this is not true. In fact, the field becomes stronger by exactly such an amount that the stopping radius remains unchanged.

The qualitative pattern of the formation and evolution of a gas disk in a binary system looks like this: suppose that the flow takes place through the inner Lagrangian point. The gas jet is twisted around the neutron star in the form of a ring. Due to turbulent viscosity, the ring spreads to the disk on a time scale

$$t_{\mathrm{r}} \approx \frac{R_{\mathrm{out}}^2}{\nu_{\mathrm{t}}} \approx \frac{R_{\mathrm{out}}^2}{\alpha v_{\mathrm{s}} H} .$$

Here, ν_{t} is the kinematic turbulent viscosity. The gas attains a radius R_{d}, where the pressure exerted by matter becomes equal to the magnetic field pressure. If $R_{\mathrm{d}} > R_{\mathrm{c}}$, the centrifugal barrier obstructs accretion and only the angular momentum of the compact star is transported over the disk. We can consider two possibilities.

1) If the total matter falling on the outer edge does manage to flow out, acquiring an additional angular momentum in the process, the problem is time-independent, the mass of the disk is constant, and only the angular velocity of the neutron star changes.

2) If the tidal forces result in an effective pumping of torque from the disk to the orbital angular momentum at the outer boundary (Paczynski 1976; Papaloizou, Pringle 1977), the mass of the disk slowly increases. The inner parts rapidly adapt themselves to these changes so that the structure of the disk at each instant of time is described by (7.48). If the disk radius becomes equal to the corotation radius, accretion sets in and a powerful source of X-ray radiation comes into action in a time of the order of t_{r}.

Thus, accumulator disks may be responsible for the formation of transient sources.

7.5 Nonstationary Disk Accretion. Model of Transient X-Ray Sources

The flow of matter described in the previous sections points towards the need to take into account the time-dependent equations of disk accretion. The theory of nonstationary disk accretion was considered by Lynden-Bell and Pringle (1974), Lightman (1974), Bath and Pringle (1981), Filipov (1984), and Lyubarskii and Shakura (1987).

It can be easily seen that the system of angular momentum transport and continuity equations

$$\frac{\partial \Sigma R^2 \omega_k}{\partial t} + \frac{1}{R}\frac{\partial}{\partial R}(R\Sigma v_r \omega_k R^2) = \frac{1}{R}\frac{\partial}{\partial R}(R^2 W_{r\varphi}) \ ,$$

$$\frac{\partial \Sigma}{\partial t} + \frac{1}{R}\frac{\partial}{\partial R}(R\Sigma v_r) = 0 \ ,$$

can be reduced by taking into account the viscous stresses

$$W_{r\varphi} = -v_t \Sigma R \frac{\partial \omega}{\partial R}$$

into a diffusion-type time-dependent equation for the surface density

$$\frac{\partial \Sigma}{\partial t} = \frac{3}{R}\frac{\partial}{\partial R}\left[R^{1/2}\frac{\partial}{\partial R}(v_t \Sigma R^{1/2}) \right] \ . \tag{7.49}$$

However, it is more convenient to transform to new variables F and k (Filipov 1984):

$$I = 2\pi W_{r\varphi}R^2 \ ,$$
$$k = \sqrt{GMR} \ , \tag{7.50}$$

where F is the frictional force applied to a unit length of the annular disk at radius R, and k is the specific Keplerian torque. In this case, we can write instead of (7.49)

$$\frac{\partial F}{\partial t} = \prod \frac{F^m}{k^n}\frac{\partial^2 F}{\partial k^2} \ , \tag{7.51}$$

where $\prod$ is a parameter depending on the scale and type of viscosity, as well as on the nature of absorption, m and n are fractions depending on the particular model. Such a model with $m = 0$ was considered by Lynden-Bell and Pringle (1974).

For the standard α-model proposed by Shakura and Syunyaev (1973) m and n had the following values: $m = 3/10$, $n = 8/10$ if the absorption is dominated by free-free transitions; $m = 4/10$, $n = 12/10$ if the Thomson scattering dominates.

The self-similar solution describing time-dependent accretion from the dead disk becomes

$$\dot{M}(t) \propto t^{-1/(n+2)} \tag{7.52}$$

after the inner boundary radius becomes smaller than the corotation radius (Filipov 1984). Accretion begins suddenly and then attenuates according to (7.52). In the disk where free-free transitions in the absorption dominate, the accretion rate (and hence the luminosity also) will decrease according to a power law $L_x \propto t^{-5/14}$. Observations show that beyond the peak the X-ray flux indeed decreases according to a power law.

The model considered here can be effective in cataclysmic variables in which the accreting compact star is a white dwarf (Smak 1971; Bath 1973; Pringle, Savonije 1979).

7.6 Relativistic Propeller

Shvartsman (1970a) was the first to consider the acceleration of charged particles in the rapidly rotating magnetosphere of a neutron star, whose size is smaller than (but close to) the radius of the light cylinder. For $R_m \approx R_l$, the electric field $E \approx \omega R_m B/c \propto B$ is strong and the maximum energy $\varepsilon \approx EeR_m \approx \mu^2 \omega^2 e/c^2 \approx 10^{13} p^{-2} \mu_{30}$ eV acquired by a particle in such a field is much higher than the rest energy of the particle. This estimate just illustrates the basic possibility of accelerating a particle to relativistic energies and is by no means an indication of the characteristic energy acquired by the particles in this case. Unfortunately, investigations of this type of relativistic "propeller" regime have not been carried out so far. The only work in this direction was carried out by Tsygan (1981) who considered the acceleration of a particle in a high-intensity electric wave. However, it is obvious that such investigations must be carried out. The relativistic propeller is a stage passed by almost every single neutron star immediately after the ejection regime (Chap. 11).

7.7 Objects That Can Become Propellers

The propeller regime is a necessary stage in the evolution of a neutron star which must pass through this stage if it has a high rotational velocity at the time of its formation.

7.7.1 Binary Systems

It was mentioned above that the existence of long-period pulsars in close binary systems directly points towards the existence of a spinning-down mechanism for the neutron star that is much more effective than spinning down according to the magnetic dipole law (which results in ejection by neutron stars). For example, in order to spin-down to a period of 500 s according to the magnetic dipole law, a neutron star requires a time $t \approx 10^{13} \mu_{30}^{-2}$ years which is much longer than the lifetime of massive stars ($\sim 10^7$ years) for all reasonable fields. Consequently, binary systems must contain propeller neutron stars.

Let us consider the astrophysical properties of such propellers. First of all, prospective propellers must be transient X-ray sources. The above theoretical analysis shows that especially favorable conditions exist when the accretion rate is relatively high and the cooling mechanisms are effective. In this case, we can expect a nonstationary two-stream or disk accretion. Apparently, for a low accretion rate from the stellar wind, a quasi-stationary atmosphere emitting radiation in the hard X-ray region is formed. However, its X-ray luminosity is low and is lower than

$$L_x \lesssim \dot{M} \frac{GM_x}{R_\mathrm{m}} \approx 10^{32} \dot{M}_{15} R_9^{-1} \text{ erg/s} \ .$$

In fact, the luminosity is lower by a factor of $(t_\mathrm{br}/t_\mathrm{r})$, where t_br is the cooling time and t_r is the time of fall.

Prospective propeller objects may include non-X-ray binary systems with relativistic components (Cherepashchuk, Aslanov 1984).

7.7.2 Single Neutron Stars

An analysis of the statistical properties of radiopulsars (Chap. 8) shows that a single neutron star is generated in our Galaxy every $15-20$ years. For about $10^6 - 10^7$ years, these neutron stars manifest themselves as radiopulsars at the ejection stage. There are about 100000 such pulsars in the Galaxy. The pulsar mechanism ultimately stops being operative. What is the fate of these stars?

Let us consider the physical conditions in the interstellar matter surrounding a single neutron star. Of course, the interstellar medium is highly inhomogeneous (Kaplan, Pikel'ner 1979): the physical conditions vary most significantly in the case under consideration, i.e., from the densest regions (where the density $\varrho_\infty \approx 10^{-22}$ g/cm^3, and the temperature $T \approx 10^2$ K) to hot and rarefied regions in which $\varrho_\infty \approx 10^{-25} - 10^{-26}$ g/cm^3 and $T \approx 10^5 - 10^6$ K. Nevertheless, it is expedient to consider moderate values of the parameters $T_\infty \approx 10^4$ K and $\varrho_\infty \approx 10^{-24}$ g/cm^3.

Table 7.1. Parameters characterizing the interaction of a single neutron star with interstellar plasma

Parameter	$\varrho_\infty = 10^{-22}$ g/cm^3 $T_\infty = 10^2$ K	$\varrho_\infty = 10^{24}$ g/cm^3 $T_\infty = 10^4$ K	$\varrho_\infty = 10^{-26}$ g/cm^3 $T_\infty = 10^6$ K
Capture radius (R_G)	2.7×10^{14} cm	2.7×10^{14} cm	1.3×10^{14} cm
$\dot{M}_\mathrm{c}$	2.2×10^{13} g/s	2.2×10^{11} g/s	5.5×10^{8} g/s
Potential luminosity $L \equiv \dot{M}_\mathrm{c} G M_x / R_x$	2.2×10^{33} erg/s	2.2×10^{31} erg/s	5.5×10^{28} erg/s
Alfvén radius (R_A)	3×10^{9} cm	10^{10} cm	$\sim 10^{11}$ cm

The values of parameters characterizing the interaction of a neutron star with the interstellar medium are presented in Table 7.1. An analysis of the relations between the characteristic quantities presented in Table 8.2 shows that almost every neutron star is inevitably transformed into the propeller regime after the completion of the ejection stage. It is possible that such neutron stars are the sources of gamma-ray bursts (Lipunov et al. 1982).

However, the model of gamma-ray burst sources proposed by Bisnovaty-Kogan et al. (1975) also cannot be ruled out. It is proposed that the bursts are a result of the decay of non-equilibrium heavy impurities which could be accumulated in the crust of the neutron star during its formation.

8. Ejecting Stars

In this chapter, we shall consider rapidly rotating neutron stars, each having its own radiation. Electromagnetic waves and relativistic particles ejected by such stars scatter the surrounding plasma without giving it a chance to accrete. The concept of "rapidity" of rotation, however, remains relative. It was shown in Chap. 4 that the critical period p_E distinguishing the ejecting stars is a function of the medium parameters, viz., the potential accretion rate $\dot{M}_c$, the magnetic field of the star, and a number of other physical quantities. The formula for R_{st} obtained in Chap. 4 is a first approximation since we did not take into account the change in the accreting plasma parameters under the action of the energy flux ejected by the neutron star. We shall consider the finer aspects of the problem in this chapter.

Neutron stars were discovered as ejecting stars or radiopulsars. Investigations of such objects have continued for the last twenty years, and a vast body of observational data has been accumulated. A large number of theoretical works have also been published. Reviews and monographs have been devoted to the investigation of radiopulsars (Ginzburg 1971; Ter Haar 1972; Usov 1977; Smith 1977; Manchester, Taylor 1978). Hence we shall not delve into the details of the theory and observations of radiopulsars but concentrate our attention on problems which have not been studied extensively before and which have a bearing on the topics described in this chapter.

8.1 Observed Characteristics of Radiopulsars

Following the discovery of radiopulsars by the Cambridge group, the quest for such sources was continued at many radio observatories in the world, including the Jodrell Bank observatory of Manchester University in England, Owens Valley Radio Observatory of the California Institute of Technology (USA), the Radiophysics Laboratory of the Scientific Research Organization, Molonglow (Australia), and the Radioastronomical Observatory of the Physical Institute of the Academy of Sciences (USSR). Over 500 radiopulsars have been discovered to data. Most of these were detected in decameter radiation at frequencies $\lesssim 400\,\mathrm{MHz}$. We shall briefly describe the main observed characteristics of the pulsars.

8.1.1 Periods and Their Variation

The periods of radiopulsars are spread over a wide range (three orders of magnitude): from the fastest (millisecond) pulsar PSR 1937+21 having a period $p = 1.56 \times 10^{-3}$ s (Backer et al. 1982) to the slowest PSR 0525+21 having a period of 3.75 s (Fig. 8.1). The high stability of the pulse arrival time (which is sustained up to the 12th decimal place) makes it possible to compare pulsars with an atomic clock. The millisecond pulsar is especially stable and movements in the solar system can be measured from the instants of pulse arrivals from this source.

The absence of radiopulsars with periods longer than 4 s and a shortage of pulsars with periods below 0.1 s are real and not due to any limitations imposed by the measuring techniques.

The high stability and low period of pulsars prove beyond doubt that this phenomenon is associated with rotating neutron stars emitting synchrotron radiation.

The periods of radiopulsars are increasing. The time scale $t_{sd} = p/\dot{p}$ of radiopulsar spinning down varies from a few thousand years to several hundred million years. For some pulsars, it has not been possible as yet to measure $\dot{p}$. The $\dot{p}-p$ diagram for radiopulsars is presented in Fig. 8.2.

The increase in the pulse arrival period is naturally explained by the spinning down of the neutron star. The energy losses corresponding to this process are given by

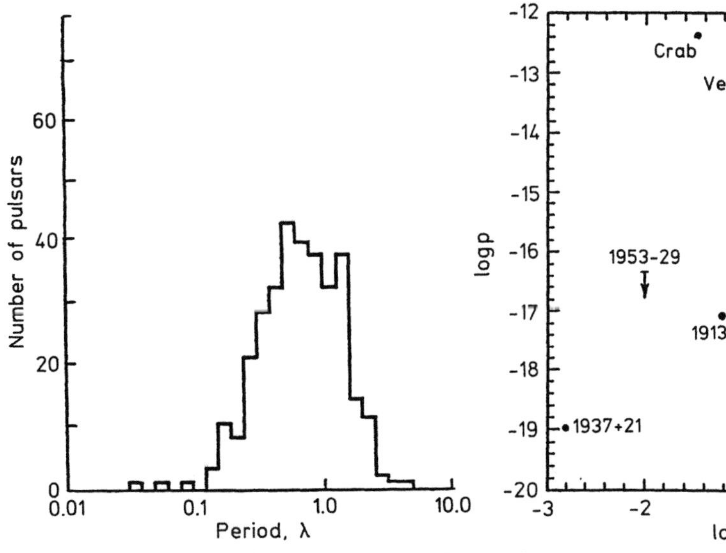

Fig. 8.1. Distribution of radiopulsars according to period (Manchester, Taylor 1978)

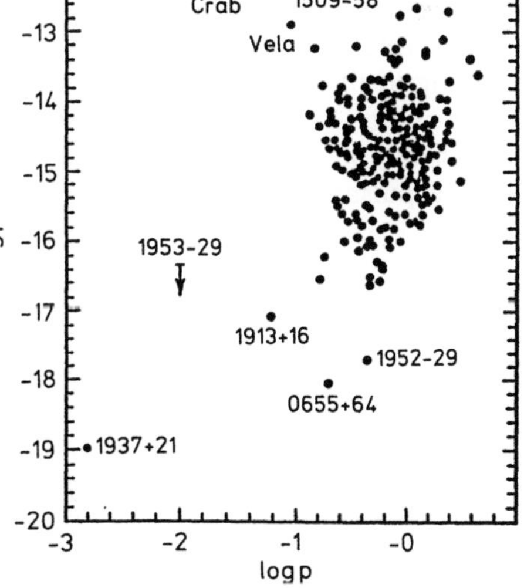

Fig. 8.2. The $\dot{p}-p$ diagram for radiopulsars (Manchester, Taylor 1978)

$$L_{\text{rot}} = \frac{d}{dt}\left(\frac{I\omega^2}{2}\right) = I\omega\dot{\omega} \ . \tag{8.1}$$

Observations reveal that the dissipation of the rotational energy of neutron stars is much higher (by a factor of $10^2 - 10^5$) than the luminosity of pulsars in the radiofrequency range. For the pulsar in the Crab nebula, $\dot{p} \approx 4.2 \times 10^{-13}$ (in dimensionless units). Dividing the pulsar period $p = 0.033$ s by $\dot{p}$, we obtain the characteristic spinning-down time of $t_{\text{sd}} \approx 2.5 \times 10^3$ years, which has the same order of magnitude as the characteristic lifetime of the neutron star (it should be recalled that according to the Chinese chronicles, this neutron star was formed in the year 1054). The power law for spin-down can be written

$$\frac{dI\omega}{dt} = -A\omega^n \ . \tag{8.2}$$

The time of spin-down from an infinitely high frequency is given by

$$t_{\text{d}} = \frac{t_{\text{sd}}}{n-1} \ . \tag{8.3}$$

For $n = 3$, the characteristic time (say, of magnetic dipole losses) is $t_{\text{d}} = t_{\text{sd}}/2$. The introduction of the factor of 2 improves the agreement between the dynamic age (determined from spinning down) and the real age of the Crab pulsar. This is an astonishing level of accuracy in astrophysics.

Measurement of the higher-order derivatives of frequency allows us, in principle, to determine the spin-down index n. It follows from (8.2) that

$$n = \frac{\omega\ddot{\omega}}{\dot{\omega}^2} \ . \tag{8.4}$$

So far, this method of determining n has given only one positive result. According to the data of Nelson et al. (1970), and Groth (1975), $n \approx 2.5$ for the Crab pulsar. Random fluctuations of $\dot{\omega}$ in other pulsars are so large that they completely wash out the secular variations in the frequency derivative $\ddot{\omega}$. Two of the pulsars (Crab and Vela) exhibited spin-up, which will be discussed below.

8.1.2 Pulse Structure

Unlike the X-ray pulsars, radiopulsars have much sharper pulses (Fig. 8.3). On the average, the relative pulse duration is of the order of just a few hundredths of the time between pulses: ~ 0.04. The pulse shape keeps on changing at random and only the profile averaged over a large number of pulses remains unchanged. Some pulsars exhibit an intermediate pulse between the main pulses. Observations of high resolution in the time show that the radioflux varies chaotically (and sometimes quasiperiodically) with time down to tens of microseconds (Hankins 1971).

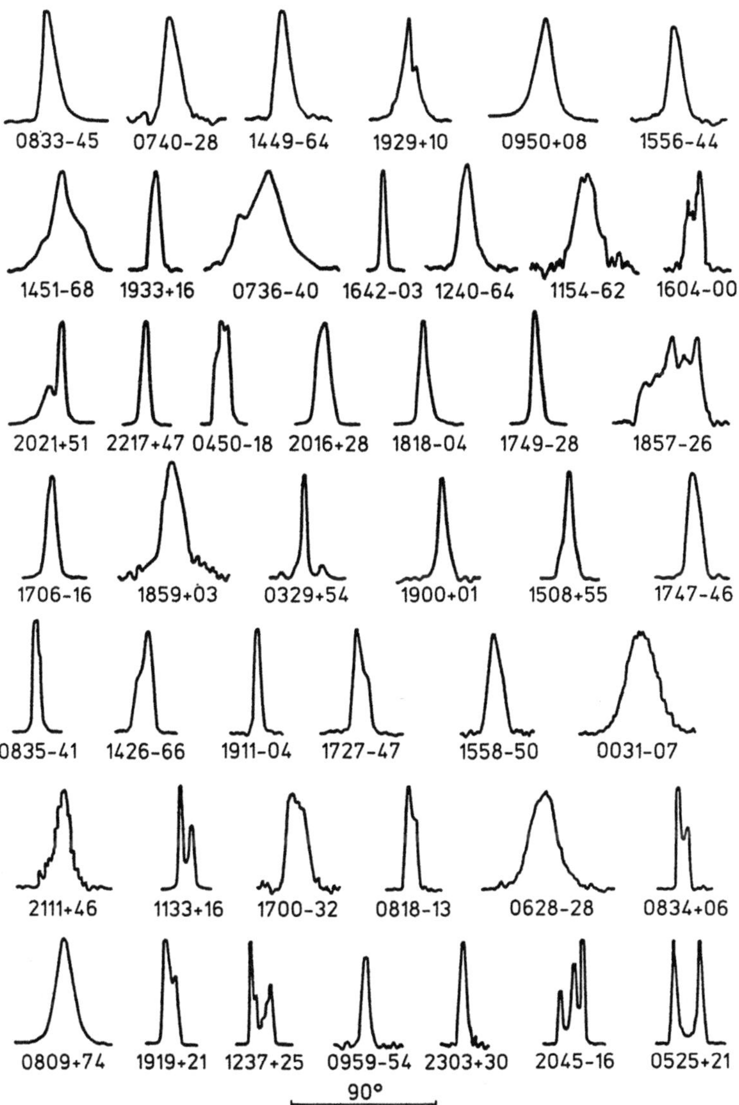

Fig. 8.3. Pulse profiles of radiopulsars (Manchester, Taylor 1978)

8.1.3 Spectrum and Luminosity

Figure 8.4 shows examples of spectra of several radiopulsars. In spite of a large diversity in the data, the following general properties can be singled out. In the frequency range between several hundred megahertz to tens of gigahertz, the spectral flux density decreases according to a power law, $I_\nu \propto \nu^{-\alpha}$. The characteristic fluxes at the pulse peaks attain values of up to ~ 100 Jy (1 Jy $= 10^{-26}$ W/cm^2 Hz). The flux averaged over a period does not exceed a

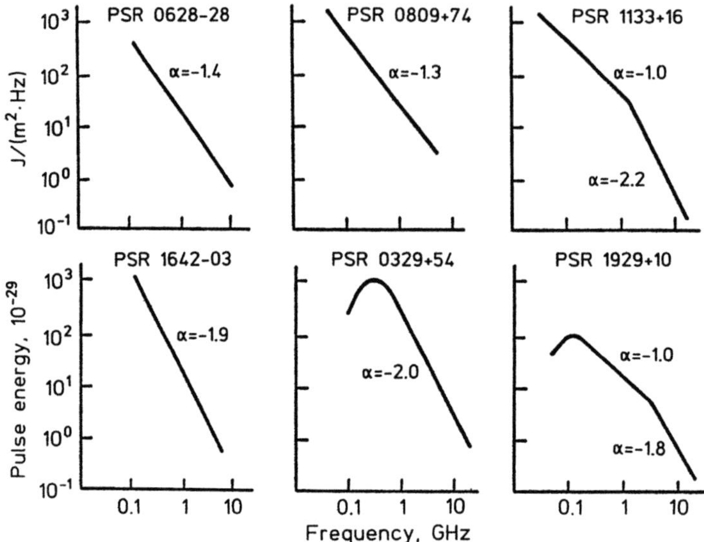

Fig. 8.4. Spectra of several radiopulsars (Manchester, Taylor 1978)

few Jy. In the long-wave range (frequency below 100 MHz), flux absorption or even a cut-off is observed.

By measuring the pulse arrival time at different frequencies, we can determine the degree of dispersion and hence estimate the distance from the radiopulsar (see below). The luminosities corresponding to the observed fluxes are of the order of $10^{30} - 10^{32}$ erg/s, and the brightness temperatures are 10^{30} K. Such a large value of the luminance temperatures undoubtedly points towards the nonthermal nature of the radiowaves emitted by the pulsars.

It should be emphasized once again that with the exception of a few rapid pulsars (like the Crab pulsar and the Vela pulsar), we observe only the radiowaves emitted by the pulsars. As a rule, the luminosity in the radiofrequency range is thousands of times lower than the rotational energy loss determined by $\dot{p}$ [see (8.1)].

Young pulsars (of about $\sim 10^4$ years) are surrounded by the remnants of supernova explosions and emit energy not only in the radiofrequency range, but also in the optical, X-ray, and gamma-ray ranges.

The luminosity of the Crab nebula in the X-ray range is 2.5×10^{37} erg/s, which is about 5% of the total rotational losses by the pulsar.

At the end of 1970s and the beginning of 1980s, attempts were made to look for compact X-ray sources in many supernova remnants at the Einstein observatory (Helfand et al. 1980). The results corresponding to 12 objects in the remnants of supernova explosions emitting synchrotron X-ray radiation (Helfand 1984) are presented in Table 8.1. Attempts at finding thermal X-rays from bright pulsars proved to be futile, but led to the establishment of quite rigorous upper limits of X-ray luminosity.

Table 8.1. Remnants emitting in the X-ray range

Remnant	L_x, erg/s	$L_{\rm rot}$, erg/s	$L_x/L_{\rm rot}$	t, years	d, kpc
Crab	2.5×10^{37}	4.6×10^{38}	0.05	10^3	1
G 29.7 − 0.3[a]	4×10^{36}	−	−	−	2
G 21.5 − 0.9[a]	2.6×10^{35}	−	−	−	3
MSH 15 − 52	2×10^{35}	2×10^{37}	0.01	2×10^3	6
3 C 58	10^{35}	−	−	−	12
G 74.9 + 1.2[a]	8×10^{34}	−	−	−	20
Vela	10^{33}	7×10^{36}	0.002	10^4	0.2
CTB 80	8×10^{32}	−	−	−	1
G 326.3 − 1.8[a]	5×10^{32}	−	−	−	0.4
PSR 1055 − 52	0.4×10^{32}	3×10^{34}	0.01	5×10^5	~0.05
PSR 0355 + 54	10^{32}	4.6×10^{34}	0.002	6×10^5	−
PSR 1642 − 07	6×10^{31}	10^{33}	0.06	3×10^6	0.6

[a] Remnants in which the compact source was not discovered

About 140 supernova remnants whose ages are between $\sim 10^{2.5}$ and $\sim 10^{4.5}$ years have been discovered so far. A review of the results of the quests for radio, optical, X-ray or gamma-ray point sources was published by Helfand and Backer (1983). The main result of these investigations is that only nine of the remnants were found to contain point sources. Of these, two (SS 433 and CTB 109) are binary X-ray systems.

Thus the radiofrequency range is the main source of information concerning radiopulsars. Radiation in this range is strongly polarized: linear polarization depends on the phase of the period and attains a value of 100% (Radhakrishnan et al. 1969; Alekseev 1971, 1973). Circular polarization reaches several tens of percent (Craft et al. 1968). The direction of the polarization plane rotates with the rotational period of the pulsar.

8.1.4 Distribution of Pulsars in Space

The distribution of radiopulsars in the sky is not isotropic; they are concentrated towards the Milky Way plane. This was observed in early investigations of radiopulsars and leaves no doubt about their galactic nature. Mainly two methods are used for estimating the distance from radiopulsars: measurement of the interstellar absorption at the 21 cm line and the dispersion measure. These two methods are completely independent. In the first, the distribution of the cold component of interstellar hydrogen and its movement in the Galaxy are assumed to be known. However, in order to find the distance from the value of the dispersion measure, we must know the density distribution of free electrons. This was indeed determined from observational data on radiopulsars.

It is well known that the electromagnetic waves propagate in plasma with a group velocity differing from the velocity of light, i.e.,

$$v_g = c\sqrt{1-(v_p/v)^2} \ , \tag{8.5}$$

where v_p is the plasma frequency [see (4.16)]. Since the pulsar radiation is discontinuous, it becomes possible to measure the time difference Δt between the arrival of the same pulse at different frequencies. At frequencies that are considerably higher than the plasma frequency ($v \gg v_p$), the difference in the arrival times is given by

$$\Delta t = t_2 - t_1 \approx \left(\frac{1}{v_2^2} - \frac{1}{v_1^2}\right) Dm \, \frac{e^2}{2m_e c} \ , \tag{8.6}$$

where Dm is the dispersion measure:

$$Dm = \int_0^d n_e \, dr \ , \tag{8.7}$$

d being the distance from the pulsar. Dm is usually measured in pc/cm^3.

Formula (8.6) effectively serves as a powerful tool for investigating the distribution of free electrons in the interstellar medium. It was used to determine the mean electron density in the vicinity of the Sun: $n_e \approx 0.03$ cm^{-3}. The error in the distance from the pulsar determined from the value of Dm is entirely due to our lack of knowledge about the density variation in a given direction.

Knowing the distance from a pulsar, we can determine its z-coordinate in the Galaxy. These results show that pulsars form a highly flattened subsystem with a thickness of several hundred parsecs. Manchester and Taylor (1978) proposed the following formula for the distribution of pulsars along the z-coordinate in the Galaxy:

$$\varrho(z) = \frac{1}{460} e^{-|z|/230 \, pc} pc^{-3} \ . \tag{8.8}$$

The normalization condition is chosen to be $\int_{-\infty}^{+\infty} \varrho(z) \, dz = 1$. This formula is applicable only in the vicinity of the Sun. An analysis of the observational data reveals that pulsars are distributed nonuniformly over the galactic disk. Apparently, their distribution is annular with the maximum density in the region of $5-8$ kpc from the center of the Galaxy (Guseinov, Yusifov 1984).

8.1.5 Spatial Velocity of Radiopulsars

Let us compare the distribution of radiopulsars along the z-coordinate with the distribution of other objects in the Galaxy. The characteristic distribution height of OB stars is ~ 80 pc, while that of the supernova explosion remnants is even lower, viz., ~ 60 pc (Lozinskaya 1986). It can be seen that the thickness of these subsystems, which are possibly the predecessors of radiopulsars, is much smaller than that of radiopulsars. It was mentioned first by Gunn and Ostriker (1970) that this difference is due to a higher dispersion in the velocities of radiopulsars (~ 100 km/s).

Indeed, measurements of intrinsic motions of radiopulsars over the sky show that some pulsars have a velocity of several hundred kilometers per second.

The tangential velocity can be used to estimate the so-called kinematic age of a pulsar:

$$t_k = \left| \frac{z}{\dot{z}} \right| . \tag{8.9}$$

Assuming that the velocity distribution is Maxwellian and the velocity direction is random, Gunn and Ostriker (1970) obtained the value $t_k \approx 10^6$ years for the kinematic age, which is at least a few times smaller than the dynamic age of pulsars determined from the spinning-down time t_d of their rotation.

However, it should be emphasized that the data on intrinsic motions are statistically incomplete and also subjected to a strong selectivity: only the velocities of pulsars moving very rapidly are measured. Hence the obtained results cannot be extended directly to the entire subsystem of radiopulsars. It was shown by Tutukov et al. (1984) that if various selection effects are taken into consideration, the velocities of most radiopulsars turn out to be much lower than 100 km/s.

8.1.6 Pulsars and Binary Systems

The stability of the pulse repetition period of radiopulsars is much higher than that of X-ray pulsars. Thus, a departure of the pulse arrival time from the value determined beforehand on the basis of several years of observations does not exceed 1 ms. This means that we can detect periodic motion around the barycenter with an amplitude of ~ 300 km. In principle, it should be possible to detect even an earth-type planet revolving around a pulsar at a distance of ~ 1 astronomical unit. However, with the exception of four cases, duality has not been observed in pulsars. The restrictions imposed by observations on the binary nature of pulsars are shown in Fig. 8.5 (Lamb, Lamb 1976). No radiopulsar is known to form a binary system with a visible component. Four radiopulsars (PSR $1913+16$, $0655+64$, $0820+02$ and $1953+29$) have companions which are not manifested in any way in the optical region and are probably degenerate stars (Sect. 8.9).

8.2 Radiopulsars as Ejecting Neutron Stars

Thus, most of the radiopulsars are single neutron stars surrounded by the interstellar medium. This is true even for the four pulsars which are parts of

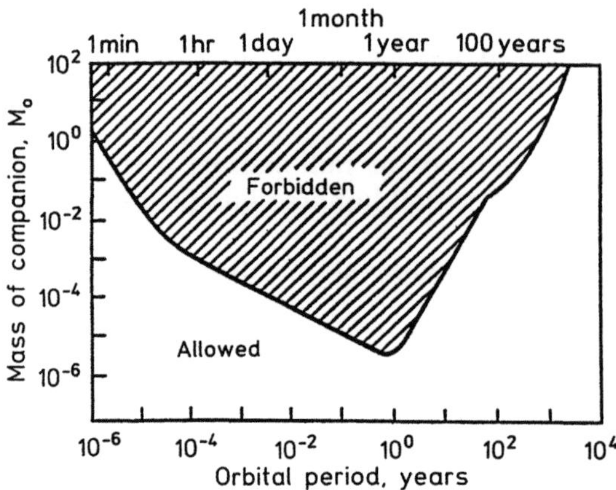

Fig. 8.5. The stability of radiopulsar periods makes it possible to impose restrictions on their binary character. Binary systems cannot have parameter values in the hatched region (Lamb, Lamb 1976)

binary systems. Obviously, their components are not in a position to supply any appreciable amounts of matter.

Let us estimate the Shvartsman radius (Chap. 4) at which the pressure of ejected radiation and relativistic particles becomes equal to that of the interstellar matter. We shall assume that rotational losses are completely in the form of low-frequency electromagnetic waves and relativistic particles. In this case, the Shvartsman radius is determined by the equality

$$\frac{L_{\text{rot}}}{4\pi R^2 c} = \varrho_\infty v_\infty^2 \ , \tag{8.10}$$

where ϱ_∞ is the density of the interstellar medium and v_∞ is the velocity of the moving star. Substituting $L_{\text{rot}} = -I\omega\dot{\omega}$ into this expression, we get

$$R_{\text{Sh}} \approx 10^{16} I_{45}^{1/2} \varrho_{-24}^{-1/2} v_6^{-1} p^{-3/2} \dot{p}_{-15}^{1/2} \ \text{cm} \ . \tag{8.11}$$

The gravitational capture radius R_G is equal to

$$R_G \approx 2.7 \times 10^{14} m_x v_6^{-2} \ \text{cm} \ . \tag{8.12}$$

It follows from Fig. 8.2 that the change in the pulsar period is $\dot{p} \gtrsim 10^{-18}$, so that $R_{\text{Sh}} > R_G$ and gravitation does not play a significant role. This means that radiopulsars are indeed ejecting neutron stars.

Assuming that the rotational losses of a pulsar are described by the magnetic dipole relation, the equality $R_{\text{Sh}} = R_G$ leads to the critical period (4.52):

$$p_E \approx 6\mu_{30}^{1/2} v_6^{1/2} \varrho_{-24}^{-1/4} m_x^{-1/2} \ \text{s} \ . \tag{8.13}$$

The periods of all known radiopulsars clearly satisfy the inequality (4.48). Moreover, the maximum periods of pulsars are quite close to the value p_E and the attenuation of radiopulsars can be attributed to the fact that the interstellar medium penetrating into the capture radius quenches the pulsar. It should be stated that this natural explanation for the absence of radiopulsars with periods exceeding $5-10\,\mathrm{s}$ is not widely accepted.

It is worth noting that p_E depends weakly on the magnetic field of a neutron star, its velocity, and the basic parameters of the interstellar matter.

While deriving (8.13), we certainly did not take into account a number of circumstances which can change the critical period manifold towards lower or higher values. For example, we must take into consideration the fact that a radiating pulsar (basically, we mean the radiation emitted by magnetic dipole waves and relativistic particles) is not isotropic (the consideration of anisotropy decreases the critical period). We also did not take into account the inverse effect of pulsar radiation on the parameters of the interstellar medium. We shall take this into consideration in the following analysis. This will increase the value of the critical period.

The hypothesis explaining the quenching of radiopulsars through the penetration of the interstellar medium within the capture radius seems to contradict the indisputable fact that the pulsars are located in a thin layer of a few hundred parsecs, where the entire interstellar matter is effectively concentrated. Conversely, there are practically no radiopulsars high above the galactic plane where the interstellar matter does not obstruct ejection. In this case, however, there is no controversy. The concentration of radiopulsars in the plane is associated with the fact that the pulsars are formed in it. It remains to be clarified why the pulsars are not "reignited" upon falling into rarer layers high above the galactic plane.

The answer to this question is the following. We recall that it is difficult to suppress ejection. But as soon as plasma penetrates beyond the capture radius and into the light cylinder, ejection begins in the rarer medium (Shvartsman 1970c). This peculiar hysteresis is explained by the fact that the pressure of accreting plasma within the capture radius increase more rapidly than the pressure of the ejected flux. In short, the transition $E \to P$ from the ejection state to the propeller state is determined by (8.13), while the reciprocal transition $P \to E$ occurs with a much shorter critical period.

8.3 Pulsar Electrodynamics and Generation
of Relativistic Particles

The above estimates show that the interstellar medium does not affect the physical situation inside the light cylinder. The analysis of the operation of radiopulsars can be started from the vacuum approximation. As a rule, a magnetic dipole rotating in vacuum is an emitter of magnetic dipole waves with a frequency equal to the rotational frequency of the dipole. However, the emergence of static electric fields in the vicinity of the rotating magnetic dipole is more interesting. This is a well-known effect and is applied to unipolar inductors (Landau, Lifshitz 1982). However, in the theory of radiopulsars it was first used by Goldreich and Julian (1969).

8.3.1 Vacuum Approximation

We shall assume the star to be an ideally conducting sphere of radius R_x rotating with a frequency ω. The magnetic field outside the sphere $R > R_x$ has the dipole structure

$$B = [3e_r(n_{\mathrm{m}} \cdot e_r) - n_{\mathrm{m}}]\mu/R^3 . \tag{8.14}$$

As before, μ is the magnetic dipole moment. Following Tsygan (1981), let us determine the electric field outside the sphere. The electric field potential Φ in a reference frame rotating together with the sphere is described by the equation (Fowley et al. 1977)

$$\Delta\Phi + \frac{2\omega n_\omega \cdot B}{c} = -4\pi\varrho_{\mathrm{e}} = 0 . \tag{8.15}$$

The electric field is given by

$$E = -\nabla\Phi . \tag{8.16}$$

Equation (8.15) is approximate. In deriving it, we omitted the term $\propto(\omega R/c)^2$, and hence this relation can be used only deep inside the light cylinder. The electric field inside an ideally conducting sphere is equal to zero: $E = 0$. Consequently, $\Phi = \Phi_0 = $ const. The solution of (8.15) with boundary conditions $\Phi|_{R = R_x} = \Phi_0$ and $\Phi|_{R \to \infty} = 0$ is

$$\Phi = \Phi_0 \frac{R_x}{R} + \frac{3(n_\omega \cdot e_r)(n_{\mathrm{m}} \cdot e_r) - (n_\omega \cdot n_{\mathrm{m}})}{3eR_x} \left(\frac{R_x}{R} - \frac{R_x^3}{R^3}\right) \omega\mu . \tag{8.17}$$

The surface charge density at the surface of the sphere is proportional to the change in the normal component of the electric field:

$$\Sigma_{\mathrm{e}} = \frac{E_{\mathrm{n}}}{4\pi}\bigg|_{R = R_x} = -\frac{1}{4\pi}\frac{\partial\Phi}{\partial R}\bigg|_{R = R_x}$$

$$= \frac{\Phi_0}{4\pi R_x} - \frac{3(n_\omega \cdot e_r)(n_{\mathrm{m}} \cdot e_r) - (n_\omega \cdot n_{\mathrm{m}})}{6\pi c R_x^2}\omega\mu . \tag{8.18}$$

The total charge on the sphere is $q = 0$. Hence we obtain

$$q = \int\Sigma_{\mathrm{e}}dS - \int\frac{\omega n_\omega \cdot B}{2\pi c}dV = 0 .$$

This leads to the potential

$$\Phi_0 = -\frac{2(n_\omega \cdot n_{\mathrm{m}})}{3cR_x}\omega\mu .$$

Finally, we obtain from (8.17) the electric field in a rotating reference frame:

$$E_{\text{rot}} = \frac{\omega\mu R_x^2}{cR^4}\{n_\omega(n_\text{m}\cdot e_r)+n_\text{m}(n_\omega\cdot e_r)+e_r[(n_\omega\cdot n_\text{m})-5(n_\omega\cdot e_r)(n_\text{m}\cdot e_r)]\}$$

$$-\frac{\omega\mu}{cR^2}\{n_\omega(n_\text{m}\cdot e_r)+n_\text{m}(n_\omega\cdot e_r)+e_r[(n_\omega\cdot n_\text{m})-3(n_\omega\cdot e_r)(n_\text{m}\cdot e_r)]\} \ . \tag{8.19}$$

In order to obtain the electric field strength in an inertial reference frame, we make use of the Lorentz transformation

$$E = E_{\text{rot}} + \frac{1}{c}[B\times v] \ , \quad v = [\omega n_\omega \times R] \ .$$

As a result, we get

$$E = \frac{R_x^2\omega\mu}{cR^4}\{n_\omega(n_\text{m}\cdot e_r)+n_\text{m}(n_\omega\cdot e_r)+e_r[(n_\omega\cdot n_\text{m})-5(n_\omega\cdot e_r)(n_\text{m}\cdot e_r)]\}$$

$$+\frac{\omega\mu}{cR^2}\{n_\omega(n_\text{m}\cdot e_r)-n_\text{m}(n_\omega\cdot e_r)\} \ .$$

This expression was obtained by Deutsch (1955). Formula (8.19) shows that the electric field around a rotating sphere is quadrupolar.

The case when the dipole axis coincides with the rotational axis of the sphere is especially important. In this case, a simple exact solution can be found for the electric field outside the sphere. The electrostatic potential is

$$\Phi = -\frac{2\mu\omega}{3c}\frac{R_x^2}{R^3}P_2(\sin\theta) \ , \tag{8.20}$$

where θ is measured from the equator of rotation and $P_2(\sin\theta)$ is the Legendre polynomial which is of pure quadrupole type.

It can be found that $E\cdot B\propto\sin^3\theta$, whence it can be seen that the electric and magnetic fields are parallel (or antiparallel) at the pole $\theta = \pm\pi/2$. The electric field strength at the pole of the star is found to be of the order of

$$E_{\|} \approx \frac{\omega R_x}{c}B_0 \approx 10^8 p^{-1}\mu_{30}R_6^{-2}\,\text{V/cm} \ . \tag{8.21}$$

As before, p is the rotational period in seconds. For $p = 1$ s and $\mu_{30} = R_6 = 1$, the electric field acting on the electrons is stronger than the gravitational field by a factor of 10^9. Apparently, such strong fields cannot exist in extended regions: the charges detached from the surface of the star are polarized and neutralize the electric field. Hence a realistic model of pulsars must take into account the presence of plasma inside the magnetosphere of a neutron star.

8.3.2 Magnetosphere in the Presence of Plasma

Goldreich and Julian (1969) proposed a model of a magnetosphere of a neutron star in which $n_{\mathrm{m}}||n_{\omega}$ and which is characterized by the presence of two essentially different regions: the region enclosed between the lines of force not extending beyond the light cylinder and the region from poles along the lines extending beyond the light cylinder (Fig. 8.6).

In the first region, plasma rotates as a rigid body with the magnetic field of the neutron star. Here, the charge density is determined from the Maxwell equation

$$\varrho_{\mathrm{e}} = \frac{1}{4\pi}\,\mathrm{div}\,E \; . \tag{8.22}$$

Neglecting the inertia of the particles, the electric field (for an ideally conducting plasma) is

$$E = -\frac{\omega}{c}\,[n_{\omega}\times R]\times B \; . \tag{8.23}$$

Substituting the charge density and the corresponding number density of the excess charges of one sign over the other sign into (8.22), we obtain

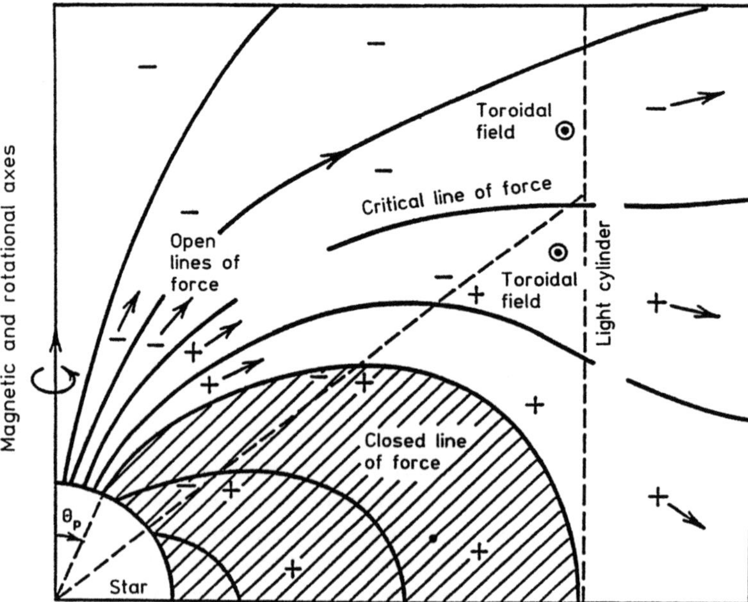

Fig. 8.6. Structure of the magnetosphere of a radiopulsar according to (Goldreich, Julian 1969)

$$\varrho_e = -\frac{\omega}{2\pi c}(\boldsymbol{n}_\omega \cdot \boldsymbol{B}) \ ,$$

$$n_e = 7 \times 10^{-2} B_\omega p^{-1} \, \text{cm}^{-3} \ ,$$

(8.24)

where $\boldsymbol{B}_\omega$ is the component of magnetic field strength parallel to the rotational axis of the star. Inside the rotating rigid part of the magnetosphere, $(\boldsymbol{E} \cdot \boldsymbol{B}) = 0$, so that no component of the electric field is directed along the magnetic lines of force and the particles cannot be accelerated. In the region of open field lines, however, there exists a flux of relativistic particles accelerated along the lines of force. Electrons flow along the lines of force close to the poles, while positively charged particles flow near the boundary with the rigidly rotating region. Initially, it was assumed that electrons are accelerated in a wide region with characteristic size of the order of the radius of the neutron star itself.

The upper estimate for the maximum energy to which a particle can be accelerated is given by the expression

$$\varepsilon_{\max} \approx eER_x \approx \frac{e\omega\mu}{cR_x} \approx 10^{17} p^{-1} \mu_{30} R_6^{-1} \, \text{eV} \ .$$

(8.25)

For a more realistic estimate, we must take into account the flexural radiation associated with the fact that even when a charge moves along a field line, it emits radiation because the line is curved. The power of flexural radiation in the ultra-relativistic case is

$$I = \frac{2e^2}{3m^4c^7} \frac{\varepsilon^4}{R_{\text{cur}}^2} \ ,$$

(8.26)

where R_{cur} is the radius of curvature of the field line. The maximum energy is apparently determined by the balance between the energy acquired per unit time due to the work done by the electric field and the energy $I = eEc$ lost by radiation. Thus we get

$$\varepsilon_{\max} \approx \left(\frac{3\mu}{2eR_x}\right)^{1/4} mc^2 \approx 10^7 mc^2 \ ,$$

(8.27)

which gives 10^{14} eV for electrons and 10^{17} eV for protons.

The relativistic particles may also be accelerated outside the light cylinder in the field of a magnetic dipole wave (Gunn, Ostriker 1971; Kegel 1971; Kulsrud et al. 1972; Tsygan 1981). An electromagnetic wave is called intense if the work done by its electric field on a particle during one oscillation period is much larger than the rest energy of the particle. It can be easily verified that a magnetic dipole wave in the vicinity of the light cylinder is an intense wave. The energy acquired during one period is $eBc/\omega (E \approx B)$. Hence the parameter characterizing the wave intensity can be written in the form

$$b = \frac{eB}{mc\omega} \approx \begin{cases} 10^8 \mu_{30} p^{-2} & \text{for electrons ;} \\ 10^{11} \mu_{30} p^{-2} & \text{for protons .} \end{cases} \tag{8.28}$$

Calculations show that in the vacuum approximation, particles may be accelerated to energies of $10^{14} - 10^{15}$ eV in the vicinity of fast pulsars. The total radiant power emitted by a pulsar is of the order of magnitude of the magnetic dipole losses in different models.

It is clear, however, that the emergence of a pulsar is ruled out in the Goldreich-Julian model. We must introduce a misalignment between the magnetic and rotational axes. The problem is complicated even in the noninertial approximation. However, analysis points towards a preservation of the qualitative pattern, especially at small angles between the magnetic axis and the rotational axis (Michel 1982).

The existence of an intermediate zone with a strong electric field can be sustained only if charges of both signs are detached from the surface of the star. However, it was mentioned by Ginzburg and Usov (1972) and by Ruderman and Sutherland (1975) that the work function of ions in a cold neutron star is too high and they will not be detached from the surface. Consequently, the electric field is mainly screened and this leaves a narrow vacuum gap in which the potential difference is

$$\Delta\Phi \approx \frac{\omega B_0 z^2}{c} \; ; \tag{8.29}$$

z being the height of the gap. It was mentioned by Sturock (1971) and by Ruderman and Sutherland (1975) that an electron-positron avalanche may be formed within the gap. Inside the gap, the electric field strength corresponding to the potential difference (8.29) is

$$E(z) = \frac{2\omega B_0}{c} z (n_{\mathrm{m}} \cdot n_\omega) \mu \omega \; . \tag{8.30}$$

The emergence of an electron-positron avalanche can be described as follows (Tsygan 1981 a). Suppose that a charged particle or a pair of particles appear in the gap (say, as a result of creation of an electron-positron pair from a gamma quantum moving in a strong magnetic field). The particle will be entrained by the electric field and accelerated to energies with a characteristic gamma-factor

$$\gamma(z) = \frac{e}{mc^2} \int_0^z E(z)\,dz = \frac{e\omega B_0}{mx^3} z^2 \cos\beta \; . \tag{8.31}$$

In turn, the charged particle will emit a (flexural radiation) quantum with a characteristic energy

$$\varepsilon = \hbar\omega(z) \approx \frac{1}{2}\frac{\hbar c}{R_{\text{cur}}}\gamma^3(z) \ . \tag{8.32}$$

For the dipole field, the radius of curvature of the field line reaching the light cylinder is

$$R_{\text{cur}} = 4R_x\sqrt{c/\omega R_x}/3 \ . \tag{8.33}$$

Gamma quanta are emitted along the path of the relativistic particles, i.e., along the lines of force. However, since the lines are curved, the quanta gradually start moving at an angle δ towards the field lines. When the condition

$$\sin\delta > \sin\delta_{\text{cr}} \approx 2mc^2/\hbar\omega \tag{8.34}$$

is satisfied, a new electron-positron pair may be created. The process is then repeated and amplified. The amplification factor is equal to the number of final particles generated from one initial particle and is given by (Tsygan 1981 a)

$$k_{\text{amp}} = \frac{2mc^2\gamma_{\text{max}}}{\hbar\omega_{\text{min}}} \approx 3\times10^5 B_{12}^{6/7} R_7^{-3/7} p^{-1/7} \ . \tag{8.35}$$

Assuming that the electron-positron pairs in the gap are close to saturation, $n_e \approx (\omega n_\omega \cdot B)/(2\pi ce)$, we find that the total flux of relativistic particles generated by the star is

$$\frac{dn}{dt} \approx k_{\text{amp}} 2\pi nc\left(\frac{R_x}{\sqrt{\omega R_x/c}}\right)^2 \text{s}^{-1} \ . \tag{8.36}$$

For a typical pulsar with a period $p = 1$ s, magnetic field $B = 10^{12}$ Oe, and $k_{\text{amp}} = 3\times10^5$, the ejection rate is $\sim 10^{36}$ particles/s.

This is the preliminary picture of acceleration of relativistic particles in a gap. Some new interesting effects have been observed in ultrastrong magnetic fields $B \geqslant 10^{13}$ Oe. For example, Shabad and Usov (1982) showed that an effective refractive index appearing in such fields causes gamma quanta to move along the magnetic field lines. This effect suppresses the electron-positron pair generation.

Before concluding this brief account of the main processes taking place in the magnetosphere of a pulsar and described in the literature, we would like to make mention of the work of Beskin et al. (1983) who obtained the solution of the self-consistent problem on the structure of the magnetosphere of a pulsar.

8.4 Mechanisms of Radiation

The energy received in the form of radiowaves at frequencies $\lesssim 400\,\text{MHz}$ ($\lambda \gtrsim 75$ cm) is an insignificant part of the total energy loss by a neutron star which was determined, by us, from the spinning down of pulsars. The only source of information concerning most of the radiopulsars is in fact a weak second order effect. This is equivalent to studying an ordinary star, say, by simply investigating its X-ray emission. Hence it should be very interesting to observe the main energy flux emitted by radiopulsars.

The manner in which practically all the energy of radiopulsars is carried away is not yet clear. It is usually assumed that this energy may be carried by relativistic particles and magnetic dipole waves. But the proportion in which the energy is distributed beween these two channels is not clear. Magnetic fields in the Galaxy muddle the trajectories of relativistic particles and we are unable to receive them from radiopulsars. Magnetic dipole waves have an extremely low frequency and it is usually assumed that it is impossible to observe them.

Indeed, electromagnetic waves can propagate in a plasma with a frequency higher than the plasma frequency $v_{\mathrm{p}} \approx 870\, n_{-2}^{1/2}\,\text{Hz}$, where $n_{-2} = n/10^{-2}\,\text{cm}^3$ is the number density of free electrons. It should be recalled that the fastest pulsar known until 1982 was in the Crab nebula and had a rotational frequency $v \approx 30\,\text{Hz}$. Under such conditions, there can be no question of observing magnetic dipole radiation. The "working" of a pulsar is reminiscent of an idle radar installation in which the entire radiation is transformed into heat.

However, when it became clear after the discovery of a millisecond pulsar that there must exist long-lived rapidly rotating neutron stars in the Galaxy, hopes were raised that magnetic dipole waves could be detected (Lipunov 1983a). The frequency of a millisecond pulsar is $v = 642\,\text{Hz}$ and approaches the plasma frequency at $n_{-2} \approx 1$.

In principle, neutron stars can rotate more rapidly (Sect. 6.13), and their frequencies may even be higher than 1 kHz. Moreover, if the magnetic field of a neutron star has a magnetic multipole moment, we can expect magnetic multipole radiation at higher frequencies. For example, the energy flux density of the magnetic dipole waves emitted by a millisecond pulsar (the distance is assumed to be equal to 2.5 kpc) is

$$F_{1937+21} \approx 8 \times 10^{16} I_{45} \Delta v_1^{-1}\ \text{W/m}^2\,\text{Hz} \tag{8.37}$$

where Δv_1 is the receiver bandwidth in kHz. This is two orders of magnitude higher than the sensitivity of modern magnetometers.

In the above relation, we have not taken into account the absorption due to free-free transitions which may considerably lower the flux being received. Hence this relation is of a rather illustrative nature. It was mentioned by Lipunov (1983a) that to observe radiation in the region of several kilohertz, we must move away from the Sun in order to avoid absorption in the solar wind. In this connection the recent discovery made by Kurth et al. (1985) on board the spaceship Voyager acquires added significance. A low-frequency electromagnetic radiation at a frequency of ~ 3 kHz was detected by the magnetometers installed on the spaceship. The origin of this radiation has not been established unambiguously, though the authors believe that the most likely source is the boundary of the heliopause formed as a result of the interaction of solar wind with the interstellar gas. Nevertheless, it would be quite interesting to investigate whether pulsars are responsible for this radiation. But let us turn to the reliably observed frequency range.

High brightness temperatures of radiopulsars can be explained by assuming a coherence radiation mechanism (Ginzburg, Zheleznyakov 1970a, b). Two types of coherence mechanisms, the aerial and maser, were considered. The aerial mechanism operates, for example, during the emission of a cluster of particles whose size is much smaller than the emitted wavelength. In this case, amplitudes are added instead of their squares, and hence the total luminosity is found to be proportional to the square of the number of particles rather than to the number of particles, as is usually the case. However, such clusters rapidly spread out due to dispersion of particle velocities and plasma instabilities (Ginzburg 1971; Ter Haar 1972).

The maser mechanism of radiation is operative in the case when a population inversion of momenta and energies is somehow created. The phasing is automatically ensured by the radiation itself. Two types of masers were considered. In the first of these, radiowaves are amplified directly (Ginzburg 1971) while masers of the other type operate at plasma frequencies and radiowaves are just a secondary product of transformation of plasma waves into electromagnetic waves (Kaplan, Tsytovich 1973).

Any radiation mechanism must be able to explain the main feature of radiation, that is, its highly directional nature. Since the pulse width is about ten times smaller than the interval between pulses, the directionality diagram must have an opening angle not exceeding $\sim 10°$.

Directionality emerges in a natural way in maser mechanisms. Two possibilities can be considered. One involves maser radiation in the vicinity of the neutron star surface (Chiu, Canuto 1971; Ginzburg et al. 1969). In the second work, it is assumed that the observed frequency of the radio pulse is close to the plasma frequency in the plasma flow. The second possibility concerns emission outside the light cylinder (see, for example, Michel 1971). In the latter case, however, it is difficult to explain the stability of the pulse shape and, moreover, the phasing of pulses in optical, X-ray and radiofrequency regions observed, for example, in the Crab nebula pulsar.

At present, only three radiopulsars are known to emit radiation outside the radiofrequency range. The most thoroughly investigated pulsar in different frequency (from radio to gamma) ranges is the Crab pulsar. The emitted optical and X-ray radiation is polarized and, significantly, has a power spectrum with an index close to that of the nebula emission spectrum itself. This prompted Shklovskii (1970) to suggest that synchrotron radiation from the same electrons serves as the mechanism of emission by both the nebula and the pulsar.

The sharp peak of optical pulses in the Crab nebula indicates that some particles emit a beam whose width is less than 10^{-2} rad.

A comparison of the Crab and Vela pulsars shows that the power of optical radiation decreases strongly with increasing pulsation period, nearly in proportion to p^{-12}. This fact has not been explained satisfactorily so far.

X-ray radiation emitted by rapidly rotating pulsars is apparently also of synchrotron type. However, the specific mechanism responsible for such radiation is not clear so far (Manchester, Taylor 1978).

8.5 Caverns Around Neutron Stars

The plasma surrounding ejecting stars is stopped by the free electrons of the wave and by relativistic particles. Let us find the shape and properties of the boundary in this case.

We shall follow the approach used by Lipunov and Prokhorov (1984). It is convenient to start with a situation in which the neutron star is a part of a binary system. Of course, we are speaking of binary systems in which the neutron star is paired with a normal star which is losing matter. Numerical simulation of the evolution of neutron stars shows that the fraction of ejecting stars in massive systems with normal components may become several tens of percent of the total number of binary systems (Chap. 11). So far, no reliable candidates have been found among the observable sources which may be such objects. However, there is no doubt that such sources must exist. In Sect. 8.9 we shall describe some sources with appropriate properties.

8.5.1 Caverns in Binary Systems

Suppose that a neutron star having a rotational magnetic luminosity L_m forms a binary system (with semi-major axis a) with a normal star losing matter in the form of stellar wind (with a loss rate $\dot{M}_0$).

Let us find the Shvartsman radius in accordance with the definition given in Chap. 4 and by equating the plasma pressure to the electromagnetic pressure P_m:

$$P_\mathrm{A} = \varrho v^2 = \frac{\dot{M}}{4\pi a^2} v_\mathrm{w} = P_\mathrm{m} = \frac{L_\mathrm{m}}{4\pi R^2 c} \, ,$$

where R is the distance from the neutron star. The Shvartsman radius is

$$R_\mathrm{sh} = a \left(\frac{L_\mathrm{m}}{\dot{M}_0 v_\mathrm{w} c} \right)^{1/2} . \tag{8.38}$$

When making estimates, it is convenient to use the fact that the outflow rate, luminosity and the stellar wind velocity of hot stars are connected through the empirical relation (see, for example, Barlow, Cohen 1977)

$$\dot{M}_0 = \alpha_\mathrm{w} \frac{L_0}{v_\mathrm{w} c} \; ,$$

where $\alpha_\mathrm{w} \approx 0.2 - 0.4$. This leads to the approximation

$$R_\mathrm{Sh} \approx \left(\frac{L_\mathrm{m}}{L_0}\right)^{1/2} a \; . \tag{8.39}$$

The luminosity of normal stars whose mass varies from a few times to several tens of times the mass of the Sun lies in the interval between 10^{34} erg/s to $10^{38} - 10^{39}$ erg/s. The rotational magnetic luminosity of neutron stars also changes in the same wide range. For aged radiopulsars, $L_\mathrm{m} \approx 10^{30}$ erg/s, while for young pulsars L_m may be as high as $10^{38} - 10^{40}$ erg/s. Formula (8.39) shows that entirely different situations may arise in different binary systems at various stages of evolution. For example, $R_\mathrm{Sh} \gtrsim a$ or $R_\mathrm{Sh} \ll a$. This means that while deriving the formula for a cavern in the general case, we must take into account the proximity of the normal star.

Let us calculate the shape of a cavern surrounding a neutron star, assuming that the relativistic wind ejected by the star is spherically symmetric. We shall also assume that $R_\mathrm{Sh} > R_\mathrm{G}$. At the cavern boundary (Fig. 8.7), the following equilibrium condition must be satisfied:

$$\frac{\dot{M} v_\mathrm{w}}{4\pi R_0^2} \cos^2 \psi + P_\mathrm{g} = \frac{L_\mathrm{m}}{4\pi R^2 c} \cos^2 \chi + \frac{\delta L_\mathrm{m} t}{3 V} \; , \tag{8.40}$$

where R_0 is the distance from the normal star, V is the volume of the cavern, and $P_\mathrm{g} = n_\mathrm{i} k_\mathrm{B} T_\mathrm{i} + n_\mathrm{e} k_\mathrm{B} T_\mathrm{e}$ is the gas pressure. The second term on the right-

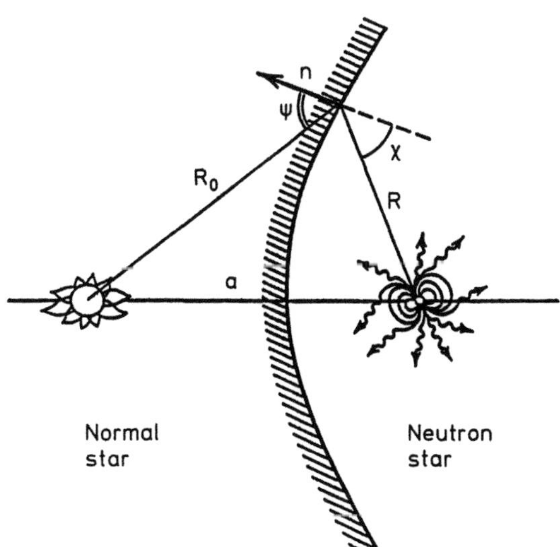

Fig. 8.7. Calculation of the shape of the cavern boundary

hand side of the equilibrium equation (8.40) describes the contribution of the pressure of magnetic dipole radiation which is reflected at the cavern walls and is accumulated inside it. The factor δ equals 0 for an open cavern and approximately 1 for a closed cavern. The statistical gas pressure P_g can be presented in the form

$$P_g = A R_0^{-n-2} \; .$$

For an isothermal stellar wind (as a rule, this is the situation commonly encountered), $n = 0$. We go to dimensionless variables:

$$r = \frac{R}{a} \; ; \quad r_0 = \frac{R_0}{a} \; ; \quad r_{Sh} = \frac{R_{Sh}}{a} \; ;$$

$$\tau = \frac{ct}{a} \; ; \quad v = \frac{V}{a^3} \; ; \quad k = \frac{A R^{-n}}{4 \pi \dot{M}_0 v_w} \; .$$

In this case, (8.40) becomes

$$\frac{\cos^2 \psi}{r_0^2} + \frac{k}{r_0^{-n-2}} = \frac{r_{Sh}^2}{r^2} \cos^2 \chi + \frac{\delta \tau}{v} \frac{4 \pi}{3} r_{Sh}^2 \; . \tag{8.41}$$

Let us consider the case of isothermal stellar wind ($n = 0$) when there is no accumulation of magnetic dipole radiation ($\delta = 0$):

$$\frac{\cos^2 \psi + k}{r_0^2} = \frac{r_{Sh}^2}{r^2} \cos^2 \chi \; . \tag{8.42}$$

It follows from symmetry considerations that at the anterior point of the cavern (i.e., the point closest to the normal star), $\cos \psi = \cos \chi = 1$. This leads to the following expression for the distance from the anterior point:

$$r_+ = \frac{r_{Sh}}{\sqrt{k+1} + r_{Sh}} \; . \tag{8.43}$$

Similarly, we can determine the distance r_- from the posterior point of the cavern. Since we get

$$\frac{k}{r_0^2} = \frac{r_{Sh}^2}{r_-^2} \quad \text{and} \quad r_- - r_0 = 1 \; , \tag{8.44}$$

$$r_- = \frac{r_{Sh}}{\sqrt{k} - r_{Sh}} \; .$$

If $r_{Sh} < \sqrt{k}$, the cavern is closed. The value of k is of the order of the ratio of the square of the velocity of sound to the outflow rate of the stellar wind, so that in real situations $k \approx 10^{-4} - 10^{-6}$. Hence closed caverns must be quite

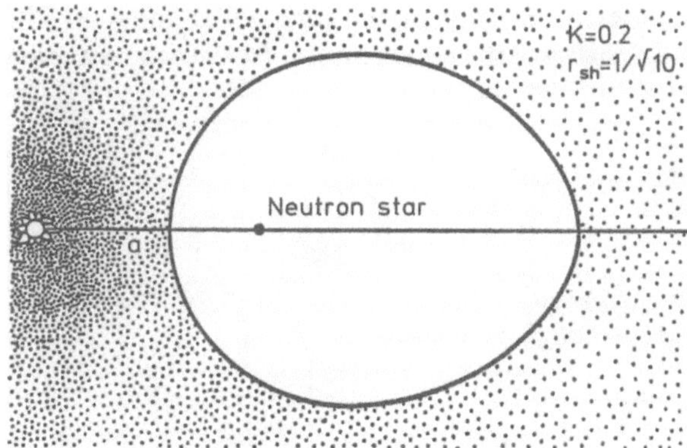

Fig. 8.8. Shape of a closed cavern

compact: $R_{\mathrm{Sh}} \lesssim (10^{-2} - 10^{-3})a$. The assumption $k = 0$ may serve as a good approximation in most cases.

It was mentioned above that electromagnetic waves whose frequency is higher than the plasma frequency ($v_{\mathrm{p}} \approx 9 \times 10^3 n_{\mathrm{e}}^{1/2}\,\mathrm{Hz}$) may propagate in a plasma. The characteristic density in the stellar wind is estimated from the continuity condition

$$n \approx 6.3 \times 10^{10}\, \dot{M}_{-6}\, a_{10}^{-2}\, v_8^{-1}\,\mathrm{cm}^{-3} \ ,$$

where $a_{10} = a/10R_\odot$. Thus the magnetic dipole radiation is reflected from the walls. Suppose that the reflection coefficient $n_{\mathrm{r}} \lesssim 1$. In this case, the above analysis will remain valid if we redefine the Shvartsman radius as

$$\tilde{r}_{\mathrm{Sh}} = r_{\mathrm{Sh}}/(1-n) \ .$$

This gives

$$\tilde{r}_+ = \frac{r_{\mathrm{Sh}}}{(1+n)\sqrt{k+1} + r_{\mathrm{Sh}}} \ .$$

It can be seen that the size of the cavern has increased. As $n_{\mathrm{r}} \to 1$, the cavern cannot remain stationary. It is either always open or pulsating quasi-periodically. In the latter case, the closed cavern expands right up to the maximum value r_{max}:

$$r_- \to r_{\mathrm{max}} = \frac{1}{(k/r_{\mathrm{Sh}}^2)^{1/3} - 1} \ .$$

The shape of the closed cavern is shown in Fig. 8.8.

8.5.2 Caverns Around a Single Neutron Star

The results obtained above can be easily extended to the case of a single neutron star interacting with the interstellar medium. We use the relation

$\dot{M}_c = \pi R_G^2 \varrho_\infty v_\infty$ (ϱ_∞ and v_∞ are the density of the interstellar medium and relative velocity respectively). In this case the equation for the cavern boundary assumes the form

$$\frac{\dot{M}_c v_\infty}{\pi R_G^2} \cos^2 \psi + P_\infty = \frac{L_m}{4\pi R^2 c} \cos^2 \chi + \frac{\delta L_m t}{3 V} \ . \tag{8.45}$$

We use the notation

$$k_1 = \frac{\pi R_G^2 P_\infty}{\dot{M} v_\infty} \ ,$$

$$R_{Sh} = \frac{1}{2} \left(\frac{L_m}{\dot{M} v_\infty c} \right)^{1/2} R_G \tag{8.46}$$

and the dimensionless parameter $r = R/R_G$. This gives

$$\cos^2 \psi + k_1 = \frac{r_{Sh}^2}{r^2} \cos^2 \chi + \frac{\delta L_m t \pi R_G^2}{3 V \dot{M} v_\infty} \ .$$

If $\delta = 0$, we get

$$\cos^2 \psi + k_1 = \left(\frac{r_{Sh}}{r} \right)^2 \cos^2 \chi \ .$$

Accordingly, the distance to the anterior and posterior points of the cavern can be presented in the form

$$r_+ = \frac{r_{Sh}}{\sqrt{1+k_1}} \ ,$$

$$r_- = \frac{r_{Sh}}{\sqrt{k_1}} \ .$$

Let us estimate the size of a cavern for a typical aged radiopulsar with $L_m \approx 10^{33}$ erg/s. The accretion rate $\dot{M}_c$ in the interstellar medium cannot be much higher than $\sim 10^{10}$ g/s. Putting $v_\infty \approx 10$ km/s in (8.46), we obtain $R_{Sh} \approx 10^2 R_G$, i.e., $R_{Sh} \approx 10^{16} - 10^{17}$ cm. Magnetic dipole radiation may accumulate inside such a huge cavern. In this case, the size of the cavern will increase: $R \propto t^{1/3}$.[1]

---

[1] Apparently, a cavern of this type was discovered around the pulsar PSR 1957+20.

8.5.3 Effect of Relativistic Wind on Accretion Flow Parameters

We shall try to take into account the rearrangement of the surrounding medium under the action of pulsar radiation in the same way as the computation of stationary shells in the propeller regime (Sect. 7.1). Such an analysis, which was carried out by Davies and Pringle (1981), is an extension of the model proposed by Rees and Gunn (1974) for the Crab nebula.

Suppose that a cavern is formed around an ejecting star (for the sake of simplicity, we shall assume that the cavern is spherical). We assume that the entire energy L_m emitted by the pulsar is dissipated at the internal wall of the cavern. Obviously, the size of the cavern has the same order of magnitude as the Shvartsman radius. Near the inner boundary of the shell, the velocity of sound is equal to the maximum possible value:

$$a_s = \frac{c}{\sqrt{3}} \ .$$

Since $a_s \gg v_{ff}$ the pressure gradient is small and the pressure in the shell is constant. Taking this into consideration, we obtain from (5.16) for the conservation of the energy flux

$$\varrho \propto R^4 \ ,$$
$$a_s \propto R^{-2} \ , \quad R_{Sh} < R < R_{out} \ . \tag{8.47}$$

The outer radius of the hot shell is determined from the condition $a_s = a_\infty$:

$$R_{out} \approx \left(\frac{c/\sqrt{3}}{a_\infty} \right)^{1/2} R_{Sh} \approx 3 \times 10^{10} p^{-2} \mu_{30} \dot{M}_{15}^{-1/2} v_8^{-3} \ \text{cm} \ .$$

It follows from the second of the solutions (8.47) that the ratio of the velocity of sound to the free-fall velocity is $a_s / v_{ff} \propto R^{-3/2}$. Consequently, as a neutron star is spun down, v_{ff} becomes of the order of a_s first at the outer boundary. Hence Davies and Pringle proposed that the shell collapses and the ejection stage is terminated when the inequality $R_{out} \lesssim R_G$ is satisfied. This condition is equivalent to the following inequality for the period:

$$p > p_E \approx 1.2 \dot{M}_{15}^{-1/4} \mu_{30}^{1/2} v_8^{-1/2} \ \text{s} \ . \tag{8.48}$$

If, however, the radius of the light cylinder is larger than the capture radius, i.e., $R_l > R_G$, the earlier condition for the termination of the ejection stage remains valid [see (4.49)]. Formula (8.48) can be used if

$$\dot{M}_{15} \lesssim 70 v_8^4 \mu_{30}^2 m_x^{-4} \ .$$

Thus, a consideration of the inverse effect of pulsar radiation on the properties of accreting plasma makes the ejection possible even when the Shvartsman radius is smaller than the capture radius. It can be seen that in this case the estimate of the critical period p_E does not change significantly. This is primarily due to the fact that the power of a pulsar, and hence the pressure of the pulsar wind, depends very strongly on its period (in proportion to p^{-4}). However, it remains unclear whether the shell will be stable when its inner radius becomes smaller than the capture radius.

Lipunov and Prokhorov (1984) proposed the mechanism for the formation of radiobursts in the case $R_{Sh} < R_G$ which takes into account the possible contribution of pressure of the accumulated magnetic dipole radiation. In this case, the process becomes time-dependent.

Let us assume that the magnetic dipole radiation reflected at the cavern walls accumulates inside it and does not allow the cavern to collapse. If the cavern radius $R < R_G$, the condition at its boundary can be written

$$\frac{L_m t}{4\pi R^3} \approx \frac{\dot{M_c}}{4\pi R^2} \sqrt{\frac{2GM_x}{R}} . \tag{8.49}$$

It can be seen that the pressure of the radiation in such a cavern rises more rapidly (as R^{-3}) than the plasma pressure (which increases as $R^{-5/2}$). The cavern is filled very quickly by radiation and begins to swell (as t^2). When its radius becomes equal to the capture radius, the gravitational force is no longer important. The size of the cavern increases slightly, it is detached from the neutron star and drifts upward with the stellar wind emanating from the normal component. Apart from the magnetic dipole radiation, the cavern may also contain relativistic particles (which are also reflected from the cavern boundary). Such caverns must produce very short radiopulses upon floating and bursting. The total energy and the characteristic radioburst recurrence time can be easily estimated by using (8.49) for $R = R_G$. The energy of a radioburst is given by

$$E_b = L_m t_b \approx 2.4 \times 10^{35} \dot{M}_{-6} v_8^{-5} m_x^3 a_{10}^{-2} \text{ erg} . \tag{8.50}$$

The recurrence time t_b and the duration δt_b of a burst are

$$t_b \approx \left(\frac{a}{R_{Sh}}\right)^2 \frac{R_G}{c} , \quad \delta t_b \approx \frac{R_G}{c} . \tag{8.51}$$

For $R_G \approx 10^{10}$ cm and $a/R_{Sh} \approx 10 - 10^{-2}$, t_b varies from a few hundred seconds to several weeks.

8.6 Change in Radiopulsar Period

8.6.1 Spin-down of Pulsars and Their Magnetic Fields

The observed increase in the pulsar period is one of the most exactly measured astronomical effects. However, the theory of this effect is qualitative rather than quantitative even today. Of course, there is no doubt about the origin of spin-down, which lies in the dissipation of the rotational energy of a magnetized neutron star. The dissipation is caused by the electromagnetic field of a rotating star. The details of this effect are rather sketchy at present. For reasons described above, many (quite different) models proposed for this effect lead to the same spin-down relation (in order of magnitude). In most cases, the spin-down of a neutron star is reduced to the magnetic dipole deceleration

$$\frac{dI\omega}{dt} = -\frac{2}{3}\frac{\mu^2\omega^3}{c^3}\sin^2\beta \ . \tag{8.52}$$

If we neglect the dependence on β on the right-hand side and consider that the quantities ω and $\dot{\omega}$ can be observed while the moment of inertia does not differ strongly from the value 10^{45} g cm^2 (Chap. 2), this law can be used to estimate the magnetic dipole moment of neutron stars. Figure 8.9 shows the histogram of the distribution of radiopulsars according to the values of their magnetic dipole moment, constructed from the data of Manchester and Taylor (1981). The main feature of this distribution is that most of the observed radiopulsars have magnetic dipole moments $\mu = 10^{30}$ Oe cm^3, which correspond to a field $B = 10^{12}$ Oe for the radius $R_x = 10$ km. However, there is a considerable spread in μ which varies from 10^{28} to $10^{31.5}$ Oe cm^3. Quite reasonable estimates of the magnetic field have been obtained in this way. The results are in excellent agreement with the estimate for the magnetic field of accreting X-ray pulsars obtained from gyrolines (Sect. 6.6). On the other hand, the magnetic fields of X-ray pulsars obtained from their periods p and period variations $\dot{p}$ (Sect. 6.7) are considerably higher in some cases.

The reason behind such a strong difference in the characteristics of neutron stars (X-ray and radiopulsars) may lie in the extremely strong selection effect. In fact, radiopulsars with high magnetic fields spin-down and attenuate more rapidly (Shakura 1975). Hence the probability of detecting highly magnetized stars among these pulsars is low. In contrast, the accreting X-ray pulsar effect requires a considerable slowing down of rotation and is more likely, the higher the magnetic field of the neutron star.

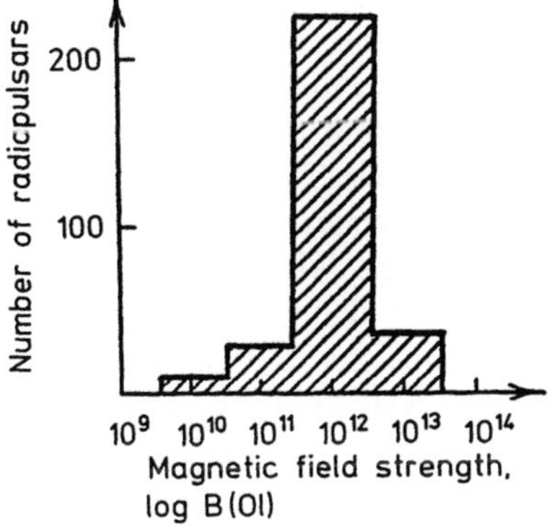

Fig. 8.9. Distribution of radiopulsars according to magnetic field values obtained from the magnetic dipole formula. The neutron star radius is taken equal to 10 km

Are there other spin-down mechanisms? It was mentioned in Sect. 8.2 that the observed spin-down index [formula (8.4)] for the Crab pulsar is $n = 2.5$ and not 3, as expected from the magnetic dipole spin-down. An explanation is required for this difference even though it is not large. It is quite likely that the decrease in the index is due to the fact that the magnetic field structure differs from purely dipole field: the lines of force are stretched by the relativistic wind. Hence the magnetic field decreases more slowly towards the light cylinder. The index n may also vary due to secular variation of the angle β (Goldreich 1970).

Apart from these "quasimagnetic dipole" mechanisms, entirely different types of spin-down mechanisms were also considered. It was observed by Ostriker and Gunn (1969) that at the initial stage of evolution or, to be more precise, at the stage of rapid rotation, the emission of gravitational waves may be largely responsible for the spin-down of neutron stars. It is well known (Landau, Lifshitz 1973) that the necessary condition for the emergence of gravitational waves is the variation of the quadrupole moment of the star with time. The quadrupole moment may appear due to the distorting effect of the magnetic field. In this case, the spinning down takes place according to the law $\dot{\omega} \propto \omega^6$. The radiant power decreases sharply with frequency so that the gravitational radiation for most pulsars is insignificant. Quadrupole moments may also arise due to the bifurcation type star instability (Chaps. 2 and 6). This effect can be observed only in rapidly rotating neutron stars with periods $p \approx 10^{-3}$ s.

Huang et al. (1983) proposed an original mechanism of dissipation of rotational energy. If follows from the Weinberg-Salam theory of electroweak interactions that a neutron undergoing acceleration must emit neutrinos. For example, a neutron rotating about an axis must emit neutrinos in the same way that an electron rotating about a field line emits light quanta (synchrotron radiation). Quantum vortices (elongated in the direction of the rotational axis of the star) emerge in the superfluid component of the neutron star (Sect. 2.3). Participating in the vortex motion, neutrons will emit neutrinos which carry both energy and angular momentum as they leave the neutron star. Since the vortex excitation is associated with the overall rotation, it is obvious that the emission of neutrinos will spin-down the neutron star. Unfortunately, the estimate of the efficiency of this mechanism is hardly reliable and contains a number of unknown parameters which strongly influence the final result. Nevertheless, the authors were able to obtain observational confirmation for their model (see also Malov 1985). However, it is doubtful that this mechanism is really responsible for the rotational losses in neutron stars. The spin-down torque in this model is found to be independent of frequency. This unpleasant property forces us to rule out the mechanism even without a detailed theoretical study. Obviously, the same mechanism must also apply to accreting X-ray pulsars. This, however, contradicts the observations. All X-ray pulsars are in the equilibrium state. But equilibrium is not possible if the spin-down torque is independent of frequency (Sect. 6.3).

8.6.2 Spin-up Episodes and Internal Structure of Neutron Stars

Soon after the discovery of glitches in the spinning down of radiopulsars (Fig. 1.3), Baym et al. (1969) proposed that this phenomenon should be used for investigating the internal structure of neutron stars. Within the framework of a simple two-component model (superfluid and normal components), the change in the angular momentum of a neutron star is described by the system (6.90) of equations. For radiopulsars, $\dot{\omega}$ is negative between glitches and constant to a high degree of accuracy. Under these conditions, the time-independent solution of (6.90) has the form

$$\dot{\omega}_s = \dot{\omega}_c \ ,$$
$$\omega_s - \omega_c = \frac{I_s}{I_c}\frac{\tau_c}{t_{sd}}\omega_c \ . \tag{8.53}$$

It should be recalled that the subscripts s and c correspond to the superfluid and normal components respectively, τ_c is the characteristic time of angular momentum exchange between the crust and the core, and $t_{sd} = p/\dot{p}$ is the spin-down time scale. It can be seen that the crust rotates a bit more slowly in the normal state. However, glitches in the process of spin-down show that such a stationary pattern is violated from time to time. For example, the increase in the frequency in the Vela pulsar was accompanied by a jump $\Delta\dot{\omega} \approx 10^{-2}\,\dot{\omega}$. Baym et al. (1969) attributed this phenomenon to the rapid rearrangement of the hard crust (starquake). The qualitative reason behind this phenomenon is quite clear: the star is slightly flattened at the poles as a result of rotation. The solid crust tries to retain its shape but the centrifugal forces are reduced as the star slows down. The load at the equatorial parts increases and the crust cracks at a certain instant of time. The moment of inertia I_c of the crust sharply decreases and its rotation becomes more rapid. However, the superfluid component, which "realizes" this only after a time τ_c, takes away the angular momentum from the crust and thus slows it down. A relaxation to the previous value of $\dot{\omega}$ occurs, but at a different frequency ω (Fig. 8.10). It   follows   from

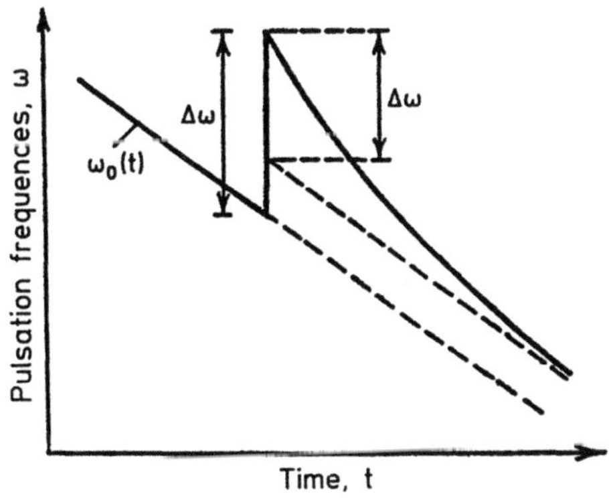

Fig. 8.10. Change in the rotational frequency of a neutron star upon an abrupt variation of the moment of inertia in the two-component model (Manchester, Taylor 1978)

(6.90) that the relative variation in the spin-down rate of the crust (which is actually observed) is approximately equal to

$$\frac{\Delta\dot{\omega}_c}{\dot{\omega}_c} \approx \frac{\Delta\omega}{\omega}\frac{t_{sd}}{\tau_c}\left(1 - \frac{\Delta I_s}{I_s}\frac{I_c}{\Delta I_c}\right) , \tag{8.54}$$

where ΔI_s and ΔI_c denote the variations in the moments of inertia of the superfluid and normal components respectively. For starquakes, $\Delta I_s \ll \Delta I_c$, and hence $\Delta\dot{\omega}_c/\dot{\omega}_c \gg \Delta\omega/\omega$. The complete solution of the system (6.90) after the glitch is described by

$$\omega_c(t) = \omega_0(t) + \Delta\omega_s[1 - Q(1 - e^{-t/\tau_d})] , \tag{8.55}$$

where

$$\tau_d = \tau_c \frac{I_s}{I} ,$$

$$Q = \frac{(\Delta\dot{\omega}_c)^2}{\Delta\dot{\omega}_c\Delta\omega} = \frac{I_s}{I}\left(1 - \frac{\Delta I_s I_c}{\Delta I_c I_s}\right) \approx \frac{I_s}{I} . \tag{8.56}$$

In principle, the above model correctly describes the glitches observed in the Crab and Vela pulsars. It can even be stated after an analysis of the observations that the relative fraction of the superfluid component in the Crab pulsar is much higher ($Q \approx 1$) than in the Vela pulsar.

However, the starquake model does not provide any explanation for the extremely high frequency at which the breakdown occurs in the period. The crust is not able to accumulate the necessary deformations in a few years (viz., the characteristic interval between jumps). Hence the idea of corequakes was put forth by Pines et al. in 1972. The core has a much higher rigidity and can accumulate higher deformation stresses. The idea of explaining small fluctuations of the period by the "precipitation" of superfluid vortices to the wall (crust) of the star (Pines, Shaham 1972) seems to offer some promise.

8.7 Evolution of Radiopulsars

Here we shall consider the following questions:

1) Which objects are the progenitors of radiopulsars?
2) How long do radiopulsars live and what rules do they obey?
3) What happens to the neutron star after the termination of the radiopulsar stage?

These questions are just a part of the single general problem of investigation of the evolution of neutron stars. The general approach to the analysis of such an evolution will be described in Chap. 11.

8.7.1 Origin and Age of Pulsars

Radiopulsars are the only reliably identified class of ejecting neutron stars so far. The fact that radiopulsars are concentrated in a relatively thin layer (of thickness ~ 400 pc) of the galactic disk implies that pulsars are genetically linked to the stars of the galactic disk. Apparently, a considerable part of generated radiopulsars have small periods. This is indicated by the observed properties of young neutron stars around which the remnants of the supernova explosions are observed (Table 8.2, Seward, Harnden 1984). The age of such objects can be determined independently in two ways: 1) from the formula $t_e = t_{sd}/2$ for the characteristic age; and 2) from the age of the supernova remnants.

It can be seen from Table 8.2 that four out of five young pulsars have periods much smaller than one second. We have specially included the X-ray pulsar in the remnant of CTB 109 in this table. This pulsar is of quite different type and is accreting. This example directly indicates that in many cases neutron stars are generated with large periods and do not cases neutron stars are generated with large periods and do not pass through the ejection stage (Lipunov, Postnov 1985). With the exception of this pulsar, the characteristic age and the age of the remnant are generally in agreement with each other.

However, the age of most of the pulsars is much higher than the life time of the supernova remnant. This is one of the reasons why only a few of the pulsars are surrounded by remnants. The age of such pulsars is estimated independently from their proper motion. As a rule, the kinematic age is lower than the characteristic age. The histogram showing the distribution of the number of pulsars according to their characteristic age (Manchester, Taylor 1978) clearly indicates that the mean age of the pulsars is several million years. A detailed analysis shows that the mean life of a pulsar is $\sim 5 \times 10^6$ years, which is in good agreement with the mean kinematic age. The characteristic time of oscillations in the z-coordinate is $\sim 10^8$ years. Hence it can be concluded that radiopulsars only move away from the galactic plane and do not manage to complete a single oscillation.

Table 8.2. Characteristics of young neutron stars associated with the remnants of supernova explosions

Parameter	Pulsar				
	Crab	(LMC) 0540−69	1509−58	Vela	SNR CTB 109
p, s	0.033	0.050	0.150	0.089	7
$\dot{p}$, 10^{-15} s/s	422	479	1540	125	−
$\dot{E}_{rot}$ 10^{38} erg/s	4.7	1.5	0.18	0.071	−
Age, 10^3 years	1.24	1.66	1.55	11.3	10
L_x of the pulsar, 10^{36} erg/s	1	2.4	0.019	10^{-5}	0.2

We can now easily estimate the birth rate of radiopulsars. The observed surface density of radiopulsars in the vicinity of the Sun is $\sim 90\,\text{kpc}^{-2}$, while the rate of their generation per unit area is $\sim 2 \times 10^{-5}\,\text{year}^{-1}\,\text{kpc}^{-2}$. Multiplying the characteristic surface area of the galactic disk by this quantity, we find that the approximate generation rate is one pulsar per century. A more exact evaluaton gives the estimate of one pulsar per thirty years. Considering that not every pulsar can be seen from the Earth because of the directionality of radiation, the estimate of pulsar birth rates becomes several times higher. On the other hand, the estimate of the birth rate depends on the accuracy with which the distance from the pulsar is determined. In turn, this accuracy depends on the assumption concerning the mean electron density in the Galaxy. Taking these uncertainties into consideration, a reasonable estimate for the radiopulsar birth rate is one pulsar every $20-40$ years.

Massive stars (stars with a mass exceeding $8-12 M_\odot$) are formed in the Galaxy at nearly the same rate. Together with the theoretical calculations for the evolution of stars, this directly points towards the genetic connection between radiopulsars and massive OB stars (Chap. 11).

The characteristic age of a large number of radiopulsars is higher than the estimated mean value of 5×10^6 years by an order of magnitude or two. (For example, the characteristic age of the pulsar PSR $1952+29$ is 4.3×10^9 years.) This may be due to the fact that the pulsars are generated not with periods much smaller than the observed periods, but rather with periods quite close to the observed ones.

Gunn and Ostriker (1970) proposed another explanation for the existence of radiopulsars with enormous characteristic age. According to them, the magnetic field of a neutron star is not constant and attenuates with a characteristic time of $\sim 10^7$ years. This also explains the attenuation of radiowaves in $10^6 - 10^7$ years. However, such a rapid attenuation of the magnetic field runs into serious theoretical difficulties. Moreover, it seems that a rapid attenuation of the magnetic field is in direct contradiction to the observations: the accreting X-ray pulsar Her X-1, which is a neutron star, has a magnetic field strength of $\sim 10^{12}$ Oe and simultaneously forms a part of a low-mass binary system having an age of $\sim 10^8$ years (see Chap. 6).

The spatial distribution and the birth rate of radiopulsars indicate that most of them are products of the evolution of massive stars. There are two ways of formation of radiopulsars.

1) The formation of a rapidly rotating neutron star soon after the collapse of the degenerate core of a single massive star or a star in a binary system, followed by the disintegration of the binary system. This is the method of formation of radiopulsars usually accepted in the literature.

2) An alternative way was proposed by Bisnovaty-Kogan and Komberg (1974) who mentioned that some accreting pulsars in binary systems have rotational periods characteristic of radiopulsars. Moreover, the periods of accreting pulsars are reduced. Consequently, following the collapse of the normal component and a breakdown of the binary system, the old neutron star

may become a radiopulsar (E-pulsar). In other words, the rapid rotation of neutron stars, which is necessary for the emergence of the radiopulsar effect, may be caused not only by a collapse, but also gradually as a result of the accretion spin-up of the neutron star in a binary system (Srinivasan, Van den Heuvel 1982; Lipunov 1982a). This idea was widely supported by Alpar et al. (1982), Joss and Rappaport (1983), and others following the discovery of a millisecond pulsar by Backer et al. (1982).

Calculations of the evolution of neutron stars in binary systems show that a large number of radiopulsars are old neutron stars that have passed the accreting pulsar stage in a binary system (Kornilov, Lipunov 1983b). Such pulsars are known as recycled pulsars and occupy certain regions on the $(\dot{p}-p)$-diagram (Kolosov et al. 1989).

Before concluding this discussion, we observe that the inclusion of binary systems as progenitors of radiopulsars removes the contradiction between the birth rate of radiopulsars and the rate of supernova explosion. The latter, determined from the observation of remnants, apparently does not exceed $1/40$ years^{-1} (Lozinskaya 1986). The situation is improved if we consider that two radiopulsars may be generated simultaneously as a result of a repeated explosion in a binary system. Moreover, the explosion of a massive star in a close binary system may not be accompanied in general by a supernova explosion (Shklovskii 1971). As a matter of fact, there is no room in a close binary system for a supergiant envelope which is necessary for channeling the explosion energy into a slow optical burst.

8.7.2 Evolution of the Radiopulsar Period

In the electrodynamic models of radiopulsars considered above, the total energy losses are close to magnetic dipole losses. It should be recalled that the exact equation for the variation of the angular momentum of the neutron star under the action of the magnetic dipole radiation has the form of (8.52).

It is convenient to write this equation in a different form by replacing the rotational frequency by the period $p = 2\pi/\omega$:

$$\dot{p} \frac{8\pi^2}{3} \frac{\mu^2 \sin^2 \beta}{c^3 Ip} \; . \tag{8.57}$$

The main evolution diagram for radiopulsars is the $\dot{p}-p$ diagram (Fig. 8.2). During the course of its evolution, each radiopulsar leaves a track on this diagram whose form depends on the variation $\dot{p}(p)$ and the behavior of physical quantities appearing on the right-hand side of (8.57). Two types of processes are usually considered: 1) dissipation of the magnetic field and 2) a slow alignment of the magnetic axis with the rotational axis of the neutron star (Macy 1974).

Obviously, for a statistical description of an ensemble of radiopulsars, we must consider the functions of their distribution over various parameters like periods (or frequencies) and magnetic dipole moments $\varphi(\omega,\mu)$. Let us discuss this question in detail (Lipunov 1987a).

The evolution of the distribution function must satisfy Liouville's equation with a nonzero right-hand side

$$\frac{\partial\varphi(\omega,\mu,t)}{\partial t} + \nabla_{\omega,\mu}\varphi(\omega,\mu,t)v_{\omega,\mu} = \psi(\omega,\mu,t) \ . \tag{8.58}$$

Here, $\nabla_{\omega,\mu}$ is a divergence operator in the space of frequencies and magnetic moments and $\psi(\omega,\mu,t)$ is a function describing the parameters of the generated stars. For the case when the magnetic dipole moment attenuates according to

$$\dot\mu = \frac{\mu}{\tau} \ , \quad \mu = \mu_0 e^{-t/\tau} \ ,$$

and the pulsars are generated with nonzero periods and identical magnetic moments

$$\psi(p,t) = v\delta(p) \ ,$$

we obtain the following self-similar solution:

$$\varphi(p) = \begin{cases} \dfrac{v\tau p}{4\pi^2 A\tau\mu_0^2 - p^2} & \text{for} \quad p \leqslant p_{\max} \ , \\ \\ 0 & \text{for} \quad p > p_{\max} \ , \end{cases} \tag{8.59}$$

where

$$p_{\max} = 2\pi[A\mu_0^2\tau(1-e^{-2t/\tau})]^{1/2} \ , \quad A = \frac{2}{3c^3 I} \ .$$

Obviously, this solution can describe only the left side (to the maximum) of the distribution of pulsars over periods. In order to attain full agreement between theory and observations, we must consider the fact that pulsars are generated with different initial periods and magnetic fields. The general solution taking this into account will be presented in Chap. 11.

8.8 Spatial Velocities of Radiopulsars

Assuming that radiopulsars are formed near the symmetry plane of the Galaxy, their observed distribution along the z-coordinate can be explained by assuming that radiopulsars have very high velocities in space. Let us estimate the spatial velocity of a radiopulsar by assuming that the mean distance of the radiopulsar from the galactic plane $\langle z \rangle = 200\,\text{pc}$ and its lifetime is $\tau = 5 \times 10^6$ years. Obviously,

$$v > v_z = \frac{\langle z \rangle}{\tau} \approx 40\,\text{km/s} \ . \tag{8.60}$$

In principle, such velocities can be assumed to be high. For example, the class of runaway stars includes those whose spatial velocities are higher than $\sim 30\,\text{km/s}$.

However, the very first measurements of the proper motion of radiopulsars led to unexpected results. The velocities determined were found to be of the order of several hundred km/s (Lyne et al. 1982), which is much higher than the estimate (8.60). This result was confirmed in subsequent investigations: the tangential velocities of some radiopulsars attain values of 500 km/s and even higher (Lyne et al. 1982). This gave rise to the general opinion that pulsars are objects moving with anomalously high velocities in space. As a matter of fact, this is true perhaps only partially. The data on pulsar velocities is still not statistical and is at the same time subjected to a number of selection effects. Suffice it to say that proper motion has so far been measured only in about 30 radiopulsars, i.e., in less than 10% of their total number. Moreover, proper motion has naturally been observed only in those pulsars which move with an anomalously high velocity. A consideration of these and other effects significantly lowers the estimate for the mean spatial velocity of radiopulsars. It is found that most of them have moderate velocities $\sim 30-40\,\text{km/s}$ (Tutukov et al. 1984). Nevertheless, the existence of superfast pulsars does require an explanation.

Three mechanisms were considered to explain the anomalously high pulsar speeds. These include: 1) asymmetric emission of magnetic dipole waves caused by a displacement of the dipole center relative to the body of the star (Tademaru, Harrison 1975); 2) anisotropy of the collapse of a normal star into a neutron star (Shklovskii 1971); and 3) decay of a binary system following the explosion of one of the stars (Blaau effect) (Gott et al. 1970).

The first two mechanisms are actually quite similar: both envisage the pumping of gravitational energy of a collapsing star into the kinetic translational energy. After all, the rotational energy of a pulsar is the result of the work done by the gravitational forces during collapse. In this case, the estimate of the velocity acquired by the neutron star contains a free parameter, that is, the anisotropy parameter (see below). Things are quite different in the case of the Blaau effect. If the explosion in the binary system takes place instantaneously, the velocity acquired by the stars flying apart is completely determined by the initial and final masses, the rotational periods, and the eccentricity. An analysis of the evolution of a massive binary system shows (Chap. 11) that before the second explosion the binary system consists of a helium star with a mass $\sim 10\,M_\odot$ and a neutron star with a mass $\sim 1\,M_\odot$. The helium star collapses, shedding 90% of its mass, and the system decomposes (it should be recalled that the necessary condition for the decay of a binary system is that it should instantaneously shed more than half its mass). The maximum velocity acquired by runaway stars is close to the initial orbital velocity but does not exceed this value. The maximum velocity of a neutron star revolving around a helium star is ~ 500 km/s. According to the ratio of the masses of these stars, the orbital velocity of the helium star is about one-tenth of this, i.e., ~ 50 km/s. Hence it is clear that the Blaau mechanism together with present concepts of the evolution of binary systems provides a natural explanation for the existence of radiopulsars with a velocity of $\lesssim 700$ km/s. It is interesting to note an important corollary of this mechanism which can be qualitatively verified. A rapidly moving neutron star must be an old star. For example, if we could detect a radiopulsar moving with a velocity of $300-500$ km/s and emitting thermal X-rays which are associated with the cooling of the neutron star and point towards its recent origin, the Blaau mechanism can be immediately discarded for such a pulsar.

No such observations have been reported so far. In general, the Blaau mechanism seems to be quite natural and fairly effective. It must, however, be accepted that most radiopulsars have moderate spatial velocities ($\lesssim 100$ km/s). After all, not all radiopulsars are generated by binary systems, and the components of a majority of binary systems are separated by large distances and do not acquire high velocities upon decomposition of the system.

The radiopulsar PSR 2111+46 is an exception (Lipunov et al. 1986). This pulsar has a proper motion $\mu'' \cos \alpha \approx (107 \pm 60) \times 10^{-3}$ and $\mu'' \cos \delta \approx 40 \times 10^{-3}$ arc seconds/year. At a distance of ~ 4.3 kpc (Manchester, Taylor 1981), this corresponds to a tangential velocity $v_t \approx 2200$ km/s. Such a rapid motion cannot be explained by the standard pattern involving the disintegration of a binary system. This leaves only the mechanisms associated with the possible anisotropy of energy liberation during collapse.

Suppose that a part of the energy is liberated anisotropically during collapse. In this case, the velocity acquired by a neutron star is determined from the law of momentum conservation

$$M_x v = \frac{E_{ej}}{v_{ej}} \, ,$$

where $E_{ej} \approx 0.1 \, Mc^2$ is the energy liberated at the time of explosion and v_{ej} is the velocity with which this energy is carried away. Calculations prove (Imshennik, Nadezhin 1982) that the main part of energy during the formation of a neutron star is carried away in the form of neutrinos. Putting $v_{ej} = c$, we obtain

$$v \approx 0.1 \beta c \ .$$

For $\beta \approx 10^{-1}$, the velocity $v \approx 3000$ km/s. Let us find out the reasons behind the emergence of such an anisotropy.

It was mentioned by Chugai (1984) that the neutrino emission must be asymmetric in the strong magnetic field of the neutron star being formed.

However, detailed calculations for such a mechanism (Dorofeer et al. 1985) show that even in superstrong magnetic fields the star acquires a velocity of under 100 km/s.

Another source of anisotropy discussed in the literature (Lipunov 1983b) may be the tidal distortion of a collapsing star in a binary system. The distortion of the shape of a star due to tidal deformation can be estimated from

$$\beta_0 \approx \frac{2\varrho^4}{q} \, ,$$

where ϱ is the radius of the collapsing star in units of the separation between the components and q is the ratio of the mass of the collapsing part of the star to the mass of the companion. The size of the collapsing part of the star is the size of the degenerate core and does not exceed 10^9 cm. It is quite obvious that in a massive binary system, where the components are separated by a distance of the order of 10^{12} cm, the tidal anisotropy is quite insignificant. The situation is quite different for low-mass binary systems with white dwarfs. The value of ϱ in this case may be several tens of percent, and hence the initial distortion of the shape of the star may be $\beta_0 \approx 10^{-4}$. If we take into account the additional increase in the anisotropy during collapse (Tsygan 1981), the anisotropy may rise to ~ 0.1 and the velocity may become as high as several thousand kilometers per second. Of course, these are very rough estimates. We must take into account the fact that the outer layers of the white dwarf do not make a significant contribution to the mass. This results in a considerable decrease in the anisotropy of energy liberation. On the other hand, the nonsymmetric ignition of matter inside a white dwarf due to a distortion of its shape may play a significant role. In short, the possible value of anisotropy is hard to predict.

8.9 Ejecting Stars in Binary Systems

It should be recalled that only four radiopulsars forming parts of binary systems had been discovered by 1985. In all four cases, the second star has not been observed in any of the electromagnetic ranges. Apparently, this is due to the fact that the invisible components are old evolved stars (degenerate dwarfs, neutron stars, black holes).

8.9.1 Radiopulsars Forming Pairs with Degenerate Stars

Soon after the discovery of the radiopulsar PSR 1913+16 in a binary system (Hulse, Taylor 1975), it became clear that such a binary system is an ideal "laboratory" for investigating relativistic effects. The following effects were observed owing to a high stability of the period of radio pulses: 1) Doppler's classical effect; 2) Doppler's transverse effect; 3) rotation of the apsidal line;

4) gravitational red shift; 5) variation of the binary system period, which was found to be in complete accord with the predictions of Einstein's general theory of relativity. Measurement of these effects made it possible to determine the masses of stars to a previously unattained degree of accuracy. According to the latest data (Taylor, Weisberg 1989), the mass of the radiopulsar is $M_x = (1.442 \pm 0.003) M_\odot$ while the mass of the second star is $M = 1.386 \pm 0.003) M_\odot$. The characteristic spin-down time is 2×10^8 years. This leads to an estimate for the dipole moment of $\mu \approx 10^{28}$ Oe·cm^3. The field strength is $B \approx 10^{10}$ Oe.

The most important question is why did the binary pulsar system PSR 1913+16 not disintegrate during the formation of the neutron star? The evolution of the binary system providing an answer to this and other questions was described by Srinivasan and Van den Heuvel (1982). This problem will be discussed in greater detail in Chap. 11. For the present we shall consider just one aspect.

It is assumed that the radiopulsar PSR 1913+16 is a former X-ray pulsar, and the rotational period of the neutron star observed now is close to the period established during accretion. It should be recalled that the equilibrium period of a pulsar in the case of disk accretion is [see (6.80)]

$$p \approx 1 \mu_{30}^{6/7} L_{37}^{-3/7} \text{ s} . \tag{8.61}$$

Putting $L_x = 10^{38}$ erg/s, which is close to the Eddington limit, we get $p \approx 0.059$ s and $\mu_{30} \approx 10^{-2}$. These values are in good agreement with the estimates obtained from the magnetic dipole formula. It should be noted, however, that the actual accretion rate may be much higher than the critical value corresponding to the Eddington limit. In this case, the equilibrium period is found to be slightly different (Chap. 9).

Why do radiopulsars not form pairs with normal stars? In the very first years following the discovery of radiopulsars, it became clear that for some reason they avoid forming binary systems. Three hypotheses were put forth to explain such a strange incompatibility:

1) Binary systems disintegrate after the first explosion which leads to the formation of a neutron star. Since it was clear from the very beginning that the lighter component explodes first in view of the mass exchange, the binary system certainly cannot lose more than half its total mass. Hence the disintegration must be associated with the anisotropy of the collapse (Shklovskii 1976).

2) Shvartsman (1971a) interpreted the absence of radiopulsars in binary systems from an entirely different position. Neutron stars remain in the binary systems, but the radiopulsar stage is much shorter than the stage for a single star due to the presence of accretion flows. Hence the probability of finding a radiopulsar in a binary system is much lower.

3) Illarionov and Syunyaev (1975) noted the fact that coherent radiowaves with a wavelength longer than 75 cm, which make it possible to observe most

of the radiopulsars, are absorbed effectively in the stellar wind of a normal star.

Suppose that a normal star loses matter in a spherically symmetric way. In this case, the optical thickness associated with the free-free absorption in the stellar wind is estimated by

$$\tau_{\text{br}} \approx 3.2 \times 10^8 \, T_4^{-3/2} \dot{M}_{-8}^2 a^{-3} v_8^{-2} \lambda_{75}^2 \; , \tag{8.62}$$

where $T_4 = T_w/10^4 \, \text{K}$ is the temperature of the stellar wind, $\lambda_{75} = \lambda/75 \, \text{cm}$ is the wavelength of the radiofrequency radiation, and a is the semi-major axis of the orbit expressed in units of solar radii. In massive binary systems with $\dot{M}_8 \approx 100$, we have $v_8 \approx T_4 \approx 1$. Hence the stellar wind is opaque even for very broad systems ($a \lesssim 10^3$).

Which of the above three mechanisms is responsible for the absence of radiopulsars in binary systems with normal stars? To answer this question correctly, we must carry out a detailed analysis of the evolution of normal and neutron stars in binary systems (Chap. 11). A simulation of such an evolution showed that all three factors are significantly responsible for the avoidance of binary systems by radiopulsars (Kornilov, Lipunov 1984). This statement will be discussed in detail in Chap. 11.

In order to understand the importance of the problem concerning, for example, the anisotropy, let us consider the following estimate. It is well known that even massive stars on the main sequence lose matter at a rate of $\dot{M}_0 \lesssim 10^{-8} \, M_\odot/\text{year}$. The formation of a young radiopulsar is most probable at the instant when the neighboring normal star is on the main sequence (after all, this is the most prolonged state of a normal star). It can be easily seen from (8.62) that for $v_8 \approx 2-3$, the stellar wind becomes transparent in the case of systems with $a \gtrsim 10^{2.5} - 10^3$, i.e., systems with periods ranging from a few years to tens of years. Such broad pairs with neutron stars are actually observed (e.g., A 0535 + 26 or X Per with a period of ~ 500 days). In such broad systems, matter in the stellar wind near the neutron star is extremely rarefied so that the phase of the ejecting pulsar may exist for quite a long time. The number of such separate systems is extremely large, and is even larger than the number of close binary systems according to statistical investigations. This leads to the inevitable conclusion that radiopulsars must be visible in pairs with normal stars. But they are nowhere to be seen! This could be understood by assuming a moderate anisotropy which is enough for the disintegration of broad binary systems after the first explosion. In order to explain the absence of radiopulsars forming pairs with the normal stars, we must assume (Kornilov, Lipunov 1984) that during formation a neutron star acquires a kick velocity of $\sim 80-100 \, \text{km/s}$ on the average. On the other hand, these calculations show that $\sim 0.1\%$ of radiopulsars must form binary systems with normal stars and must be visible as radiopulsars. Hence it can be hoped that such pulsars will indeed be observed in near future. This would be important not only for confirming the theory of evolution of neutron stars, but also for understanding processes occurring in the stellar wind of normal stars. After all, a radiopulsar revolving in the semitransparent stellar wind of a normal star is a sort of probe scanning an inhomogeneous plasma by pulsed radiation (Lipunov, Prokhorov 1984). In this work, three periodic effects which could be observed in radiopulsars in binary systems were computed: 1) absorption of radio waves (light curve); 2) variation of the pulse arrival time (dispersion measure curve); and 3) Faraday rotation of the polarization plane in the magnetic field of stellar wind.

These calculations did not take into account the refraction of radiowaves. In most cases, this refraction can be neglected. However, in cases where it is significant, interesting details appear on the light curve of the radiopulsar (Lipunov, Prokhorov 1987).

8.9.2 "Reflection" Effect

It seems at first glance that a binary system is not the best place where an ejecting neutron star could exist: the dense stellar wind absorbs coherent radio waves, and reduces the lifetime of the star at the ejection stage. This is indeed so. However, there is another important factor which favorably (from the point of view of observation) distinguishes a neutron star in a binary system from a single neutron star. It should be recalled that most (99.99%) of the energy dissipated by single radiopulsars in the form of relativistic particles and low-frequency electromagnetic waves is not observable. In the case of a binary system, however, the normal star may capture a considerable part of the relativistic wind ($\geq (R_0/a)^2/4$, where R_0 is the radius of the normal star) and convert it into a form in which it can be detected on the Earth. This gives rise to an interesting analog of the classical reflection or heating effect (Lipunov 1980a; Lipunov, Prokhorov 1984). The main mechanisms of energy conversion are the synchrotron radiation of relativistic particles in the magnetic field of a normal star and nuclear transformation of ultrahigh-energy particles colliding with the stellar wind matter of the same star.

The total power of the energy of relativistic particles transformed in this way is estimated from

$$L_{\text{ref}} \approx \frac{1}{6} \left(\frac{R_{\text{eff}}}{a} \right)^2 \frac{\mu^2 \omega^4}{c^3} \approx 10^{31} \varrho_{\text{eff}}^2 \mu_{30}^2 p^{-4} \text{ erg/s} ,$$

where $\varrho_{\text{eff}} = R_{\text{eff}}/a$ is the characteristic size of the capture region expressed in units of the semi-major axis ($R_{\text{eff}} \geq R_0$). For close binary systems, $\varrho_{\text{eff}} \approx 0.1 - 0.3$ and the transformation coefficient may be several percent.

Apart from the reflection effect, transient floating and bursting of caverns, or radioburster phenomenon, may also be observed (Sect. 8.5).

8.9.3 Observational Evidence of the Existence of Ejecting Stars in Binary Systems

The first probable candidates in this class of objects are the sources of variable nonthermal radio waves, which are identified with binary stellar systems. These include sources of the type Cyg X-3, Cir X-1, LSI$+61°$. The model of a young radiopulsar in the binary system Cyg X-3 was first proposed by Basko et al. (1974). All three sources are characterized by variable nonthermal radio waves (Nicolson 1984; Molnar et al. 1984). A strict periodicity is observed for the systems Cir X-1 and LSI$+61°$: the radiation comes in the form of radiobursts with the orbital period of the binary system.

The discovery of periodic radiation of ultrahigh energies ($\gtrsim 10^{15}$ eV) from the source Cyg X-3 is quite interesting (Stepanyan 1984). This fact has been confirmed in a number of observations and poses a number of astrophysical and physical problems (see, for example, Oyama et al. 1986).

The communication by Abbott et al. (1984) on the discovery of nonthermal radio waves from hot OB and WR stars has aroused considerable interest. According to the authors, a considerable part ($\sim 10\%$) of single OB stars emit such radiation. In principle, this does not contradict the results of numerical simulation of the evolution of neutron stars (Chap. 11), according to which a majority of neutron stars in binary systems must be in the ejection stage. In a number of cases, the variability of radio waves can be attributed to the variable absorption of radio waves from a neutron star in the stellar wind of the normal component (Lipunov, Prokhorov 1984, 1987).

9. Supercritical Regimes

Each massive close binary system sooner or later enters a stage of intense mass exchange, when the flow rate assumes large values, $\sim 10^{-4}-10^{-6} M_\odot$/year. If all the matter were to fall on the surface of the neutron star, its luminosity would be $\sim 10^{40}-10^{42}$ erg/s, which is much higher than the Eddington limit (3.27). This inevitably means that in the life of a neutron star in a close binary system there are times when we must take into consideration the radiation pressure emerging as a result of accretion. It was agreed in Chap. 4 to call the regime of interaction of a neutron star with the surrounding matter the "supercritical regime" if the energy liberated at the stopping radius exceeds the Eddington luminosity limit:

$$\dot{M}_c \frac{GM_x}{R_{st}} \geqslant L_{Ed} = \frac{4\pi GM_x c}{\kappa} \ . \tag{9.1}$$

A supercritical regime is possible not only in massive binary systems. An analysis of the mass exchange processes shows that the supercritical regime may set in even in low-mass binary systems where $M \lesssim (8-10)M_\odot$ (Chap. 11).

Three states of a neutron star are possible in the supercritical regime: the superejector state SE, the superpropeller state SP, and the superaccretor state SA. It can now be stated without doubt that neutron stars of at least one of these three types must exist in the Galaxy. However, no such object has been identified reliably in observations so far. This is mainly due to the fact that a neutron star in the supercritical regime must be surrounded by an optically thick envelope. The most interesting phenomena that one could imagine cannot be observed from a distance, which considerably hampers the identification of the object. Secondly, the intense mass exchange takes place very rapidly (on the galactic scale) and hence there are not many supercritical neutron stars in the Galaxy.

Interest in supercritical accretion rose sharply after the discovery of the source SS 433. However, the number of investigations on the subject is not large (less than 10). The complexity of the problems associated with these investigations does not raise hopes of a rapid and final solution of the entire problem. Nevertheless, this will not prevent us from pointing out the main characteristic features of supercritical regimes and trying to look for them in some observed objects.

Everywhere below, we shall assume that away from the stopping radius the accretion regime is close to the disk regime. This is associated with the fact that large mass flows to the neutron star are formed at the instant when the normal star fills its Roche lobe and matter flows through inner Lagrangian point. The angular momentum of matter captured by the neutron star in this case is always large enough to form an accretion disk.

The following brief review of the results on supercritical regimes is based mainly on the works by Shakura and Syunyaev (1973); Zel'dovich et al. (1972); Basko and Syunyaev (1976); Lipunov (1982b); and Lipunov and Shakura (1982).

9.1 Superaccretor

9.1.1 Accretion Pattern

We shall assume that the disk accretion regime is attained and the accretion rate of matter captured by the star is higher than the critical rate (4.28),

$$\dot{M}_c \gtrsim \dot{M}_{cr} \approx 5 \times 10^{-7} \mu_{30}^{4/9} m_x^{-1/9} M_\odot/\text{year} \ . \tag{9.2}$$

In this case, the spherization radius R_s at which the disk luminosity becomes equal to the Eddington limit is larger than the stopping radius (Fig. 9.1). In the supercritical accretion regime proposed by Shakura and Syunyaev (1973), matter begins to flow out of the disk at $R \lesssim R_s$ so that the accretion rate of matter becomes a function of the radius:

$$\dot{M}(R) = \dot{M}_{cr} \frac{R}{R_s} \ . \tag{9.3}$$

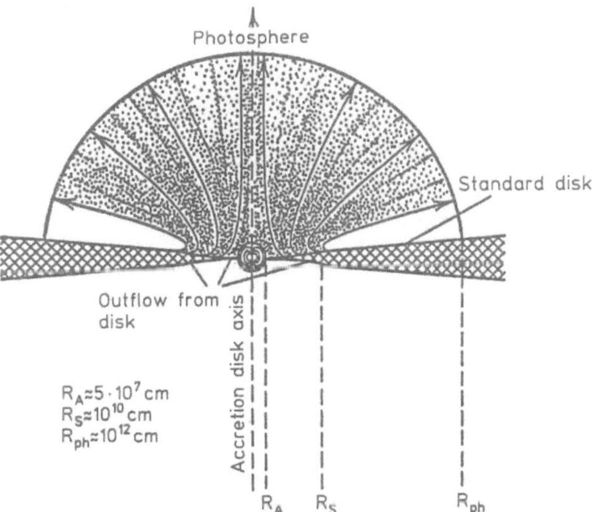

Fig. 9.1. Qualitative pattern of flow of matter around a superaccretor

It was shown in Chap. 4 that the Alfvén radius is determined from

$$R_{\mathrm{A}} = \left(\frac{\mu^4 G M_x}{8 L_{\mathrm{Ed}}^2}\right)^{1/9} \approx 4.6 \times 10^7 \mu_{30}^{4/9} m_x^{-1/9} \ \mathrm{cm} \ . \tag{9.4}$$

Near $R \approx R_{\mathrm{A}}$, the magnetic field pressure becomes comparable with the dynamic pressure of the plasma and the disk disintegrates. If the star rotates so slowly that the corotation radius is larger than the Alfvén radius, all the matter reaching the magnetosphere will fall on the poles of the neutron star. Such a regime was conditionally called the superaccretor regime SA. The condition $R_{\mathrm{c}} \geqslant R_{\mathrm{A}}$ can be written as an inequality for the rotational period of the neutron star:

$$p \geqslant p_{\mathrm{A}} \approx 0.17 \mu_{30}^{2/3} m_x^{-2/3} \ \mathrm{s}. \tag{9.5}$$

If the fall of matter from the neutron star on the surface were steady, the luminosity of the neutron star would considerably exceed the Eddington limit (Lipunov 1982b):

$$L_{\mathrm{NS}} = \frac{R_{\mathrm{A}}}{R_x} L_{\mathrm{Ed}} \approx 46 L_{\mathrm{Ed}} \mu_{30}^{4/9} m_x^{-1/9} R_6^{-1} \ . \tag{9.6}$$

It was mentioned by Zel'dovich et al. (1972), Ruffini and Wilson (1971), and Basko and Syunyaev (1976) that in this case the larger part of the energy will be carried away by neutrinos (see below).

How does such an object look to an observer from the outside? Let us estimate the optical thickness of matter flowing from the disk on account of Thomson scattering. The outflow velocity will be of the order of the parabolic velocity at the spherization radius:

$$v_{\mathrm{sh}} \approx \sqrt{2 G M_x / R_{\mathrm{s}}} \approx 10^8 \dot{M}_{-4}^{-1/2} m_x^{1/2} \ \mathrm{cm/s} \ . \tag{9.7}$$

Integrating over the flow, we obtain the photosphere radius

$$R_{\mathrm{ph}} = \frac{\dot{M} \sigma_{\mathrm{T}}}{4 \pi v_{\mathrm{w}} m_{\mathrm{p}}} \approx 2 \times 10^{12} \dot{M}_{-4}^{3/2} m_x^{-1/2} \ \mathrm{cm} \ , \tag{9.8}$$

and the temperature

$$T_{\mathrm{ph}} \approx 1.5 \times 10^4 \left(\frac{L}{L_{\mathrm{Ed}}}\right)^{1/4} \dot{M}_{-4}^{-3/4} m_x^{1/4} \mathrm{K} \ . \tag{9.9}$$

It is obvious that to a remote observer, the supercritical accretion disk will look like a supergiant star with an intense loss of matter. All hard radiation will be absorbed by the optically thick outflowing shell.

Let us now recall that a considerably supercritical flux of matter will fall down on the surface of the neutron star. In this case, it is quite natural to expect that the regime may become highly nonstationary. The characteristic variability time will then be given by (Lipunov 1982b)

$$t_{ff} \approx \frac{R_A^{3/2}}{\sqrt{2GM_x}} \approx 6 \times 10^{-2} R_8^{3/2} m_x^{-1} \text{ s} \tag{9.10}$$

and the characteristic frequency by

$$\nu_{ff} \approx 30 \mu_{30}^{-2/3} m_x^{-1/3} \text{ Hz} . \tag{9.11}$$

In this case, a part of the matter is shed from the polar column at a velocity of the order of the parabolic velocity on the surface of the neutron star:

$$v_j \approx 1.2 \times 10^{10} m_x^{1/2} R_6^{-1/2} \text{ cm/s} .$$

The jet collimation angle is of the order of the angular opening of the polar column [formula (6.45)]:

$$\varepsilon_j \approx 16.0° \mu_{30}^{-2/9} m_x^{1/18} R_6^{1/2} . \tag{9.12}$$

It should be recalled that, as before, R_6 is the radius of the neutron star expressed in units of 10^6 cm. Of course, the most favorable condition for observing such jets would be the alignment of the magnetic axis and the rotational axis of the neutron star.

9.1.2 Neutrino Pulsar

Thus a superaccretor is a powerful source of neutrinos. It is remarkable that the neutrino radiation can be modulated by the rotational period of the neutron star.

It should be recalled that the effective generation of neutrinos sets in at temperatures of 5×10^9 K (Sect. 6.2). The confinement of such a hot plasma requires a very high magnetic field strength ($\lesssim 10^{13}$ Oe) near the poles of the neutron star. In this case, the main process will be the annihilation of electron-positron pairs: $e^+ + e^- \rightarrow \nu_e + \tilde{\nu}_e$.

The anisotropy of neutrino radiation is due to two reasons, the absorption of neutrinos by the neutron star itself, and the nonsymmetric momentum distribution of neutrinos formed in a strong magnetic field.

Let us first consider the absorption in the neutron star. For a cold neutron star, the neutrino absorption will be determined mainly by three processes (Shapiro, Teukolsky 1985), scattering by free nucleons, absorption by nucleons, and partial scattering by electrons. The characteristic neutrino energy ε_ν at temperatures of $\sim 5 \times 10^9$ K (Sect. 6.2) is of the order of 1 MeV. The absorption cross section for a neutrino with such an energy is of the order of the fundamental cross section determined by the weak interaction constant σ_0:

$$\sigma_0 = \frac{4}{\pi}\left(\frac{\hbar}{m_e c}\right)^{-4}\left(\frac{G_F}{m_e c^2}\right)^2 \approx 1.76\times10^{-44}\,\mathrm{cm}^2 \ .$$

In this case, the optical thickness of the neutron star is found to be larger than unity

$$\tau_\nu \approx \frac{\sigma_0 \varrho R_x}{m_p} \approx 10\varrho_{15}R_6 \ ,$$

where $\varrho_{15} = \varrho/10^{15}\,\mathrm{g/cm}^3$ is the mean density of the neutron star. In the course of its rotation, the star will periodically obstruct from the observer the bands where neutrino generation takes place.

The other factor responsible for the directionality of neutrino radiation is the anisotropy of neutrino generation in a strong magnetic field (see, for example, Ternov et al. 1982). The combined action of these two mechanisms and the rotation of the neutron star are responsible for the phenomenon of neutrino pulsars.

The neutrino luminosity of a pulsar has the order of magnitude $L_\nu \approx 6\times10^{39}\mu_{30}^{4/9}m_x^{8/9}R_6^{-1}$ erg/s. The corresponding energy flux and the number of particles on the Earth are given by

$$F_\nu \approx 5\times10^{-5}\mu_{30}^{4/9}m_x^{8/9}R_6^{-1}d_1^{-2}\,\mathrm{erg/cm}^2\,\mathrm{s} \ ,$$

$$N_\nu \approx 30\mu_{30}^{4/9}m_x^{8/9}R_6^{-1}d_1^{-2}\,\mathrm{cm}^{-2}\,\mathrm{s}^{-1} \ ,$$

where d_1 is the distance from the source in kpc. In comparison, the solar neutrino flux with this energy is of the order of $10^{11}\,\mathrm{cm}^{-2}\,\mathrm{s}^{-1}$. And yet these objects are the most powerful constant sources of neutrinos in the Galaxy. The period of a neutrino pulsar is estimated by the expression $p\approx p_A$.

Of course, the discovery of neutrino pulsars is still only a wish. This, however, does not prevent us from conjuring up an image of such objects. The phenomenon of neutrino pulsars offers a unique opportunity of probing the interiors of neutron stars; neutrino radiation could play the part of a sort of X-ray camera.

9.1.3 Spin-up and Spin-down

The change in the angular momentum of a neutron star in the supercritical disk accretion regime is described by the equation

$$\frac{dI\omega}{dt} = \dot{M}_{cr}\sqrt{GMR_A} - \kappa_t\frac{\mu^2}{R_c^3} \ .$$

The neutron star tends to its equilibrium period which is found to be of the order of the critical period p_A:

$$p_{eq} \approx \frac{1}{\kappa_t^{1/2}}p_A \approx 0.26\mu_{30}^{2/3}m_x^{-2/3}\kappa_t^{-1/2}\,\mathrm{s} \ .$$

If the initial period of the neutron star is $p \gg p_{eq}$, the relaxation time to the equilibrium state is approximately equal to $t_{rel} \approx 300 I_{45} m_x^{1/3} \mu_{30}^{-4/3}$ years. This time is much smaller than the lifetime of the binary system at the stage of intense mass exchange $(\sim 10^4 - 10^5$ years). Consequently, the neutron star always manages to attain equilibrium.

The mass of a superaccretor increases rapidly, thus making the collapse of a neutron star into a black hole quite likely. The characteristic mass-doubling time is given by

$$t_M = \frac{M}{\dot{M}_{cr}} \approx 2 \times 10^6 m_x^{10/9} R_6 \mu_{30}^{-4/9} \text{ years} \ .$$

It can be seen that the higher the magnetic dipole moment of the star, the more probable the collapse (Lipunov 1982b). For $\mu_{30} \approx 1 - 10$ and a supercritical accretion period $\sim 10^4 - 10^5$ years, the mass of the neutron star manages to increase by $1 - 10\%$. If the initial masses of neutron stars are distributed uniformly in the interval $M_{Ch} < M_x < M_{OV}$, about $1 - 10\%$ of the neutron stars will be transformed into black holes at the supercritical accretion stage.

9.2 Superejectors and Superpropellers

In principle, a neutron star may retain its initial rapid rotation up to the instant when the Roche lobe is filled by the normal star. Consequently, it may turn up either in the ejection stage, if the stopping radius R_{st} is larger than the radius of the light cylinder (superejector), or in the propeller stage, if the stopping radius is smaller than the radius of the light cylinder but larger than the corotation radius R_c (superpropeller). Outwardly, this situation will not differ significantly from the one considered in the last section. The outflowing optically thick envelope will completely absorb hard X-rays as well as the relativistic particles. However, the situation will be quite different in the inner parts of the disk. At the stopping radius, matter will come to halt and a sort of bubble filled with static and free electromagnetic fields and relativistic particles will be formed within the envelope. So far, this situation has not been investigated at all. Attempts have been made to construct a qualitative picture of jets from such a bubble however.

We would·like to mention that in the course of evolution, a part of neutron stars clearly passes through the superejection and superpropeller stages (Chap. 11). This points towards the need for theoretical investigations of such exotic objects.

9.3 Is SS 433 a Superaccretor?

The unique spectral properties of SS 433 discovered by Margon et al. (1979) drew the attention of researchers and initiated a large number of observational

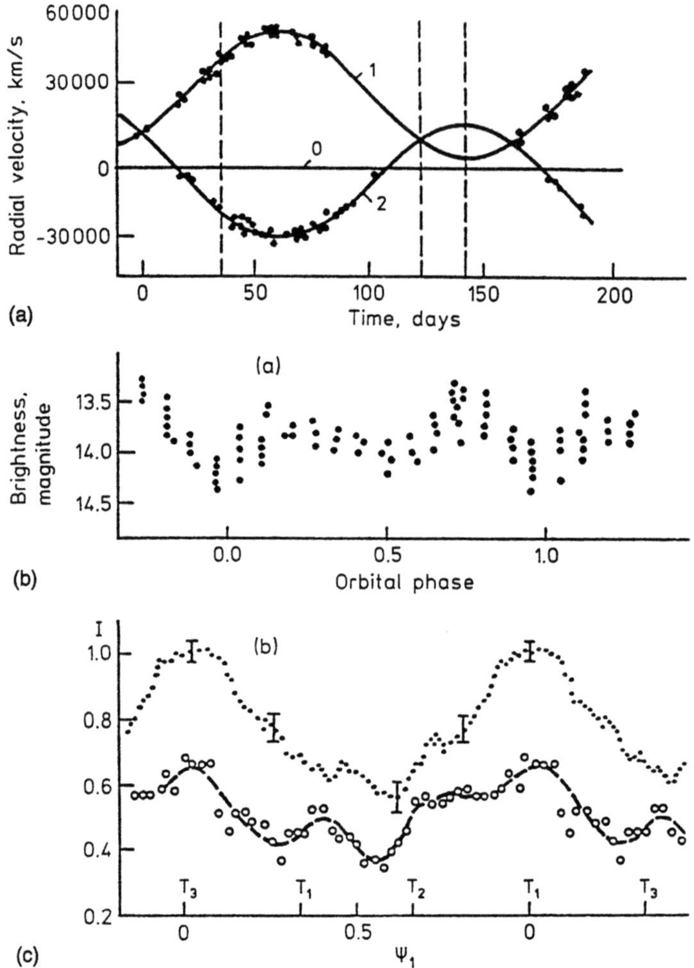

Fig. 9.2a–c. Spectral and photometric properties of SS 433. (a) Curves of radial velocities of relativistic emission, (b) orbital curve of brightness, (c) precession curve of brightness

and theoretical works (Fig. 9.2a). As a result of spectroscopic and photometric observations, it was established that SS 433 is a close binary system with a period of ~13.1 days (Crampton et al. 1980; Gladyshev et al. 1979; Cherepashchuk 1981) (Fig. 9.2b), ejecting two gas jets in opposite directions with a velocity $v_j \approx 80000$ km/s. Apparently, one of the components of the binary system is an OB-supergiant losing mass at a rate of ~$10^{-4} M_\odot$ per year (Cherepashchuk 1981; Cherepashchuk et al. 1982), while the second component is an unusual object with an optical luminosity of $10^{39} - 10^{40}$ erg/s.

It is astonishing to note that the direction of the jets changes with a period of 164 days. This second period is called the precession period in the literature.

It should be noted that both the binarity of SS 433 and the jet precession were predicted soon after the discovery of relativistic emission lines (the former was predicted by Shklovskii in 1981 and the latter by Milgrom in 1979).

The "narrowness" of relativistic emission points towards a small jet collimation angle of $\sim 4°$. The total brightness of SS 433 correlates with the position of the relativistic jets. At the instant when the jets are directed at us (actually one of them is directed away from us), the system has maximum brightness (Fig. 9.2c).

A large number of theoretical works presenting different models of SS 433 have appeared. We believe that the entire body of observational data available to date allows us to give preference to those models assuming that the supercritical accretion to the relativistic star is mainly responsible for the phenomena taking place in this system (see, however, Kundt 1981, 1985). Among all the models proposed so far, even models of this type are quite numerous (see the review by Margon 1984).

But which of the relativistic stars (a neutron star or a black hole) is present in SS 433? The mass function obtained by Crampton et al. (1980) from the 4686 Å line of He II is 10.1 $M_\odot$. This speaks in favor of the black hole. However, the ratio of the component masses has not yet been determined exactly and it is too early to draw a final conclusion.

Before going into the details of the model, let us show how the model of the supercritical disk explains the observed phenomena in SS 433. The following arguments were put forth by Lipunov and Shakura (1982). Firstly, we note that in spite of the enormous optical luminosity of the peculiar object, its X-ray luminosity is extremely low ($\sim 10^{35}$ erg/s). However, this is just what one would naturally expect in the case of supercritical accretion described above. Suppose that the rate of accretion to a relativistic star is $\dot{M} \simeq 10^{-4} M_\odot$/year. In this case, we obtain from (9.8) the value of $\sim 10^{12}$ cm for the radius of the photosphere of matter flowing out of the disk. This is comparable with the size of the binary system. According to (9.7), we find that the outflow velocity $v_{\rm sh} \approx 10^8$ cm/s. This is also in agreement with the observed pattern of steady-state emission.

In view of a partial dissipation of the kinetic energy of jets, it can be expected quite naturally that the temperature of the envelope must be higher at the jet propulsion regions. In other words, the envelope flowing out of the disk is a quasispherical star with two hot spots precessing with a period of 164 days. This picture explains the increase in the brightness of SS 433 at the instant when the jets are directed at us (Fig. 9.2a). Moreover, it explains the exact coincidence of the brightness maximum with the instant T_3. If the photometric data could be attributed to the precession of external parts of the disk, it would be impossible to understand such a coincidence. Indeed, let us imagine that the photometric behavior is due to different orientations of the outer parts of the disk. In this case, the position of the inner parts collimating the jets would lag behind by the time of radial motion of matter in the disk:

$$t_r \approx \frac{T}{2\pi\alpha} \left(\frac{R}{H}\right)^2 ,$$

where α is the turbulence parameter, T the period of the binary system, and R/H the ratio of the outer radius of the disk to its thickness. It can be expected that $R/H \approx 10$. Since $\alpha < 1$, we find that $t_r \gtrsim 200$ days. It would be hard to understand why the maximum brightness of SS 433 is observed at the instant of maximum separation between relativistic emissions. The light curve synthesized with the precession period by using the model in which the normal star fills the Roche lobe and the neighboring component is a quasispherical star with two hot spots, is in complete agreement with the observational data.

It was assumed in the works of Lipunov et al. (1982) and van den Heuvel et al. (1983) that a superaccreting neutron star is a relativistic star. In this model, the appearance of relativistic jets may be associated with the ejection of matter from the polar column (Lipunov, Shakura 1982), or with the collimation of the outflowing matter by the internal parts of the accretion disk (Shakura, Syunyaev 1973). In the former case, the angle and velocity of jets can even be calculated [see (9.12)].

Apart from the superaccretor model, the ejector model and the propeller model (Shklovskii 1979) were also proposed.

9.4 Other Candidates

SS 433 is not the only galactic object from which relativistic jets are observed. The first object in this subgroup was the brightest source on the X-ray sky, the source Sco X-1. It was borne out by radio observations that in addition to Sco X-1, which itself is a nonstationary radiosource, two weak radiosources are situated symmetrically about it (Hjellming, Wade 1971). This phenomenon is similar to that observed in quasars and active galactic nuclei.

Let us try to find the common properties of SS 433 and Sco X-1. Both are varying radiosources and both exhibit relativistic jets [true, it is a cold gas moving at a velocity of about $80\,000$ km/s in the case of SS 433, while for Sco X-1 it is apparently a cloud of relativistic particles which practically does not move (Fomalont et al. 1983)]. Both sources are at the Eddington luminosity limit for stars $(1 - 10 M_\odot)$.

The difference between the two sources lies in their optical components. The optical component of SS 433 is a massive OB-star while for Sco X-1 it is a low-mass star of a late spectral type. Could this difference be decisive? The low-mass star which fills the Roche lobe flows out much more slowly (by a factor of about 1000) than the star in SS 433. Can it be stated that Sco X-1 is a superaccretor but, unlike Sco X-1, it has an accretion rate that does not exceed the critical limit $\dot{M}_{cr}$ [formula (9.2)] as strongly (say, only by a factor of ten)? For $\dot{M}_c \approx 10^{-7} M_\odot$/year, the photosphere radius of the outflowing matter R_{ph} is about 10^8 cm [see (9.8)]. Obviously, all radiation will lie in the X-ray range (as is actually observed). The outflow velocity of the envelope is $v_{sh} \approx 30\,000$ km/s [formula (9.7)]. The ejection of matter at such a high velocity inevitably leads to the formation of shock waves and the generation of relativistic particles and, consequently, to the emission of synchrotron radio waves.

It is interesting to note that many bright X-ray bulge sources are also radiosources (Hjellming 1978). It is even more astonishing that, as a rule, noisars (Sect. 6.14) are sources of nonthermal radio waves (Bradt et al. 1979). Are these sources weak (in the sense of a slight increase over $\dot{M}_c$) superaccretors or do they go randomly from the accretion to the superaccretion stage? In this case, the emergence of "noise" can be attributed to random pulsations in the matter flowing out of the disk in the vicinity of the X-ray photosphere. The characteristic pulsation time is estimated as

$$t_n \approx \frac{R_{ph}}{v_{ph}} \approx \frac{R_{ph}}{v_{sh}} \approx 2 \times 10^{-2} \dot{M}_{-7}^2 m_x^{-1} \text{ s} \ .$$

This corresponds to a frequency $v_n \approx 50 \dot{M}_{-7}^{-2}$ Hz.

Let us estimate the optical thickness from Thomson scattering in the outflowing envelope,

$$\tau \approx \frac{\kappa_T \dot{M}}{4\pi v_{sh} R_{in}} \approx \frac{R_{ph}}{R_{in}} \ . \tag{9.13}$$

Putting the inner radius equal to the spherization radius $R_{in} \approx R_s \approx 10^7 \dot{M}_{-7}$ cm, we obtain an optical thickness inside the envelope of $\tau \approx 10 \dot{M}^{1/2} m_x^{-1/2}$. Assuming that the X-ray luminosity $L_x \propto e^{-\tau}$, we find from (9.13) that $L_x \propto e^{-v^{-1/4}}$. The higher the pulsation frequency, the higher the luminosity.

10. Stars with an Anomalously Low Value of Gravimagnetic Parameter

It should be recalled that the gravimagnetic parameter is defined as the parameter $y = \dot{M}/\mu^2$. In this chapter, we shall discuss neutron stars which either have a high magnetic field, or are surrounded by highly rarefied matter. In both these cases, y was found to be small. The analysis of the interaction of such neutron stars with the surrounding matter and of possible astrophysical manifestations is contained in only two or three works. The inclusion of this less investigated field in a separate chapter mainly reflects the author's desire to draw the attention of those engaged in the investigation of neutron stars to this problem.

10.1 Georotators

The condition of slow rotation of a magnetized star is necessary, but not sufficient, for the existence of accretion. If the magnetosphere radius R_m is larger than the gravitational capture radius R_G, gravitation at the magnetosphere boundary will be insignificant and accretion of matter will be impossible (Illarionov, Syunyaev 1975; Lipunov 1982c). This is the situation observed in the vicinity of the Earth's magnetosphere. That is why this regime is called the georotator regime.

Let us discuss the structure of the magnetosphere of the georotator. Two situations are possible. If the velocity of a star relative to the medium is essentially supersonic, i.e., if $v \gg a_s$, the magnetosphere will be similar to the Earth's magnetosphere (Fig. 10.1 a). Conversely, if the neutron star moves slowly ($v \ll a_s$) and at the same time the capture radius is found to be much smaller than the magnetosphere radius ($R_G < R_m$), the magnetosphere boundary is described by the solution first obtained by Cole and Huth (1959) and Medgley and Davis (1962) (Sect. 5.5). It should be recalled that these solutions correspond to a constant plasma pressure $P = $ const (zero gravity!).

Investigation of the stability of the boundary of such magnetospheres shows that they are stable in the ideal MHD approximation (Akasofu, Chapman 1972). This can be easily verified by using the variational principle described in Chap. 6.

Let us now verify the real possibility of the emergence of G-type neutron stars (georotators). The condition $R_A > R_G$ can be written in the form

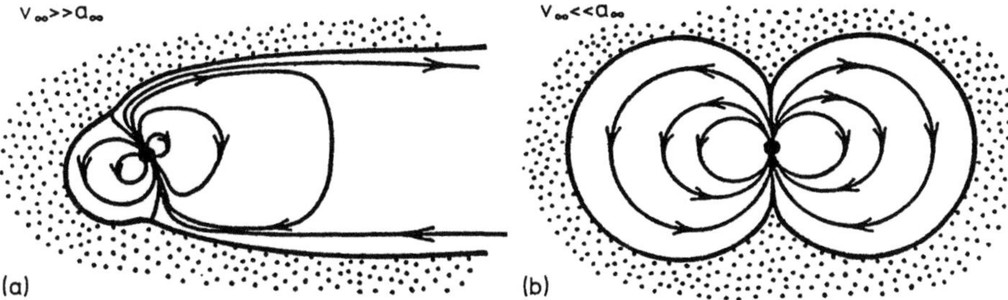

Fig. 10.1. The magnetosphere of a georotator for (**a**) rapid, (**b**) slow motion of the neutron star

$$y < y_G \approx \frac{v_\infty^7}{16(GM_x)^4} \approx 2 \times 10^{-50}\, v_8^7 m^{-4} \tag{10.1}$$

or as a condition characterizing the accretion rate of matter formally falling into the capture radius,

$$\dot{M}_c < \dot{M}_G \approx 3 \times 10^{-16}\, v_8^7 m_x^{-4} \mu_{30}^2 M_\odot/\text{year} \ . \tag{10.2}$$

Near the stars outflowing at a high speed ($v_8 \approx 2-3$) and for $\mu_{30} \approx 10-100$, the emergence of geosimilar magnetospheres is a real possibility. For example, this may be one of the reasons behind the absence of X-ray sources paired with WR-stars (Lipunov 1982 e).

In the case of isolated neutron stars, it is convenient to carry out the analysis by using (3.39):

$$\dot{M}_c \approx 7 \times 10^7\, v_7^{-3} \varrho_{-24} m_x^2 \text{g/s} \ , \tag{10.3}$$

where ϱ_{-24} is the density of the interstellar medium. If follows from (10.2) that the georotator regime is achieved when the neutron star velocity satisfies the condition

$$v_\infty > 300 \varrho_{-24}^{1/10} m_x^{3/5} \mu_{30}^{-1/5} \text{km/s} \ . \tag{10.4}$$

It should be noted that higher spatial velocities are also encountered in radiopulsars. Such neutron stars will be transformed into georotators as a result of spin-down. The evolutionary track can be presented in the form of the chain $E \rightarrow P \rightarrow G$. The probability of the emergence of such magnetospheres is especially high in the case of isolated white dwarfs.

In order to investigate the physical processes emerging in the magnetospheres of G-type stars, the experience gained in investigation of the terrestrial magnetosphere may turn out to be quite useful [see, for example, the monograph by Akasofu and Chapman (1972)].

10.2 Binary Magnetic Systems (Magnetors)

A qualitatively new regime may occur in a close binary system for a star with a small value of the gravimagnetic parameter. For a strong magnetic field or a low outflowing rate of the adjacent star, the Alfvén radius may turn out to be larger than not only the gravitational capture radius but also the separation between the stars. In this case, the adjacent star will be surrounded by the magnetosphere of the compact star (Fig. 4.9). Such a regime was first considered during an analysis of the nature of accretion in the white dwarf system AM Her (Mitrofanov et al. 1977). Such binary systems with white dwarfs are now called polars.

Unlike systems with white dwarfs, whose magnetic dipole moment can attain values up to $\sim 10^{35}\,\mathrm{Oe\,cm^3}$, the emergence of magnetic systems with neutron stars requires very specific conditions. In massive binary systems, such a phenomenon is not possible in general (even for neutron stars with anomalously strong magnetic fields: $\mu \approx 10^{32}\,\mathrm{Oe\,cm^3}$). The emergence of neutron-star magnetors is possible only in supercompact binary systems with degenerate components (after all, the Alfvén radius of even an isolated neutron star does not exceed $\sim 10^{10}\,\mathrm{cm}$). Such an exotic situation was analyzed by Nulsen and Fabian (1984) who studied the electrodynamics of a system formed by two neutron stars.

Investigations of binary magnetic systems were carried out by Chiapetti et al. (1980); Lamb et al. (1983), Campbell (1983); Andronov (1984, 1986); and Lamb et al. (1985). Some of the effects emerging in binary magnetic systems are: (1) falling of matter on just one of the poles of the accreting star; (2) the emergence of a magnetic valve near the inner Lagrangian point; and (3) induction of currents on the companion and magnetic synchronization.

11. Evolution of Stars

In this chapter, we shall analyze the evolution of neutron stars. The astrophysical manifestation of a neutron star is mainly determined by the nature of its interaction with the surroundings. This gives rise to a new concept of stellar evolution, involving a slow variation of the regime of interaction of the magnetized star. However, the neutron stars are the final product in the evolution of normal stars. Hence before considering the evolution of neutron stars, we shall briefly describe the main results of the theory of evolution of normal stars.

11.1 Normal Stars

The theory of evolution of normal stars has been discussed in a large number of books and papers (see, for example, Zel'dovich, Novikov 1971; Yungel'son, Masevich 1983). We shall first consider the evolution of single stars.

11.1.1 Single Stars

The evolution of a young star begins at the instant of nuclear burning of hydrogen. This is the most prolonged stage in stellar evolution. The lifetime of the star at the hydrogen burning stage is estimated by the following approximation formula (Yungel'son, Masevich 1983):

$$\log t_{\mathrm{H}} \approx 9.9 - 3.8 \log m_0 + \log^2 m_0 \, , \tag{11.1}$$

where m_0 is the mass of the normal star expressed in units of mass of the Sun. Formula (11.1) describes quite correctly the results of numerical computations for $1 \lesssim m_0 \lesssim 10^2$. To quote just two numerical values, we can state that the lifetime of a star whose mass is equal to the mass of the Sun is 10 billion years and that of a star having a mass $10\,M_\odot$ is 50 million years. The larger the mass of the star, the faster its evolution. As hydrogen burns at the center of the star, the latter shifts slightly upwards along the main sequence. The variation of its luminosity and temperature are associated with a slow variation of the chemical composition of the core of the star as hydrogen is gradually converted into helium until all of it burns out. This happens first at the center of the star where the density and temperature are maximum. A helium core is thus formed, and the temperature in this core is not sufficient for the burning of helium. Around the core, however, hydrogen continues to burn in a spherical layer. The helium core is thus confined inside a source of energy and a constant temperature is established in it.

Upon the appearance of a layered source, the star leaves the main sequence. The mass of the helium core gradually increases and a nuclear reaction involving the combustion of helium begins in the core at a time depending on the mass of the star. This reaction involves the fusion of three helium nuclei (3α-reaction). The increased luminosity of the star leads to the emergence of convection in the envelope surrounding the core.

In general, the temperature at the center of the helium core in a low-mass star is so low that nuclear reactions do not occur in it. However, fusion in massive stars continues right up to the formation of iron group elements. It is assumed that the convective envelope of the stars of moderate mass is cast away in the form of a planetary nebula while the residual core is transformed into a white dwarf of a particular chemical composition. Calculations show that in stars with an initial mass $\sim 8-10\,M_\odot$, a degenerate core Mg-O-Ne is formed, and ultimately leads to the formation of a corresponding white dwarf. The ancestor-descendent diagram presented in Fig. 11.1 shows the ultimate fate of stars of different masses. The important point for the subsequent analysis is that depending on a critical value of the initial mass, stars can be divided into two categories: if the mass of a star is below this critical value, the end product of its evolution will be a white dwarf. Otherwise, the mass of the core

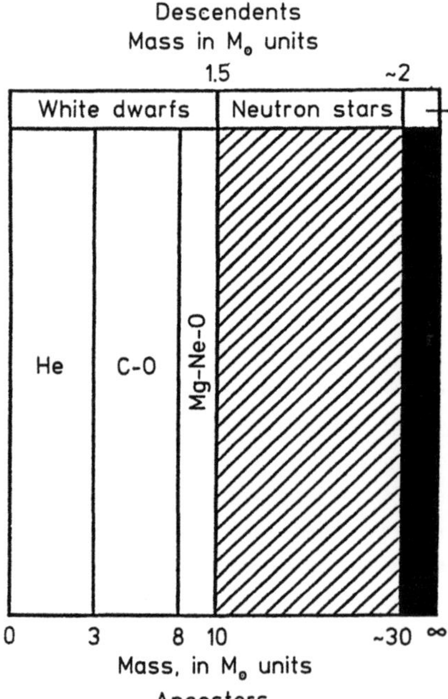

Fig. 11.1. The "ancestor-descendent" diagram. The initial mass of the ancestor star is shown on the lower axis. The mass of the degenerate remnant and its chemical composition (white dwarfs, neutron stars, black holes) are shown on top. The fact that some ancestors may not produce descendents (complete dispersion of a white dwarf) has not been taken into consideration

exceeds the Chandrasekhar limit and a neutron star or even a black hole may be formed as the end product of the star evolution. Stars capable of producing neutron stars or black holes are called massive. The critical mass value is not known to a high degree of accuracy and varies between 8 and $12\,M_\odot$. For the sake of convenience of subsequent analysis, we shall call a star massive if its mass exceeds $10\,M_\odot$. Since we are primarily interested in the emergence of neutron stars, we shall mainly consider massive stars in the following analysis. The lifetime of a massive star after its departure from the main sequence in the region of the supergiants is about $0.1\,t_{\rm H}$ [see (11.1)]. This time is mainly spent on the burning of helium and carbon while the heavier elements burn at a catastrophically fast speed (within a few minutes). For the discussion below, it is significant that as a star leaves the main sequence, the rate at which it loses its mass increases. For massive stars, it increases from $\lesssim 10^{-8}\,M_\odot$ per year on the main sequence to $\sim 10^{-5}-10^{-6}\,M_\odot$ per year at the blue supergiant stage.

Hence in stars with a mass $\gtrsim 10\,M_\odot$, a core consisting of iron group elements is formed after the burning of carbon followed by neon, oxygen and silicon. It is important that the mass of the core exceed the Chandrasekhar limit (for iron, $M_{\rm cr} \approx 1.2\,M_\odot$). It was shown in Chap. 2 that the loss of core stability is associated with the fall of the effective polytropic exponent γ to below its critical value $\gamma = 4/3$. The direct reasons behind the decrease in the adiabatic exponent are cooling due to neutrino emission, GTR effects and, the most important of all, the neutronization of matter. The hydrodynamic calculations for collapse are presented in the work by Imshennik and Nadezhin (1982). We only note here that apart from the difficulties inherent in the idealized spherical-symmetry problem on collapse, a significant role is played by the rotational effects and magnetic fields (Bisnovaty-Kogan 1970).

11.1.2 Binary Stars

Qualitatively new evolutionary effects are observed in binary systems. First indications of the different path followed by the stellar evolution in binary systems came from Parenago (1950) who implicated the Algol paradox. In the β-Persei system, the star with a smaller mass (subgiant) has considerably surpassed its more massive companion, a star belonging to the main sequence, in evolution. There are no reasons to doubt the statement that both stars were formed at the same time. However, a sharp contradiction with the main results of the theory of evolution of single stars is observed: the more massive a star, the sooner it leaves the main sequence. This contradiction was resolved by Crawford (1955) who assumed that the masses of stars are not conserved in the course of their evolution; an initially lighter star may subsequently become more massive. Mass exchange in binary systems was discussed in the pioneering works of Paczynski (1965) and Snezhko (1967). The current picture of the evolution of close binary stars (which is not a dogma and should rather be treated as a guideline) was created in the 1960s and 1970s, thanks to the excellent works of Paczynski (1971), Van den Heuvel and Heise (1972), Tutukov and Yungel'son (1973), and many other authors. According to the widely accepted definition, a close binary system is one in which a mass exchange between the components takes place. Following Van den Heuvel (1977), let us briefly consider the evolution of a massive close binary system. A significant advantage of this picture of evolution lies in the isolation of special evolutionary stages of a normal star in a binary system. In order to denote these stages, we shall use the classification proposed by Kornilov and Lipunov (1983).

I Stage. In the initial stages of evolution, the size of the stars is much smaller than the critical Roche lobe size and hence the stars practically do not feel one

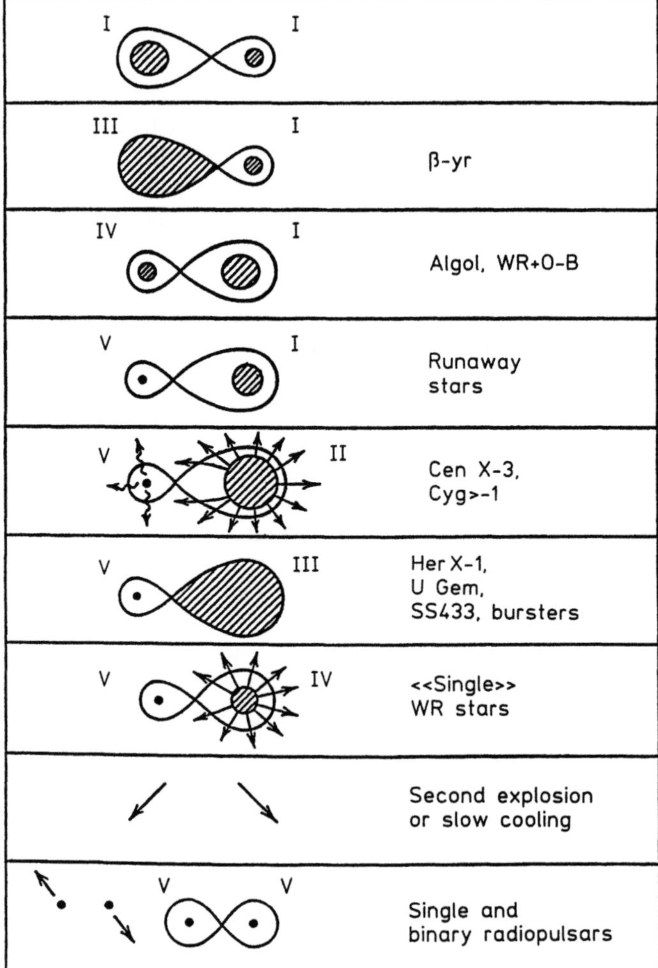

Fig. 11.2. Evolution of a close binary system

another's forces (Fig. 11.2). The duration Δt_I of this stage is approximately equal to the time t_H in which the nuclear burning of hydrogen occurs.

II Stage. The more massive of the two stars is the first to leave the main sequence and go up into the region of blue supergiants. As before, the star does not fill the Roche lobe. The duration of this stage is determined by the burning time in the layer source and is about one-tenth of t_H. Of course, the duration of this stage also depends on the separation between the stars.

III Stage. At a certain instant, the star fills the Roche lobe and begins to flow out to the companion. The outflow rate depends considerably on the ratio of the mass components, as well as on the distance between the stars. For a large ratio of the masses, the flow takes place on the thermal time scale,

$$t_{\text{th}} \approx \frac{GM^2}{RL} \approx 3 \times 10^7 m^2 \text{ years} . \tag{11.2}$$

After the masses of the stars equalize, the mass-exchange rate is determined by the nuclear time scale rather than the thermal one. It should be emphasized that a common envelope may be formed in the third stage of the star evolution (Paczynski 1976). The appearance of this envelope is associated with the fact that the thermal time of the lighter star is much larger than the thermal time of the outflowing star and for a large mass ratio the accreting star simply does not manage to attain the state of thermal equilibrium and receive the entire

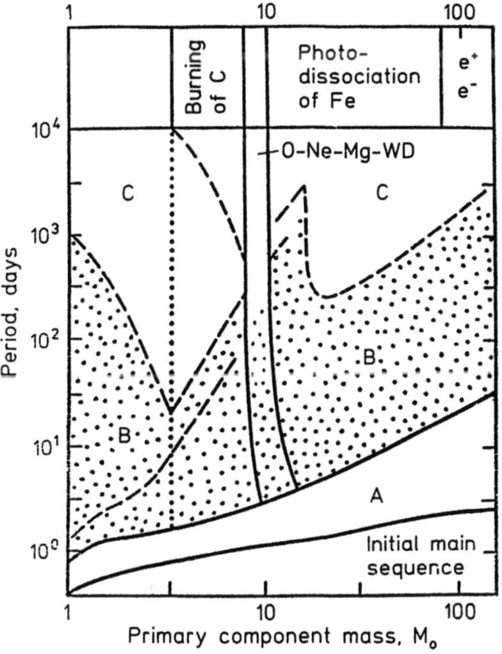

Fig. 11.3. Three types of close binary systems. The ordinate axis shows the logarithm of the binary system period. The diagram indicates the reactions taking place in the star at the instant when it completely fills the Roche lobe. The process responsible for the explosion of the star is mentioned at the top (Van den Heuvel 1983)

outflowing matter from the more massive component. The Roche lobe gets fill-ed later, the larger the period or the semi-major axis of the binary system. Hence short-period systems fill the Roche lobe even at the stage of hydrogen burning, longer-period systems fill the Roche lobe first at the stage of burning of the layered helium source, then at the helium-burning stage, and so on. Accordingly, binary systems are divided into three types, A, B and C. Sometimes a fourth type D is also introduced to formally describe systems in which there is no mass exchange. The evolution of such systems is essentially similar to that of single stars. The classification of the binary systems is presented in Fig. 11.3. The mass-exchange rate, and hence the exchange time, depends considerably on the type of the system. For systems of type A, the exchange occurs on the nuclear time scale of hydrogen burning. The exchange rate is at least ten times higher in systems of type B, while systems of type C exchange mass on the thermal time scale.

IV Stage. In the process of mass exchange, the envelope of a star flows over to the companion either completely (conservative exchange) or partially (non-conservative exchange). The mass of the remnant of the initially more massive component is roughly estimated after the exchange according to the formula (Tutukov 1980).

$$m \text{ (after exchange)} \approx 0.1\, m^{1.4} \quad \text{(before exchange)} . \tag{11.3}$$

The following relations are satisfied in the case of conservative exchange:

$$M_2 = M_{2,0} + M_{1,0} - M_1 \ ,$$
$$a_1 = a_0 \left(\frac{M_{1,0} M_{2,0}}{M_1 M_2} \right)^2 , \tag{11.4}$$

where a_0 and a_1 stand for the semi-major axis before and after the exchange. After the exchange, the star is left with a helium core of mass M_1. According to Paczynski (1965), Wolf-Rayet stars are just such remnants. In massive binary systems, $M_1 \approx 8\, M_\odot$. The lifetime of a helium star is determined by the nuclear burning time of helium (Tutukov 1980)

$$\Delta t_{\text{IV}} \approx t_{\text{He}} \approx 3 \times 10^6 m_1^{-0.7} \text{ years} . \tag{11.5}$$

V Stage. After the burning out of helium and heavier elements, an iron core is formed with a mass larger than the Chandrasekhar limit and collapses. In this case, a neutron star with mass $1.5 - 2 M_\odot$ is formed while the remaining mass is expelled from the binary system. As a result of a rapid expulsion of matter, the orbit becomes eccentric and the semi-major axis changes (even in the case of a spherical explosion). It should be recalled that the necessary condition for the disintegration of a system is that it should lose more than half its mass. Apparently, in the conservative picture considered by us, such a situation never arises since the lighter star is the one that explodes.

Thus, a binary system is formed by a normal star with mass $M_0 = M_2$ and a neutron star with mass m_x. In this case, the semi-major axis of the system is given by

$$a = a_1 \frac{1+q_*}{1+2q_*-q} \; ,$$

(11.6)

where $q_* = M_1/M_2$; $q = M_x/M_2$. This is followed by the onset of the second stage of evolution of the binary system: the initially lighter star now goes through all five stages described above.

The state of a neutron star is determined by its rotational velocity, magnetic field and, finally, by its potential accretion rate. Obviously, the latter is determined by the nature of flow from the normal star. Hence we shall briefly describe the approximate relations for the parameters of matter flowing from the adjacent star at different stages.

I Stage. The star loses matter in the form of spherically symmetric stellar wind. The flow rate is described approximately by the empirical relation (Chap. 3):

$$\dot{M}_0 = \alpha_w \frac{L_0}{v_\infty c} \; ,$$

(11.7)

where L_0 is the luminosity of the star, v_∞ is the stellar wind velocity at infinity, and $\alpha_w \simeq 0.2$ (Barlow, Cohen 1977). The characteristic values depend strongly on the mass of the star and vary between 10^{-9} and $10^{-6} M_\odot$ per year. The velocity at infinity is $v_\infty \approx 3 v_p$, where $v_p = \sqrt{2GM_0/R_0}$ is the parabolic velocity on the surface of the normal star. The empirical stellar wind velocity is normally written

$$v = v_\infty \sqrt{1-R_0/R} \; .$$

(11.8)

II Stage. The characteristics of the stellar wind in this stage are described by the same relations as in stage I.

III Stage. The flow of matter is in the form of a gas jet through the inner Lagrangian point. The flow rate and velocity are described by the approximate relations

$$\dot{M}_0 \approx \frac{M_0}{\Delta t_{III}}$$

(11.9)

$$v_\infty \approx v_{orb} \; ,$$

where v_{orb} is the orbital velocity in the binary system.

IV Stage. In order to describe the outflow from the helium star, we can use the stellar wind picture with the following parameters:

$$\dot{M}_0 \approx \frac{M_0}{t_{\text{He}}} \approx 3 \times 10^{-7} m_0^{1.7} M_\odot / \text{year} \ . \tag{11.10}$$

By the time of the second explosion, the binary system has the opposite mass ratio and the more massive star is the one to explode. The system disintegrates and two isolated neutron stars are formed.

Thus, we have considered in general terms the evolution of normal stars in binary systems. Owing to the change in roles, the system does not disintegrate after the first explosion. A natural consequence of this is the emergence of binary systems with relativistic components. The normal component in such a system successively passes through four stages. However, the paths of evolution of its companion, that is, the neutron star, are more diverse.

11.2 Evolution of Neutron Stars

The evolution of a neutron star involves a slow variation of the regimes of its interaction with the surrounding medium. Such an approach to the evolution was developed in the 1970s by Shvartsman (1970), Illarionov and Syunyaev (1975), Bisnovaty-Kogan and Komberg (1974), Shakura (1975), Wickramas-inghe and Whelan (1975), Lipunov and Shakura (1976), Savonije and Van den Heuvel (1977), and others. These works mainly considered three regimes: ejection, propeller and accretion. The classification of neutron stars was completed in the early 1980s and first computations concerning the evolution of neutron stars in binary systems were made by taking into account the evolution of the normal star (Lipunov 1982a,b; Kornilov, Lipunov 1983a,b; Lipunov 1984).

The complete classification of neutron stars includes eight types (Chap. 4): E, P, A, SE, SP, SA, G, and M. The appearance of a neutron star at any stage is mainly determined by three parameters, the dipole magnetic moment μ, the angular velocity ω (or the rotational period $p = 2\pi/\omega$), and the potential accretion rate $\dot{M}_c$ (or the potential luminosity $L = \dot{M}_c GM_x/R_x \approx 0.1\dot{M}_c c^2$). The evolution of the neutron star can be treated as a motion in three-dimensional space formed by the parameters μ, $\dot{M}_c$ and ω. Actually, there exists an additional parameter, the velocity v_∞ of the neutron star, and hence the space in which the neutron star moves has generally more than three dimensions. But the main evolutionary factors are just these three parameters, the most significant of which is the angular velocity of the star.

11.2.1 Evolution Equation

An analysis of the nature of interaction of a magnetized star with the surrounding plasma considered in Chaps. 4−10 allows us to write the approximate evolution equation for the angular momentum of a neutron star in the universal form (Lipunov 1982a)

Table 11.1. Parameters of the evolution equation of a neutron star

Parameter	Regime					
	E, SE	P, SP	A	SA	G	M
$\dot{M}$	0	0	$\dot{M}_c$	$\dot{M}_c \dfrac{R_A}{R_s}$	0	$\dot{M}_c$
κ_t	$\sim 2/3$	$\lesssim 1/3$	$\sim 1/3$	$\sim 1/3$	$\sim 1/3$	$\sim 1/3$
R_t	R_1	R_m	R_c	R_c	R_A	a

$$\frac{dI\omega}{dt} = \dot{M}k_{su} - \kappa_t \frac{\mu^2}{R_t^3} \, , \tag{11.11}$$

where k_{su} is the specific angular momentum applied to the accretion matter. This quantity is given by

$$k_{su} = \begin{cases} \sqrt{GM_x R_d} & \text{disk accretion} \\ \eta_t \Omega R_G^2 & \text{accretion without disk} \end{cases} \tag{11.12}$$

where R_d is the radius of the inner edge of the disk, Ω is the rotational frequency of the binary system, and $\eta_t \approx 1/4$ (Illarionov, Syunyaev 1975). The values of the dimensionless factor κ_t, characteristic radius R_t, and the accretion rate $\dot{M}$ in different regimes are presented in Table 11.1.

The evolution equation (11.11) is approximate. In particular, the situation concerning propellers and superpropellers is not clear (Chap. 7). In Table 11.1, R_m is the size of the magnetosphere whose value at the propeller stage is not known accurately, and may differ significantly from the standard expression for the Alfvén radius.

In the general case, the right-hand side of (11.11) is a certain function of the frequency ω and time. In many cases, the moment of inertia I may be taken out of the derivative. The evolution equation then assumes the form of a first-order linear differential equation

$$\frac{d\omega}{dt} = F(\omega, t) \, . \tag{11.13}$$

It is convenient to introduce the scalar potential $V(\omega)$ (Chap. 7; Lipunov 1987 a):

$$F(\omega) = -\nabla_\omega V \, . \tag{11.14}$$

In this case, the evolution of the magnetized star acquires a simple geometrical interpretation: the star tends to evolve to a state corresponding to the minimum potential $V(\omega)$.

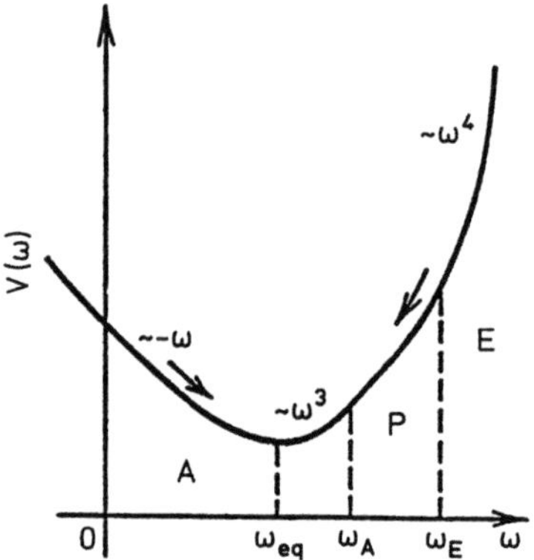

Fig. 11.4. Scalar potential describing the evolution of a magnetized neutron star in a binary system

Suppose that all the quantities on the right-hand side of the evolution equation are either independent of time or vary very slowly (more slowly than the rotational frequency). In this case, the potential for ejecting and accreting stars can be presented in the form

$$V(\omega) = A_1 \omega^4 + \text{const} \qquad\qquad \text{for E and SE ,} \qquad (11.15)$$

$$V(\omega) = \begin{cases} -A_2\omega + A_3\omega^3 + \text{const}, & \omega \geqslant 0 \\ -A_2\omega - A_3\omega^3 + \text{const}, & \omega < 0 \end{cases} \qquad \text{for A and SA ,} \qquad (11.16)$$

where A_1, A_2 and A_3 are positive constants. For propellers and super-propellers, we can expect the power dependence on frequency

$$V(\omega) = A_4 \omega^n + \text{const} \qquad\qquad\qquad (11.17)$$

for P and SP.

Figure 11.4 shows the qualitative behavior of the potential for different regimes of interaction with the surrounding plasma.

The neutron star tends to assume a state corresponding to the minimum $V(\omega)$. For accretors and superaccretors, the accretion of matter with an angular momentum leads to a minimum corresponding to a nonzero equilibrium frequency $\omega_{eq} = 2\pi/p_{eq}$. Such a situation is realized in binary systems where the accreting matter always has an angular momentum (Chap. 6).

The possible dissipation of the magnetic field of the neutron star or the process of alignment of the magnetic axis with the rotational axis may be taken into account through the equation

$$\frac{d\mu}{dt} = -\frac{\mu}{\tau} , \qquad\qquad\qquad (11.18)$$

where τ is the characteristic time of dissipation of the magnetic field.

The solution of the evolution equation (11.11) at the ejection stage taking dissipation into account is presented in Chap. 8. The general solution of the evolution equation taking into account the possible dissipation of the magnetic field at the accretion stage is described by Lipunov and Postnov (1987).

11.2.2 Statistical Description of the Ensemble of Neutron Stars

About 20 accreting neutron stars have been discovered so far, while the number of ejecting stars exceeds three hundred. In order to describe the properties of these objects, we must introduce the distribution function and write its evolution equation (Lipunov 1987a).

Let us consider an ensemble of stars, each of which is described by a certain set of parameters: $x_1, x_2, \ldots, x_i$. Let the function $\varphi(x_1, x_2, \ldots, x_i)$ describe the probability with which a randomly chosen star will have parameters in the interval $(x_1, x_1 + dx_1; x_2, x_2 + dx_2; \ldots; x_i + dx_i)$. The number dn of neutron stars whose parameters lie in this interval is given by

$$dn = \varphi(x_1, x_2, \ldots, x_i; t)\, dx_1 dx_2 \ldots dx_i dt \ . \tag{11.19}$$

The function φ must be normalized to the total number N of stars which in the general case can be a function of time:

$$\int\limits_{x_1} \int\limits_{x_2} \ldots \int\limits_{x_i} \varphi(x_1, x_2, \ldots, x_i; t)\, dx_1 dx_2 \ldots dx_i = N(t) \ . \tag{11.20}$$

In the course of evolution, the parameters of each star change in such a way that the distribution function obeys Liouville's equation with a nonzero right-hand side (see, for example, Zel'dovich, Myshkis 1973):

$$\frac{\partial \varphi}{\partial t} + \operatorname{div} \varphi \dot{x} = \psi(x, t) \ , \tag{11.21}$$

where x is a vector in the i-dimensional space, while the divergence is taken over all coordinates $x_1, x_2, \ldots, x_i$; and $\psi(x_1, x_2, \ldots, x_i, t)$ is a function describing the probability of generation of a neutron star with these parameters. It was shown in Chap. 4 that a magnetized neutron star is characterized mainly by three quantities: ω, μ, and $\dot{M}_c$ or ω, μ, and y. Hence the case $i = 3$ is of maximum interest: $x_1 = \omega, x_2 = \mu, x_3 = y$ (recall that y is the gravimagnetic parameter):

$$\frac{\partial \varphi}{\partial t} + \frac{\partial \varphi \dot{\omega}}{\partial \omega} + \frac{\partial \varphi \dot{\mu}}{\partial \mu} + \frac{\partial \varphi \dot{y}}{\partial y} = \psi(\omega, \mu, y, t) \ . \tag{11.22}$$

This equation can be reduced to a homogeneous partial differential equation with coefficients that are independent of the required function φ. The solution can be presented in the implicit form

$$\dot{\mathscr{F}}(\varphi, \omega, \mu, t) = 0 \ . \tag{11.23}$$

The equation for $\mathscr{F}$ is

$$\frac{\partial \mathscr{F}}{\partial t} + \dot{\omega}\left(\frac{\partial \mathscr{F}}{\partial \omega}\right) + \dot{\mu}\left(\frac{\partial \mathscr{F}}{\partial \mu}\right) + \left(\psi - \varphi\frac{\partial \dot{\omega}}{\partial \omega} - \varphi\frac{\partial \dot{\mu}}{\mu}\right)\frac{\partial \mathscr{F}}{\partial \varphi} = 0 \ . \tag{11.24}$$

For the sake of simplicity, we have omitted possible dependence on y. In many interesting cases, the evolution equation can be written

$$\dot{\omega} = F(\omega, \mu_0, t) \ , \tag{11.25}$$

where μ_0 is the initial magnetic moment of the neutron star. The solution of this equation is

$$\omega = F_1(\omega_0, t_0, \omega, \mu_0, t) \ . \tag{11.26}$$

The general solution of (11.22) can be presented as (Zel'dovich, Myshkis 1973)

$$\varphi(\omega, t) = \int_{\mu_1}^{\mu_2} d\mu_0 \int_0^t dt_0 \int_0^\infty \psi(\omega_0, \mu_0, t_0)\,\delta\,[\omega - F_1(\omega_0, t_0, \omega, \mu_0, t)]\,d\omega_0 \ . \tag{11.27}$$

Some particular cases of the solution (11.27) are given in Chap. 8 and by Lipunov (1987a).

Before concluding this section, let us write the equation for the function φ in the one-dimensional case, when only one parameter changes, i.e., the frequency ω:

$$\frac{d\varphi}{dt} + \varphi\frac{\partial^2 V}{\partial \omega^2} = \psi \ . \tag{11.28}$$

As before, V is the scalar potential.

This analytical consideration is useful for studying neutron stars at stages when their evolution is determined by intrinsic parameters, say, at the ejection and superejection stage. In actual practice, a neutron star passes through a whole set of states in which the external conditions are significant and can also vary with time. In this case, it is difficult to seek the analytical solution of an equation of the type (11.27), but it is easier to solve it numerically. The need for numerical solutions becomes especially clear in the analysis of the evolution of neutron stars in binary systems where the external conditions determined by the normal star vary in a complex manner. The method and the results of numerical computations by the Monte Carlo technique will be described below.

11.3 Neutron Star Tracks

Thus, the problem on the evolution of neutron stars must be solved by taking into account the evolution of a normal star. This question was discussed qualitatively by Bisnovaty-Kogan and Komberg (1974); Van den Heuvel (1977); Lipunov (1982a). We begin with the qualitative analysis presented in the last of these papers.

The most convenient way of a qualitative and quantitative analysis of the evolution of a neutron star is with the help of the $p-L$ diagram (Chap. 4). It should be recalled that L is just the potential luminosity of the neutron star. This quantity is equal to the real luminosity only at the accretion stage. Let us consider three characteristic tracks of neutron stars, assuming that all of them have the same magnetic fields.

Single neutron stars are the simplest on this diagram. To a first approximation, we can neglect the changes in the parameters of the surrounding medium. We shall assume that stars are generated with very small periods. In this case, the track of a single star is a vertical straight line (Fig. 11.5a). The neutron star successively passes the ejection and propeller stages, and then goes to the accretion stage or the georotator stage:

$$E \rightarrow P \begin{array}{c} \nearrow A \\ \searrow G \end{array} . \tag{11.29}$$

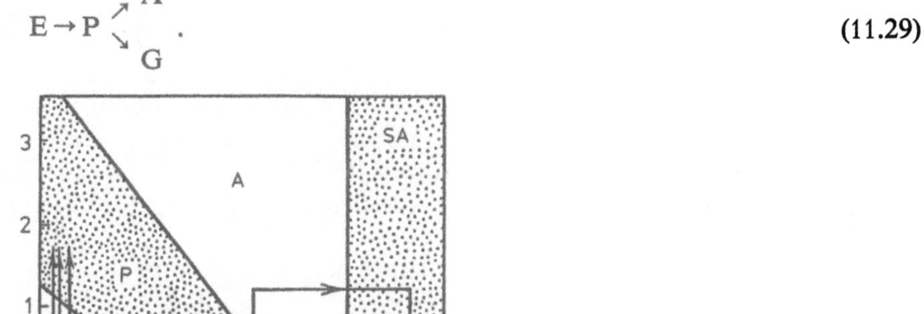

Fig. 11.5. Tracks of neutron stars on a (qualitative) $p-L$ diagram. (a) Track of a single neutron star, (b) track of a neutron star in a binary system, (c) track of a neutron star in a binary system formed at the instant when its companion fills the Roche lobe

The georotator stage is attained by a rapidly moving star [see (10.4)]. Of course, more complicated cases like the one involving a passage through dense molecular clouds are also possible.

The evolution of a neutron star in a binary system is always more complex. As a rule, the neutron star is generated at the instant when the companion lies on the main sequence (Fig. 11.5b). During the first $10^5 - 10^6$ years, the star is in the ejection stage, although it does not manifest itself as a radiopulsar since its pulse radiation is absorbed in the stellar wind of the normal star. The period of the neutron star increases in accordance with the magnetic dipole losses. After this, matter penetrates into the light cylinder and the neutron star passes first into the propeller stage and then the accretion stage. By this time, the normal star leaves the main sequence and the stellar wind is intensified. This results in the generation of a bright X-ray pulsar. The period of the neutron star stabilizes around its equilibrium value. Finally, the normal star fills the Roche lobe and the accretion rate suddenly increases; the neutron star moves first to the right and then vertically down on the $p-L$ diagram (Fig. 11.5c). In other words, the neutron star moves into the superaccretion stage SA. Its period tends to a new equilibrium value (in a normal magnetic field, the value of this period is a few tenths of a second). After the mass exchange, only the helium core of the normal star is left (Wolf-Rayet star), a separate system is formed, and the neutron star reverts to the propeller regime. Accretion is hampered by rapid rotation. This is probably the reason behind the absence of X-ray pulsars in pairs with Wolf-Rayet stars (Lipunov 1982e). Since the helium star does not have a long life ($\sim 10^5$ years), the neutron star does not have time to spin-down considerably: after the explosion of the normal star, the system disintegrates and the neutron star becomes an ejecting star, that is, a radiopulsar. The formation of radiopulsars from old neutron stars that have passed the accretion stage was first considered by Bisnovaty-Kogan and Komberg (1974).

The "loop-shaped" track discussed above can be written in the form

$$E \to P \to A \to SA \to P \to E \to \dots . \tag{11.30}$$

Figure 11.5 shows another version of the evolutionary track for a neutron star formed in the process of mass exchange in the binary system.

The overall lifetime of a neutron star in a binary system depends on the lifetime of the normal star and the parameters of the binary system. However, the rate of transition from one regime to another is proportional to the magnetic field of the neutron star.

11.4 Numerical Simulation of the Joint Evolution of Normal and Neutron Stars

To analyze the properties of an ensemble of neutron stars in the Galaxy, a special numerical program is designed to simulate the evolution of massive binary systems (Kornilov, Lipunov 1983b, 1984). In essence, this is a rough model of the real galaxy. Calculations are carried out by using the Monte Carlo method. A binary system formed by two normal stars is chosen. The instant of its formation is drawn at random, and so are its parameters which are distributed in accordance with the currently established empirical laws. The binary system then begins to evolve in accordance with the pattern described above. Each star gradually passes through the stages I → II → III → IV → V. The parameters of the star and of the matter flowing from it are assumed to be constant within each stage. The duration of a stage is calculated from the approximation formulas given in Sect. 11.2. After the appearance of a neutron star in the binary system, the neutron star evolution scheme is introduced in accordance with the approximate equation (11.11). The state of a binary system is described by a two-dimensional classification (Kornilov, Lipunov 1983a). For example, the state IIA means that we are dealing with a binary system in which the normal star is in stage II of the supergiant which does not fill the Roche lobe, while the neutron star is at the accretion stage. A typical representative of such systems is the classical massive binary system with the X-ray pulsar Cen X-3.

The evolutionary state of a binary system is defined at the current instant of time. The evolution of a large number ($\sim 10^4 - 10^5$) of binary systems is drawn at random in this way. This is how the real situation in our galaxy is simulated. It should be emphasized that the number of possible types of massive binary systems with neutron stars is much larger than the number of binary systems observed at present. Actually, only one type (IIA) of X-ray pulsar forming a binary system with a massive OB-star that does not fill its Roche lobe has been identified so far. Hence the results of such investigations are of a rather prognostic nature. Next, we describe in detail the method and the results of calculations.

11.4.1 Computational Method

The method of calculating the joint evolution of a neutron star and the normal component in the binary system for a statistical comparison with the observed characteristics (and also for predicting the hitherto unidentified stages of evolution of massive binaries) is based on the calculation of the evolution of a large number of binary systems with randomly chosen parameters from the instant $t = -t_0$ of the formation of the binary system to the present instant $t = 0$. Since we do not know the distribution of neutron stars according to their mass m_x, it is assumed that the mass distribution of the stars is uniform in the interval $m_{Ch} < m_x < m_{OV}$ ($m_{Ch} \approx 1.4$; $m_{OV} \approx 1.7$). The values of the mass components $M_{1,0}$ and $M_{2,0}$ and of the semi-major axis of the binary system at the initial instant of time are chosen in accordance with the random law (Salpeter's function):

$$\varphi(M_{1,0}) \sim M_{1,0}^{-2.35} \quad \text{for} \quad 10M_\odot \leqslant M_{1,0} \leqslant 100M_\odot \; . \tag{11.31}$$

In order to calculate the probability of distribution over the mass ratios $q_0 = M_{2,0}/M_{1,0}$, we use the law $\varphi(q_0) \propto q_0^2 (q_0 \leqslant 1)$, while the distribution over the semi-axes is calculated from $\varphi(a_0) \propto a_0^{-1}$ for $a_{\min} \leq a_0 \leq a_{\max}$. The value of $a_{\min}$ is chosen from the condition that the semi-major axis of the system must exceed the sum of the radii of the stars in the main sequence. The value of $a_{\max}$ is $\approx 2 \times 10^3$ (Tutukov 1980). The behavior of $\varphi(a_0)$ is in good agreement with the observed binary systems. It is assumed that the semi-major axis evolves in accordance with the formulas described in Sect. 11.2. The distribution of magnetic moments of neutron stars is chosen in the form

$$\varphi(\mu_0) \sim \mu_0^{-1} \, , \quad \mu_{\min} \leqslant \mu_0 \leqslant \mu_{\max} \, . \tag{11.32}$$

Thus, for a random set of quantities t_0, μ_0, $M_{1,0}$, a_0, q_0 and m_x, the evolution of both components is calculated and the class of the system and its parameters, as well as those of the neutron star, are determined at $t = 0$. In order to obtain a statistically reliable result, the evolution of $\sim 10^4 - 10^5$ systems is calculated in one numerical experiment.

11.4.2 Evolutionary Tracks

Figures 11.6 and 11.7 show examples of the evolutionary tracks of neutron stars with different initial parameters (Table 11.2). As expected, tracks in the form of a loop are the ones encountered most frequently. As a rule, the sequence of evolutionary states for a star with a magnetic moment $\mu_{30} \approx 1$ is

$$\text{IE} \rightarrow \text{IP} \rightarrow \text{IIP} \rightarrow \text{IIA} \rightarrow \text{IIISA} \rightarrow \text{IVP} \rightarrow \text{VE} \, . \tag{11.33}$$

However, other sequences are also possible. Since the initial distribution function $\varphi(q_0)$ has a peak near $q_0 = 1$, it is quite possible that the neutron star

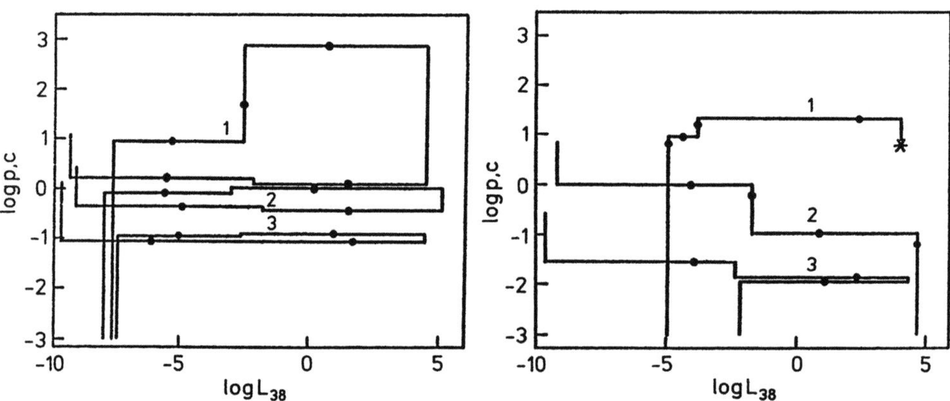

Fig. 11.6. Tracks of neutron stars obtained in the numerical experiment (Table 11.2 for explanation)

Fig. 11.7. As in Fig. 11.6

Table 11.2. Parameters of stars and sequences of stages for tracks shown in Figs. 11.6 and 11.7

Figure	Track	Initial masses of normal stars, $M_\odot$		Masses of neutron and normal stars, $M_\odot$		Magnetic moment of neutron star 10^{30} Oe cm^3	Sequence of evolutionary states
		M_1	M_2	M_x	M_0		
11.6	1	22.7	16.0	1.48	30.7	256	IE → IIP → IIA → → IIISA → IVP → VE
11.6	2	30	23	1.57	42	39	IE → IIP → IIISA → → IVP → VE
11.6	3	16.5	11.7	1.42	23.0	0.2	IE → IIP → IIISA → → IVP → VE
11.7	1	11.0	10.7	1.69	18.9	98.5	IE → IP → IIP → IIA → → IIISA → IIIBH
11.7	2	18.2	18.1	1.52	30.6	50.5	IIISE → IIISP → IVE → → IVP → VE
11.7	3	16.9	16.8	1.45	28.5	0.35	IIE → IIISP → IVE → → VE

may appear in stage II, III, or even IV instead of stage I. This situation is illustrated in Fig. 11.7. Generally, no characteristic loop is observed in this case.

The following regularities in the evolution of a neutron star were discovered (or confirmed):

1) The evolution of a star with a large magnetic moment from one state to another is more rapid.

2) Looped tracks are the ones encountered most frequently.

3) In view of the statistical closeness of the initial masses of the components, neutron stars may appear at later stages of evolution (II, III, or even IV) of a normal star.

4) The collapse of a neutron star is most likely at the III stage of evolution of a normal star, the probability of collapse being higher for larger magnetic moments.

5) Most of the neutron stars at stage I do not pass through the accretion stage A since neutron stars with large magnetic moments have a rapid evolution and fall into the stage G, i.e., are characterized by the sequence IE → IP → IG → ..., while stars with small magnetic moments do not manage to spin-down to the accretion stage in view of the low outflowing rate of the normal star at this stage.

6) For the same reasons, the emergence of stage IVA (an X-ray pulsar paired with a helium star) is unlikely, although the main reason in this case is the high outflow velocity v_w.

7) No noticeable evolution takes place in neutron stars with a very small magnetic moment ($\mu_{30} \lesssim 0.1$), and hence the phenomenon of X-ray pulsars is not observed in this case.

8) Usually, neutron stars with standard magnetic moments pass twice through the ejecting pulsar stage (IE and VE) in the course of their evolution. Most of them have small periods (less than 5 s) and may appear as radiopulsars in stage V.

11.4.3 Simulation of X-Ray Pulsars (Stage IIA) and the Choice of Optimal Parameters

It was mentioned above that only one stage IIA, that is, the stage of an accreting X-ray pulsar in a pair with a supergiant star, is the only stage to have been identified reliably with the observed objects. The total number of such objects is not large at present (about 20), and we cannot hope to exactly determine all the unknown quantities appearing in the evolution equation of a neutron star. However, we can estimate a number of parameters which strongly influence the evolution of a neutron star.

The most important quantities here are the parameter κ_t at the stage P and the upper limit $\mu_{\max}$ in the distribution of neutron stars according to magnetic moments. By varying these parameters, we were able to attain the best agreement between the computed positions of X-ray pulsars and those observed on the $p-L$ diagram. Figure 11.8 shows the computed and observed $p-L$ diagram for $\kappa_t = 10^{-2}$ and $\mu_{\max} = 10^3$, which shows that the characteristics of about 100 X-ray pulsars simulated by using the accepted parameters are close to the observed values.

These distributions show, for example, that the choice of the quantity $\mu_{\max} = 10$ contradicts the well-known observational fact that most pulsars have periods longe than 10^2 s. On the other hand, the choice $\mu_{\max} > 10^3$ does not influence the distribution of pulsars on the diagram since they normally belong to the class G in view of a strong magnetic field. The following parameters were finally chosen: $\mu_{\max} = 10^3$ and $\kappa_t = 10^{-2}$ at the stage P.

These values of the parameters were used to compute the distribution of parameters of binary systems simulated with the above-mentioned parameters.

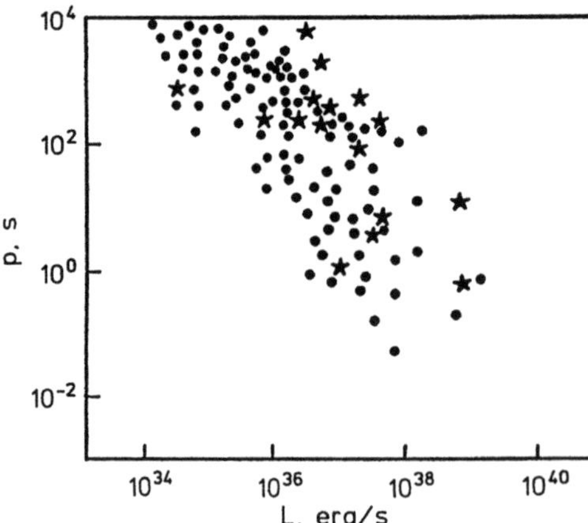

Fig. 11.8. The $p-L$ diagram for X-ray pulsars. The points show 100 "artificial" pulsars, and the stars indicate the observed pulsars

Good agreement was obtained with the observed masses of the optical components of the X-ray pulsars (for example, with the peak observed in the vicinity of $15-20\,M_\odot$). It is interesting to note that a large number of binary systems have periods $T \gtrsim 100$ days. Naturally, a larger body of statistical data (i.e., a larger number of X-ray pulsars) is required for determining more accurately the magnetic field distribution of neutron stars. In this connection, hopes are being pinned on the possibility of discovering X-ray pulsars in the nearest galaxies.

11.4.4 Abundance of Different Types of Systems in the Galaxy

The method of computation used here allows us to determine the abundance of different classes of massive binary systems with neutron stars.

Calculations were made in such a way that the number of systems with X-ray pulsars (IIA) was equal to 100. Thus all the remaining numbers are normalized to the number of X-ray pulsars. Table 11.3 contains the results of numerical experiments (see also Fig. 11.9).

Three circumstances are worth noting in this case.

1) The duration of stage III was taken to be equal to the thermal period for a normal star. In actual practice, however, the lifetime of the system at this stage may be much smaller (perhaps by an order of magnitude).

2) The number of systems at stage IV is an upper estimate since we have not taken into consideration the possibility of a neutron star being "swallowed" at the stage III by the optical component.[1]

3) The number of systems at stage V is presented only for the sake of illustration, since the time $t_{\max} = 15$ million years chosen by us is much smaller than the lifetime of a neutron star at stage IV. However, this restriction is not so important for neutron stars with a period smaller than a few seconds.

Table 11.3. Abundance of various types of massive binary systems with neutron stars and black holes. Results of numerical experiment

State of the neutron star	State of the normal star			
	I	II	III	IV
E	1700	30	0	3
P	600	40	0	45
A	0	100	0	2
SP	0	0	2	0
SA	0	0	19	0
G	20	35	0	25
BH	0	0	1	15

---

[1] The investigation of stars with neutron cores was carried out by Bisnovaty-Kogan and Lamzin (1984).

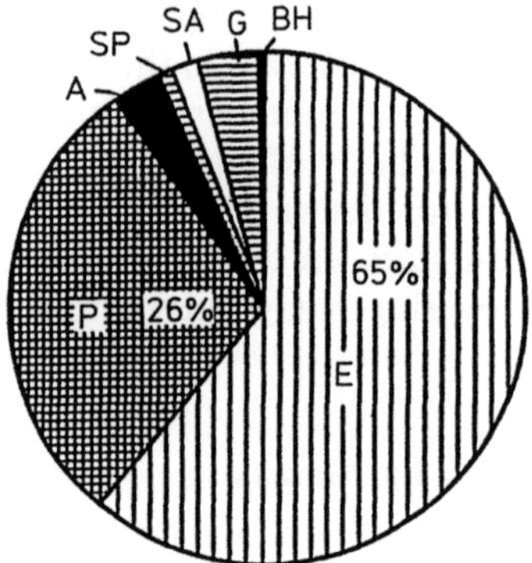

Fig. 11.9. Abundance of different types of neutron stars in massive binary systems

The results of calculations presented in Table 11.3 reveal the following regularities: (1) most of the binary systems with neutron stars are in stages IE and IP, and the number of such systems is larger than the number of X-ray pulsars by more than an order of magnitude, (2) there are no X-ray pulsars in stage I; (3) the number of X-ray pulsars at stage IV is small, which is in agreement with the fact that they are not observed in pairs with WR stars; (4) the total number of neutron stars in IIE and IIP states is comparable with the number of IIA type systems; (5) since the total number of binary systems undergoing complete evolution to stage V is about 4000, while there are about 600 black holes formed from neutron stars, the probability of collapse of neutron stars is estimated at about 10%.

We hope that the results presented in Table 11.3 will stimulate the quest of the above-mentioned binary systems with neutron stars and black holes.

11.4.5 Physical Characteristics of Neutron Stars at Various Stages of Evolution

The above calculations show that the following observational regularities in type IIA X-ray pulsars can be easily explained: (1) the luminosity of the observed X-ray pulsars varies from $\sim 10^{35}$ to $(5-8)\times 10^{38}$ erg/s; (2) most X-ray pulsars have long periods, ≥ 100 s; (3) the number of X-ray pulsars is about 1% of the total number of massive OB-stars in the galaxy.

Let us consider some important features of other types of systems.

1) The number of ejecting (type E) neutron stars with a period shorter than $1-0.1$ s (i.e., neutron stars capable of emitting not only radio radiation, but also X-rays and gamma radiation) is so large that they are quite likely to be observed in the near future in close binary systems. It should be recalled that pulsed radiation cannot be observed in view of its absorption in the stellar wind (radio bursts may be expected from these systems in some cases).

2) A large number of type VE neutron stars are known to have periods shorter than 5 s. This means that some of the observed radiopulsars are just neutron stars of this type.

3) The obtained characteristics of neutron stars and of systems of I-IVP type indicate that further theoretical investigations of the propeller stage must be carried out in order to find their possible astrophysical manifestations. Weakly pulsating X-ray (and, perhaps, gamma) radiation can be expected from such stars due to the release of rotational energy at the magnetosphere of the neutron star.

11.4.6 Two Types of Radiopulsars

About 500 radiopulsars are known to exist at present. At the moment of their formation, all of them are usually assumed to be single neutron stars which behave like ejecting objects. However, calculations show (Fig. 11.10) that some of the observed radiopulsars may be neutron stars that have gone through all the evolutionary stages in a binary system (including the accreting X-ray pulsar stage). Calculations made it possible to determine (or rather, to predict) the parameters of radiopulsars formed in this way. It should be recalled that according to our classification, such radiopulsars belong to the class VE. The specific evolution of binary systems and neutron stars leads to a certain correlation between some observed parameters of VE type pulsars. Indeed, since a neutron star in the supercritical accretion stage assumes an equilibrium rotational period determined by its intrinsic parameters (magnetic moment and mass) only, and since a neutron star does not manage to spin-down significantly at stage IV, the radiopulsar formed at stage V is characterized by a well defined relation between p and $\dot{p}$.

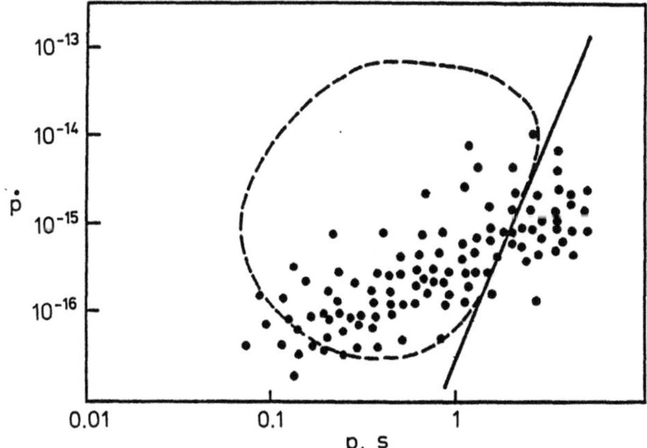

Fig. 11.10. The $\dot{p}-p$ diagram for radiopulsars. The *points* show the artificial radiopulsars emerging from binary systems. The *dashed line* separates the region of observed pulsars

11.5 Possible Candidates

It follows from Table 11.3 that statistically, it is possible to observe 18 types of binary systems with neutron stars, of which only one (IIA) has been reliably identified so far. This should be supplemented by the class IIIA to which the pulsar Her X-1 and other sources with low-mass normal components belong. In our numerical experiment, sources of this type did not appear because of the fact that we did not take into account the possibility of loss of matter in a binary system during the first mass exchange. Such a phenomenon must take place in systems with an initially large mass ratio, and leads to the formation of a low-mass binary system from the original massive binary. However, a very small fraction of such systems exists, since the probability of the formation of a massive binary with a large mass ratio rapidly falls (as q^2). Nevertheless, in view of the fact that X-ray systems of this type live much longer than massive binary systems, the probability of their observation is quite high. Hence we should also include such systems in subsequent calculations. We shall now discuss some of the observed objects which cannot be treated as candidates for other types of binary systems.

11.5.1 "Runaway" Stars

The most likely candidates for inclusion in the classes IE and IP (or IIE and IIP) are the so-called runaway stars. The high escape velocities of such stars are apparently associated with the kick momentum resulting from the collapse (and the ejection of a part of the substance) of one of the components in the pair. Since the less massive component is the one to explode first, the system does not disintegrate. The most real possibility of confirming these concepts would be the discovery of the low-mass companions of these stars. We can also expect a small periodicity in radial velocity variations in the $10-30$ km/s interval. The expected periods of revolution of such systems lie in a wide time range, from a few days to tens of years. We cannot expect strong X-ray radiation from such stars since the neutron star does not accrete (class E, P and, perhaps, G). However, at the level of $10^{30} - 10^{34}$ erg/s, the possibility of detecting weak pulsating thermal X-rays (class P) or strong pulsating nonthermal X-rays (class E) cannot be ruled out (in the latter case, we can expect radiobursts).

The star 68 Cyg, belonging to the class of shell-stars (Esipov et al. 1982) has been discovered to be a binary system. The absence of strong X-ray radiation from this star indicates that the neutron star is in one of the nonaccreting states E or P.

11.5.2 The SS 433 Object

Thanks to photometric (Cherepashchuk 1981) and spectral (Crampton, Hutchings 1981) observations, it has been reliably established at present that SS 433 is a massive binary system, with an unusual companion. Apparently, the normal star fills its Roche lobe and flows out on the thermal time scale at a rate of $\sim 10^{-4} M_\odot$/year. Consequently, it belongs to class III. The nature of the second companion is not known so well, and its mass has been estimated quite roughly, between ~ 0.5 and $\sim 5 M_\odot$. This does not allow us to associate reliably the second component with a neutron star or a black hole. According to the model worked out by Lipunov and Shakura (1982), the unusual object in SS 433 is a neutron star experiencing supercritical disk accretion, so that the binary system SS 433 belongs to the class IIISA. In our numerical experiments, objects of this class are obtained naturally and the parameters of simulated neutron stars are close to the parameters obtained for the SS 433 model. Calculations show that there must be about 20 such objects in the Galaxy. In this case, the duration of stage III was taken to be equal to the thermal time of the normal star. In actual practice, however, the duration of this stage may be much shorter, and hence the number of such systems may be smaller.

The astrophysical manifestations of neutron stars in the IIISA regime are described in Chap. 9. A neutron star in the SA regime is observationally indistinguishable from the normal star (the hard radiation is absorbed in the dense shell) with a high accretion rate. In addition, relativistic or subrelativistic bursts of matter may be observed.

11.5.3 "Single" Wolf-Rayet Stars

Strong arguments have appeared recently in favor of the binary nature of single WR stars. The mass function indicates that the invisible companion has a small mass ($\sim 1 - 3 M_\odot$). The existence of WR stars in pairs with relativistic companions follows from the evolution of massive binary systems (Tutukov, Yungel'son 1973). The decisive factor in this case would be the discovery of the high-intensity X-ray radiation from these stars. However, it was shown by Lipunov (1982e) (and later confirmed by computations) that the phenomenon of X-ray pulsars is unlikely to be observed in such systems. It can be seen from Table 11.3 that most of such systems must be in the stages IVP, IVE, and IVG (a part of these systems may belong to the class IVBH).

11.5.4 Collapse Anisotropy

In the version of evolution considered above, the collapse of a normal star is spherically symmetric. Hence the binary system does not disintegrate after the first explosion. It was mentioned above that this leads to a contradiction. As a matter of fact, radiopulsars must be observed in quite diverse systems where the density of the stellar wind near the neutron star is low. However, no radiopulsar has been observed so far in a binary system with a normal star. This paradox can be eliminated by assuming that either diverse pairs disintegrate, say, as a result of an anisotropic collapse, or neutron stars in quite diverse system are not formed for some reason.

The problem concerning the absence of radiopulsar observation in binary systems with normal stars was raised by Kornilov and Lipunov (1984), who

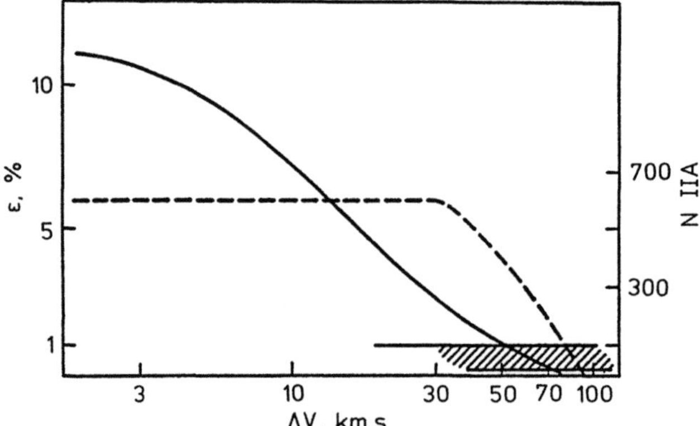

Fig. 11.11. Dependence of the number N_A of X-ray pulsars and the relative number ε of visible radiopulsars in binary systems with normal components on kick velocity as a result of anisotropic collapse (numerical experiment)

tried to associate the controversy with the collapse anisotropy. The analysis is based on simple considerations. Suppose that as a result of callapse, a star acquires a kick velocity Δv in a random direction. If the velocity Δv were too high, even very close systems would disintegrate. This is in contrast with the existence of X-ray pulsars in massive systems which number about 20. Conversely, if $\Delta v \approx 0$, radiopulsars should be observed in diverse pairs. The numerical model described above was used to calculate the number N_A of X-ray pulsars (of type IIA system) and the ratio of the number of radiopulsars in binary systems (i.e., systems in which the stellar wind is transparent to radio waves) to the number ε of single radiopulsars. According to the observations, $N_A \approx 20$ and $\varepsilon \lesssim 1/300$. Figure 11.11 shows the results of computations of N_A and ε for different values of Δv. It can be seen that an agreement with the observational data is obtained only if $\Delta v \approx 80-90$ km/s.

11.5.5 Other Numerical Models

We believe that the numerical scheme considered above is a powerful tool for verification of evolutionary patterns. The necessity of using such a numerical scheme becomes especially clear during an analysis of the evolution of low-mass systems in which there may be as many as 200 (!) different states (Lipunov, Postnov 1987). This is due to the fact that the evolution of the neutron star under consideration is supplemented by the evolution of a white dwarf as well (Lipunov, Postnov 1988).

Appendix

Magnetohydrodynamic Instabilities

Rayleigh-Taylor (RT) Instability

Imagine a vessel filled with two liquids having densities ϱ_1 and ϱ_2. It is obvious that if the heavier liquid is on top of the lighter one ($\varrho_1 < \varrho_2$), the situation will be unstable: a slight shaking of the vessel will send the heavier liquid down, expelling the lighter one to the top. This is quite clear from energy considerations: the potential energy of the vessel in a gravitational field is lower if the heavier liquid is at the bottom. The instability of the interface between the heavier and the lighter liquids is called the Rayleigh-Taylor instability, and is frequently encountered in everyday life. The fat floats on the surface of borshch (a Ukrainian soup) thanks to this instability.

The dynamics of the RT instability can be investigated by specifying a small perturbation on the boundary and solving the linear equations of hydrodynamics. We can not only verify the instability of the interface, but also determine the time of growth of this instability. Any small perturbation will increase exponentially, $\propto e^{\Gamma t}$ where Γ is called the instability increment. Exponential growth is characteristic of the initial linear stage of instability. When the perturbation of the boundary becomes large, its growth becomes slower.

The RT instability increment for an incompressible liquid can be estimated from the following considerations which explain the phenomenon from a mechanical point of view. Suppose that a drop of heavy liquid of diameter λ is immersed in the lighter liquid. This drop is subjected to gravity and the Archimedean force, and its equation of motion is

$$\varrho_2 \lambda^3 \frac{d^2 z}{dt^2} = \varrho_2 \lambda^3 g - \varrho_1 \lambda^3 g \ , \tag{A.1}$$

where g is the acceleration due to gravity and z is the depth to which the drop descends. Equation (A.1) shows that the drop moves with an effective acceleration

$$g_{\mathrm{ef}} = g(1 - \varrho_1/\varrho_2) \ . \tag{A.2}$$

The instability increment is the reciprocal of the time in which the drop descends to a depth Δz equal to the diameter λ of the drop. Obviously,

$$\Delta z = \frac{g_{\mathrm{ef}} t^2}{2} \ . \tag{A.3}$$

Substituting $\Delta z = \lambda$, we find that the RT-instability increment is given by (Lamb 1947)

$$\Gamma_{\mathrm{RT}} = (k_\lambda g_{\mathrm{ef}})^{1/2} \ , \tag{A.4}$$

where $k_\lambda = 2\pi/\lambda$. Although (A.4) is derived from a qualitative analysis, it coincides with the exact value of the RT-instability increment. Using (A.2), we obtain

$$\Gamma_{\mathrm{RT}} = (k_\lambda g)^{1/2}\sqrt{1 - \varrho_1/\varrho_2} \ . \tag{A.5}$$

Let us now replace the lighter liquid by a uniform magnetic field. If the boundary is perturbed in such a way that the magnetic field lines are pushed apart but not bent, we arrive at a situation completely analogous to the one considered above. For such perturbations, the magnetic field can be replaced by a liquid whose density and pressure are given by (Northrop 1956)

$$\varrho_{\mathrm{m}} = \frac{B^2}{4\pi c^2} \ , \quad P_{\mathrm{m}} = \frac{B^2}{8\pi} \ . \tag{A.6}$$

Substituting $\varrho_1 = \varrho_{\mathrm{m}}$ into (A.5), we obtain

$$\Gamma_{\mathrm{RT}} = (k_\lambda g)^{1/2}\sqrt{1 - (v_{\mathrm{A}}/c)^2} \approx \sqrt{k_\lambda g} \ . \tag{A.7}$$

In the approximation considered here, we neglect the term $\propto (v_{\mathrm{A}}/c)^2$, where v_{A} is the Alfvén velocity

$$v_{\mathrm{A}}^2 = \frac{B^2}{4\pi \varrho_2} \ . \tag{A.8}$$

Thus, this situation is always unstable. However, perturbation leads to a bending of lines of force and the instability is stabilized at small wavelengths λ. This happens because the lines of force tend to straighten out.

If $\boldsymbol{k}$ is the perturbation wave vector directed at an arbitrary angle to the magnetic field, the equation for the emergence of the instability becomes (Chandrasekhar 1966)

$$k_\lambda \varrho g > \frac{(\boldsymbol{k} \cdot \boldsymbol{B})^2}{4\pi} \ . \tag{A.9}$$

This inequality can be derived as follows. We calculate the force acting on a cylindrical volume element which is still connected with the boundary but has descended to a depth Δz. The equations for the incompressible liquid are

$$\Delta v = 0 \qquad\qquad \text{continuity equation ;}$$

$$\varrho\,\frac{\partial v}{\partial t} + \nabla\,\delta P = 0 \qquad \text{Euler's equation ,}$$

where δP is the pressure perturbation. The boundary is specified by $\Delta z = \eta\sin(k\cdot x)$. Obviously, δP is an analytical function, since $\Delta\delta P = 0$. We can write δP as $\delta P = \delta P(0)\,\mathrm{e}^{-kz}\sin(k\cdot x) = \eta\varrho g\,\mathrm{e}^{-kz}\sin(k\cdot x)$. In this case, the force characterizing the buoyancy of a magnetic tube is given by $-k_\lambda\eta\varrho g\Delta z\sin(k\cdot x)$. The resultant force acting on the tube is

$$F = \frac{(k\cdot B)^2}{4\pi}\,\Delta z\eta\,\sin(k\cdot x) - k_\lambda\eta\varrho g\Delta z\sin(k\cdot x)\ .$$

The tube will descend if the inequality (A.9) is satisfied.

Additional stabilization of the RT instability occurs in the case when the plasma-field boundary is curved. Let d be the displacement in the direction perpendicular to the boundary characterized by a unit normal vector n_s. The instability will develop if the gravitational force exceeds the magnetic field pressure gradient caused by the curvature of the field lines:

$$dn_s\varrho g > dn_s\nabla\left(\frac{B^2}{8\pi}\right) = dk_c\,\frac{B^2}{4\pi}\ , \qquad\qquad (A.10)$$

where k_c is the curvature of the field line at the boundary. Note that if the boundary is concave towards the magnetic field, the field facilitates the growth of instability.

In the general case, when the plasma is inhomogeneous and the magnetic field is nonuniform, the stability is verified with the help of the energy principle (Bernstein et al. 1958). According to this principle, the plasma-field boundary will be stable if the energy variation δW_s due to the deformation of the boundary is positive:

$$\delta W_s = -\tfrac{1}{2}\int_S (n_s d)^2 n_s\nabla(P_0 - B^2/8\pi)\,ds\ . \qquad\qquad (A.11)$$

Using the equalities $\nabla P = \varrho g$ and $\nabla B^2 = 2k_c B^2 n_s$, we obtain from (A.11) a result of the type (A.9). The boundary will be unstable if

$$\varrho n\cdot g > k_c\,\frac{B^2}{4\pi}\ . \qquad\qquad (A.12)$$

While perturbing the field lines, we have so far assumed them to be free. In actual practice, a situation is possible in which the field lines are fixed at the ends. This results in an additional stabilization of the boundary. In order to take this circumstance into consideration, we must supplement the integral (A.11) by the energy variation of the vacuum part occupied by the magnetic field (if the lines of force are not fixed at the ends, the energy of the field does not change). Let us suppose that the lines of force are fastened to ideally conducting plates. Obviously, the displacement d of the boundary in this case cannot be arbitrary ($d = 0$ at the ends). If l is the distance between the plates, stabilization may set in even for $k \perp B_0$ if

$$\frac{\pi B_0^2}{4l^2} > \varrho g k_\lambda \ . \tag{A.13}$$

Finally, there are two more effects (Aly 1985) which retard the growth of an RT instability. Although viscosity does not change the boundary instability condition (A.12), it checks the growth of instability for perturbations with a wave number

$$k_\lambda > k_* = (g/v_k^2)^{1/3} \ , \tag{A.14}$$

where v_k is the kinematic viscosity. For such perturbations, the increment is multiplied by a factor $(k_*/k_\lambda)^{3/2}/2$. In this case, we must consider the fact that the penetration of a magnetic field into a plasma changes its viscosity.

The second restriction arises when we take into account the finite nature of the perturbation wavelength. The RT instability stabilizes if the wave number satisfies

$$k_\lambda > k_L \approx g^{1/2} R_L^{-4/3} \omega_i^{2/3} \ , \tag{A.15}$$

where R_L is the Larmor radius of the ion and ω_i is the Larmor frequency in the field penetrating the plasma.

At the nonlinear stage, cigar-shaped plasma clusters elongated along the field are formed and move downwards by pushing the field lines apart. Two-dimensional calculations of the classical RT instability were made by Wang and Nepveu (1983) and by Wang et al. (1984). According to these authors, shortwave modes are the most effective in mass transfer. The sizes of individual clusters are controlled by effective viscosity.

Commutation Instability

The boundary between the plasma and the field can be unstable even in the case when there is no gravity. It was mentioned above [see (A.10)] that if the boundary is concave towards the magnetic field, an additional instability may emerge. If we interchange the plasma and the magnetic field, the field lines become convex and the magnetic field energy decreases. Such a commutation instability has the same increment as the RT instability:

$$\Gamma_{in} = (k_\lambda g_{ef})^{1/2} \ , \tag{A.16}$$

where

$$g_{\text{ef}} = \left(\nabla P - \nabla \frac{B^2}{8\pi}\right) \bigg/ \varrho \; . \tag{A.17}$$

Commutation instability is also observed for a plane boundary when the magnetic field strength decreases as we move away from the boundary.

Kelvin-Helmholtz Instability. A qualitatively new effect is observed when two liquids move relative to each other. It is found that, irrespective of their densities, the interface of such objects is unstable. Such an instability was first observed by Kelvin and Helmholtz (Lamb 1947) in fluid mechanics, and by Kruskal and Schwarzschild (1954) in magnetohydrodynamics.

The Kelvin-Helmholtz (KH) instability is responsible for the ripples on water and for the fluttering of sails and flags. Its emergence can be explained in an elementary manner by considering two liquids filling a volume. Suppose that the interface between the stationary and moving liquids is initially perfectly planar and the pressure in both liquids is identical. We slightly perturb the interface e.g., creating capes and bays along a shoreline. The liquid is accelerated when flowing a round a cape and, in accordance with Bernoulli's integral, its pressure decreases. The situation is reversed at the bays. As a result, the "shore" becomes more and more corrugated.

We can also argue in the following manner. Flowing past an arbitrary boundary, the liquid experiences a centrifugal acceleration

$$\frac{d^2(\Delta z)}{dt^2} = \frac{v^2}{r_c} \approx v^2 \Delta z k_\lambda^2 \tag{A.18}$$

where, as before, Δz is the depth measured from the boundary, and r_c is the radius of curvature. For the curve $\Delta z = (\Delta z)_0 \sin (k_\lambda x)$ the radius of curvature $r_c \approx 1/\Delta z k_\lambda^2$ at the most protruding point in the flow. Consequently,

$$\frac{d^2(\Delta z)}{dt^2} \approx v^2 \Delta z k_\lambda^2 \; . \tag{A.19}$$

The solution of this equation has the form $\Delta z \propto e^{\Gamma t}$, where

$$\Gamma_{\text{KH}} = k_\lambda v \; . \tag{A.20}$$

If the liquids have different densities and are situated in a gravitational field, the stability condition becomes

$$\varrho_1 \varrho_2 [(v_1 - v_2) \cdot k]^2 \leqslant (\varrho_1 + \varrho_2)(\varrho_1 - \varrho_2) k_\lambda g \; , \tag{A.21}$$

where v_1 and v_2 are the velocities of the liquids. Obviously, if the upper liquid is denser ($\varrho_2 > \varrho_1$), the boundary will always be unstable. However, if the heavier liquid is at the bottom, rather long waves will be stabilized by the force of gravity, like waves on water.

In order to find the instability increment, we need to solve the linearized equation of fluid dynamics, and the solution is sought in the form $f(z) \exp [i(k_\lambda \cdot r - \omega t)]$, where r is the coordinate in the plane of the unperturbed boundary. The requirements of the continuity of the boundary and the smallness of the perturbation at infinity lead to the following result. Since $f(z) = e^{-k_\lambda |z|}$, the perturbation attenuates as we move away from the boundary at distances of the order of the wavelength. The frequency ω is associated with the wave number k_λ through the dispersion relation

$$\omega = \frac{\varrho_1(k_\lambda \cdot v_1) + \varrho_2(k_\lambda \cdot v_2)}{\varrho_1 + \varrho_2}$$

$$\pm i \left[\frac{\varrho_1 \varrho_2}{\varrho_1 + \varrho_2} [(v_2 - v_1) \cdot k_\lambda]^2 - k_\lambda g \frac{\varrho_2 - \varrho_1}{\varrho_2 + \varrho_1} \right]^{1/2} . \tag{A.22}$$

If the density of the upper liquid is much lower than that of the liquid at the bottom, i.e., $\varrho_2 \ll \varrho_1$, unstable waves travel with the heavier liquid 1, the rate of their build-up is $\Gamma_{KH} \approx k v_0 \sqrt{\varrho_2/\varrho_1}$, where $v_0 = v_2 - v_1$, is quite small.

If the relative velocity of the liquids is comparable with the velocity of sound a_s, we must take into consideration their compressibility. The cases of sub- and supercritical flows differ quite significantly (Landau, Lifshitz 1953; Plesset, Hsieh 1964). In the supersonic regime, the instability is suppressed outside the shock cone.

Let us now consider plasma instabilities. Suppose that half the space is occupied by a uniform magnetic field B, while the second half is occupied by a uniform plasma moving at a velocity v. This problem was first considered by Northrop (1956). It was mentioned above that if the field lines are not bent by the perturbation, we can replace, in the incompressible liquid approximation, the magnetic field by a liquid whose density and pressure are given by (A.6). We put $\varrho_2 = \varrho_m = B^2/4\pi c^2$, $g = 0$ in (A.22). The increment of the KH instability growth is given by

$$\Gamma_{KH} = \frac{k_\lambda v_A v}{c} . \tag{A.23}$$

It is interesting to study the change in the instability increment if we take into account the finite size of the flow. This is the situation encountered during disk accretion onto a magnetized neutron star.

Suppose that the plane flow of an incompressible liquid is confined on both sides by a magnetic field. If the direction of flow v and the direction of the perturbation k are parallel and both are perpendicular to the direction of the magnetic field, we can use the hydrodynamic analog of the problem (Lipunov 1978c). Such a problem in fluid mechanics was considered by Rayleigh (Lamb 1947). Generalizing the dispersion relation to the case of plasma and the field, we obtain

$$(\omega - k_\lambda v)^2 + \omega^2 \left(\frac{v_A}{c}\right)^2 \coth(k_\lambda H) = 0 \; . \tag{A.24}$$

The solution of this equation has the form

$$\omega = \frac{k_\lambda v}{1 + (v_A/c)^2 \coth(k_\lambda H)} \left(1 \pm i \frac{v_A}{c} \sqrt{\coth(k_\lambda H)}\right) \; . \tag{A.25}$$

We consider the case when both flow boundaries are perturbed in phase. The plus sign corresponds to the increasing mode, and we obtain an expression for the instability increment,

$$\Gamma_{KH} = \frac{k_\lambda v v_A}{c} \cdot \frac{\sqrt{\coth(k_\lambda H)}}{1 + (v_A/c)^2 \coth(k_\lambda H)} \; . \tag{A.26}$$

For long waves $k_\lambda \to 0$, the asymptotic form can be presented as

$$\Gamma_{KH}(k_\lambda \to 0) \approx \frac{k_\lambda^{3/2} v c}{v_A} \sqrt{H} \; . \tag{A.27}$$

For short waves, Northrop's formula (A.23) serves as the asymptotic form. It can be seen that the instability increment decreases with increasing wavelength.[1]

Let us now turn to the case of an incompressible semi-infinite flow of plasma while taking into consideration the force of gravity. In this case, the instability increment is given by

$$\Gamma_{KH} = \left(\frac{k_\lambda^2 v^2}{c^2} - k_\lambda g/v_A^2\right)^{1/2} v_A \; . \tag{A.28}$$

It can be seen that only very short wave modes will be unstable:

$$\lambda < \lambda_g = 2\pi \left(\frac{v_A v}{c}\right)^2 g \; . \tag{A.29}$$

Wang and Welter (1982) generalized Northrop's problem to the case of plasma motion at an arbitrary angle to the magnetic field direction. They showed that even a weak field inside a plasma directed at an angle to the field in vacuum stabilizes the boundary.

---

[1] It is interesting to note that although $\Gamma_{KH} \to 0$ as $k_\lambda \to 0$, an unstable solution exists for an infinitely thin medium. It was shown by Rayleigh that in this case the perturbation increases in direct proportion to t (Lamb 1947).

References

Abbott DC, Bieging JH, Churchwell E (1984) *Astrophys. J.* **280**:671
Abramowicz MA, Calvani M, Nobili L (1980) *Astrophys. J.* **242**:772
Akasofu S, Chapman S (1972) *Solar Terrestrial Physics*, (Pergamon, Oxford)
Alekseev YuI (1971) *Astronom. Tsirkul.*, No. **655**:1
Alekseev YuI (1973) *Radiofizika* **16**:762
Alpar MA, Shaham J (1985) *IAU Circ. No. 4046*
Alpar MA, Cheng AF, Ruderman MA, Shaham J (1982) *Nature* **300**:728
Aly JJ (1980) *Astron. and Astrophys.* **86**:192
Aly JJ (1984) *Astrophys. J.* **283**:349
Aly JJ (1985a) In: *Proc. of the 1984 Taos Workshop*
Aly JJ (1985b) In: *Proc. Mediterranean School on Plasma Astrophysics*, Columbari, Crete
Ambartsumyan VA, Saakyan GS (1960) *Astron. Zh.* **37**:193
Amnuel' PR, Guseinov OK (1968) *Izv. Akad. Nauk. Azer.* SSR, ser. Fiz.-Tekh Mat. Nauk, No. **3**:70
Amnuel' PR, Guseinov OK (1969) *Astronom. Tsirkul.* No. **524**:3
Amnuel' PR, Guseinov OK (1972) *Astrofizika* **8**:107
Andronov IL (1984) *Astrofizika* **20**:165
Andronov IL (1986) *Astronom. Zh.* **63**:274
Anzer U, Börner G (1980) *Astron. Astrophys.* **83**:133
Arnett WD, Bowers RL (1977) *Astrophys. J. Suppl.* **33**:415
Arons J, Lea SM (1976a) *Astrophys. J.* **207**:914
Arons J, Lea SM (1976b) *Astrophys. J.* **210**:792
Arons J, Lea SM (1980) *Astrophys. J.* **235**:1016
Baade W, Zwicky F (1934) *Phys. Rev.* **45**:138
Baan WA (1977) *Ann. N.Y. Acad. Sci.* **302**:244
Backer DC, Kulkarni SR, Heiles C, Davies M, Goss WM (1982) *Nature* **300**:615
Bahcall JN, Ostriker JP (1975) *Nature* **256**:23
Bahcall JN, Ostriker JP (1977) *Nature* **256**:23
Bahcall JN, Wolf RA (1965) *Phys. Rev. B* **140**:1452
Bahcall JN, Wolf RA (1976) *Astrophys. J.* **209**:214
Bardeen J, Petterson JA (1975) *Astrophys. J. Lett.* **195**:L65
Bardeen JM, Press WH, Teukolsky SA (1972) *Astrophys. J.* **178**:347
Barlow M, Cohen M (1977) *Astrophys. J.* **213**:737
Basko MM, Syunyaev RA, Titarchuk LG (1974) *Astron. Astrophys.* **31**:249
Basko MM, Syunyaev RA (1975) *Astron. Astrophys.* **42**:311
Basko MM, Syunyaev RA (1976) *Mon. Not. Roy. Astron. Soc.* **175**:395
Basko MM (1977) *Astron. Zh.* **54**:1051
Basko MM (1980) *Astron. Astrophys.* **87**:330
Bath GT (1973) *Nature* **246**:84
Bath GT, Pringle JE (1981) *Mon. Not. Roy. Astron. Soc.* **194**:967
Baym G, Pethick C (1979) *Ann. Rev. Astron. Astrophys.* **17**:415
Baym G, Pethick C, Pines D, Ruderman M (1969) *Nature* **224**:872
Beaudet G, Petrosian V, Salpeter EE (1967) *Astrophys. J.* **150**:979
Bernstein IB, Frieman EA, Kruskal MD, Kulsrud RM (1958) *Proc. Roy. Soc.* A **244**:17
Beskin VS, Gurevich AV, Istomin YN (1983) *Zh. Eksp. Teor. Fiz.* **85**:401
Bisnovaty-Kogan GS (1970) *Astron. Zh.* **47**:813 (In Russian); Engl. translation *Sov. Astron.* **14**(4)

312 References

Bisnovaty-Kogan GS (1973) *Astron. Zh.* **50**:902
Bisnovaty-Kogan GS, Blinnikov SI (1980) *Mon. Not. Roy. Astron. Soc.* **191**:711
Bisnovaty-Kogan GS, Komberg BV (1974) *Astron. Zh.* **51**:373
Bisnovaty-Kogan GS, Lamzin SA (1984) *Astron. Zh.* **61**:323
Bisnovaty-Kogan GS, Fridman AM (1969) *Astron. Zh.* **46**:721
Bisnovaty-Kogan GS, Imshennik VS, Nadezhin DK, Chechetkin VM (1975) *Astrophys. Sp. Sci.* **35**:3
Bisnovaty-Kogan GS, Kazhdan YM, Klypin AA, Lutskii AE, Shakura NI (1979) *Astron. Zh.* **56**:359
Bondi H (1952) *Mon. Not. Roy. Astron. Soc.* **112**:195
Bondi H, Hoyle F (1944) *Mon. Not. Roy. Astron. Soc.* **104**:273
Börner G (1979) *Preprint Max-Planck-Institut, Garching,* Fed. Rep. Germany, No. 193
Bradt H, Doxsey R, Jernigan J (1979) *Adv. Space Res. Expl.* **3**:3
Brecher K, Caporaso G (1976) *Nature* **259**:377
Brecher K, Caporaso G (1977) *Ann. N.Y. Acad. Sci.* **302**:471
Brown JC, Boyle CB (1984) *Astron. Astrophys.* **141**:369
Buff J, McCray R (1974) *Astrophys. J.* **183**:147
Camenzind M (1982) In: *Proc. of the Workshop held at the Max-Planck-Institut Garching,* Fed. Rep. Germany
Campbell CG (1983) *Mon. Not. Roy. Astron. Soc.* **205**:1031
Canuto V (1977) *Ann. N.Y. Acad. Sci.* **302**:514
Chandrasekhar S (1931) *Astrophys. J.* **74**:81
Chandrasekhar S (1943) *Astrophys. J.* **97**:255
Chandrasekhar S (1966) *Hydrodynamic and Hydromagnetic Stability* (Clarendon, Oxford)
Chandrasekhar S (1969) *Ellipsoidal Figures of Equilibrium* (Yale Univ. Press, New Haven)
Chandrasekhar S (1970) *Astrophys. J.* **161**:561
Chapman S, Ferraro VC (1931) *Terr. Magn. Atmos. Elec.* **36**:77
Cherepashchuk AM (1981) *Mon. Not. Roy. Astron. Soc.* **194**:761
Cherepashchuk AM, Aslanov AA (1984) *Astron. Space Sci.* **102**:97
Cherepashchuk AM, Aslanov AA, Kornilov VG (1982) *Astron. Zh.* **59**:1157
Chiapetti L, Tanzi EG, Treves A (1980) *Space Sci. Rev.* **27**:3
Chiu H-Y, Salpeter EE (1964) *Phys. Rev. Letters* **12**:413
Chiu HY, Canuto V (1971) *Astrophys. J.* **163**:577
Chugai NN (1984) *Pis'ma Astron. Zh.* **10**:210
Cole JD, Huth JH (1959) *Phys. Fluids* **2**:624
Cominsky LR, Ossman W, Lewin WHG (1983) *Astrophys. J.* **270**:226
Cominsky LR, Wood KS (1984) *Astrophys. J.* **283**:765
Cominsky L, Simmons J, Bowyer S (1985) *Astrophys. J.* **298**:581
Craft HD, Camella JM, Drake FD (1968) *Nature* **218**:1122
Crampton D, Hutchings JB (1981) *Astrophys. J.* **251**:604
Crampton D, Cowley AP, Hutchings JB (1980) *Astrophys. J. Lett.* **235**:L131
Crawford JA (1955) *Astrophys. J.* **121**:71
Davidsen A, Malina R, Bowyer S (1976) In: *X-Ray Binaries,* NASA Report SP-389, 691
Davidson K (1973) *Astrophys. J.* **246**:1
Davidson K, Ostriker JP (1973) *Astrophys. J.* **179**:585
Davies RE, Pringle JE (1980) *Mon. Not. Roy. Astron. Soc.* **191**:599
Davies RE, Pringle JE (1981) *Mon. Not. Roy. Astron. Soc.* **196**:209
Davies RE, Fabian AC, Pringle JE (1979) *Mon. Not. Roy. Astron. Soc.* **186**:779
Deutsch AJ (1955) *Ann. Astrophys.* **18**:1
Dolginov AZ, Gnedin YN, Silant'ev NA (1979) *Propagation and Polarization of Radiation in Space* (in Russian) (Nauka, Moscow)
Dombrovskii VA (1954) *Dokl. Akad. Nauk SSSR* **94**:1021
Dorofeev OF, Rodionov VN, Ternov IM (1985) *Pis'ma Astron. Zh.* **11**:302 (in Russian)
Downs GS (1981) *Astrophys. J.* **249**:687
Dowthwaite JC, Harrison AB, Kirkman IW, McCray HJ, Orford KJ, Turver KE, Walmsley M (1984) *Nature* **309**:691

Dyson F, Ter Haar D (1973) *Neutron Stars and Pulsars* (Russian translation) (Mir, Moscow)
Eadie G, Peacock A, Pounds KA, Watson M, Jackson JC, Hunt R (1975) *Mon. Not. Roy. Astron. Soc.* **172**:35
Eddington AS (1935) *Observatory* **58**:37
Eichler D, Vestrand WT (1984) *Nature* **307**:613
Elsner RF, Lamb FK (1977) *Astrophys. J.* **215**:897
Ergma EV (1982) Itogi Nauk. Tekh. *Astronomiya* **21**:130
Ergma EV, Tutukov AV (1980a) *Astron. Astrophys.* **84**:123
Ergma EV, Tutukov AV (1980b) In: *Proc. IAU Symp. No. 88* edited by Plavec MJ, Popper DM, Ulrich RK (Reidel, Dordrecht)
Esipov VF, Klement'eva AY, Kovalenko AV, Lozinskaya TA, Lyutyi VM, Sitnik TG, Udal'tsov VA (1982) *Astron. Zh.* **59**:965 (in Russian) Engl. translation Sov. Astron. **26** (5)
Fabian AC (1975) *Mon. Not. Roy. Astron. Soc.* **173**:161
Fabian AC, Pringle JE, Webbink RF (1975) *Nature* **255**:208
Filipov LG (1984) *Adv. Space Res.* **3**:305
Fomalont EB, Geldzahler BJ, Hjellming RM, Wade CM (1983) *Astrophys. J.* **275**:802
Forman WC, Jones C, Cominsky L, Julien P, Murray S, Peters G, Tananbaum H, Giacconi R (1978) *Astrophys. J. Suppl.* **38**:357
Fowler RH (1926) *Mon. Not. Roy. Astron. Soc.* **87**:114
Fowley WH, Arons J, Sharlemann ET (1977) *Astrophys. J.* **217**:227
Friedman JL, Ipser JR, Parker L (1985) *Astrophys. J.* **292**:111
Friedman JL, Imamura JN, Dunsen RH, Parker L (1988) *Nature* **336**:560
Fruchter AS, Stineberg DR, Taylor JH (1988) *Nature* **333**:237
Ghosh P, Lamb FK (1978) *Astrophys. J. Lett.* **223**:183
Ghosh P, Lamb FK (1979a) *Astrophys. J.* **232**:256
Ghosh P, Lamb FK (1979b) *Astrophys. J.* **234**:296
Giacconi R (ed) (1981) *Astrophys. and Space Sci. Libr.* Vol. 87 (Kluwer, Dordrecht)
Giacconi R, Gursky H, Paolini FR, Rossi BB (1962) *Phys. Rev. Lett.* **9**:439
Ginzburg VL (1964) *Dokl. Akad. Nauk* SSSR **156**:43
Ginzburg VL (1971) *Usp. Fiz. Nauk* **103**:393
Ginzburg VL, Kirzhnits DA (1964) *Zh. Eksp. Teor. Fiz.* **47**:2006
Ginzburg VL, Usov VV (1972) *Pis'ma Zh. Eksp. Teor. Fiz.* **15**:280
Ginzburg VL, Zheleznyakov VV (1970a, b) *Comm. Astrophys. Space Phys.* **2**:167
Ginzburg VL, Zheleznyakov VV, Zaitsev VV (1969) *Astrophys. Space Sci.* **4**:464
Gladyshev SA, Kurochkin NE, Novikov ID, Cherepashchuk AM (1979) *Astron. Tsyrkul.* No. **1086**:1
Glass IS, Feast MW (1973) *Nature, Phys. Sci.* **245**:39
Gnedin, YN, Syunyaev RA (1974) *Astron. Astrophys.* **36**:379
Gnusareva VS, Lipunov VM (1985) *Astron. Zh.* **62**:1107
Gold T (1968) *Nature* **218**:731
Gold T (1969) *Nature* **221**:25
Goldreich P (1970) *Astrophys. J. Lett.* **160**:L11
Goldreich P, Julian WH (1969) *Astrophys. J.* **157**:869
Goldreich P, Peal SJ (1968) *Ann. Rev. Astron. Astrophys.* **6**:287
Gorbatskii VG (1965) *Trudy Astron. Observ. Leningrad State Univ.* **22**:16
Gorbatskii VG (1974) *Nova-like and Nova Stars* (in Russian) (Nauka, Moscow)
Gorbatskii VG (1977) *Cosmic Gasdynamics* (in Russian) (Nauka, Moscow)
Gott JR, Gunn JE, Ostriker JP (1970) *Astrophys. J. Lett.* **160**:L91
Grindlay JE (1981) In: *Astrophys. and Space Sci. Libr.*, Vol. 87, ed. by R. Giacconi (Kluwer, Dordrecht)
Grindlay JE, Gursky H, Schnopper H, Parsignault DR, Heise J, Brinkman AC, Schrijver J (1976) *Astrophys. J. Lett.* **205**:L127
Groth EJ (1975) *Astrophys. J. Suppl.* **29**:285
Gunn JE, Ostriker JP (1970) *Astrophys. J.* **160**:979
Gunn JE, Ostriker JP (1971) *Astrophys. J.* **165**:523

Gurevich LE (1953) In: *Proc. II Conf. on Cosmogonic Problems* (Izd. Akad. Nauk Moscow USSR)

Guseinov OK (1970) *Astron. Zh.* **47**:1143

Guseinov OK, Yusifov IM (1984) *Astron. Zh.* **61**:708

Haken H (1978) *Synergetics* (Springer, Berlin-Heidelberg)

Hankins TH (1971) *Astrophys. J.* **169**:487

Hayakawa S, Matsuoka M (1964) *Progr. Theor. Phys. Suppl.*, No. **30**:204

Helfand DJ (1981) In: *Proc. Intl. Astron. Union Symp. 95, Pulsars*, edited by Sieber W, Wielebinsky R (Reidel, Dordrecht)

Helfand DJ (1984) *Adv. Space Res.* **3**:29

Helfand DJ, Becker RH (1983) *Nature* **302**:688

Helfand DJ, Chanan GA, Novick R (1980) *Nature* **283**:337

Hjellming RM, Wade CM (1971) *Astrophys. J. Lett.* **164**:L1

Hjellming RM (1978) In: *Proc. of the International School of Physics "Enrico Fermi"*, course LXV, edited by Giacconi R and Ruffini R, *Physics and Astrophysics of Neutron Stars and Black Holes* (North-Holland, Amsterdam)

Holloway N, Kundt W, Wang Y-M (1978) *Astron. Astrophys.* **70**:L23

Hoyle F, Lyttleton RA (1939) *Proc. Camb. Phil. Soc.* **35**:592

Hoyle FJ, Narlikar JV, Wheeler JA (1964) *Nature* **203**:914

Huang J-H, Huang K-L, Peng Q-H (1983) *Astron. Astrophys.* **117**:205

Hulse RA, Taylor JH (1975a) *Astrophys. J. Lett.* **195**:L51

Hulse RA, Taylor JH (1975b) *Astrophys. J. Lett.* **201**:L55

Hunt R (1971) *Mon. Not. Roy. Astron. Soc.* **154**:141

Illarionov AF, Syunyaev RA (1975) *Astron. Astrophys.* **39**:185

Imshennik VS, Nadezhin DK (1982) *Itogi Nauki Tekhniki, Astronomiya* **21**:63

Inoue H (1975) *Publ. Astr. Soc. Jpn* **27**:311

Inoue H, Hoshi R (1975) *Progr. Theor. Phys.* **54**:415

Jackson J (1965) *Classical Electrodynamics* (Mir Moscow) [Russian translation of *Classical Electrodynamics* (Wiley, New York 1962)]

Jaroszynski M, Abramowicz MA, Paczynski B (1980) *Acta Astron.* **30**:1

Joss PC (1977) *Nature* **270**:310

Joss PC (1978) *Astrophys. J. Lett.* **225**:L123

Joss PC, Rappaport SA (1979) *Astron. Astrophys.* **71**:217

Joss PC, Rappaport SA (1983) *Nature* **304**:419

Joss PC, Avni Y, Rappaport SA (1978) *Astrophys. J.* **221**:645

Kaplan SA (1949) *Zh. Eksp. Teor. Fiz.* **19**:951

Kaplan SA, Pikel'her SB (1979) *Physics of the Interstellar Medium* (in Russian) (Nauka, Moscow)

Kaplan SA, Tsytovich VN (1973) *Nature, Phys. Sci.* **241**:122

Kardashev NS (1964) *Astron. Zh.* **41**:807

Kegel WH (1971) *Astron. Astrophys.* **12**:452

Klebesadel RW, Strong IB, Olson RA (1973) *Astrophys. J. Lett.* **182**:L85

Kolosov DE, Lipunov VM, Postnov KA, Prokhorov ME (1989) *Astron. Astrophys.* **215**:L21

Kolykhalov PN, Syunyaev RA (1979) *Pis'ma Astron. Zh.* **5**:338

Kompaneets AS (1956) *Zh. Eksp. Teor. Fiz.* **31**:876

Korn GA, Korn TM (1968) *Handbook of Mathematics for Scientists and Engineers* (McGraw-Hill, New York)

Kornilov VG, Lipunov VM (1983a) *Astron. Zh.* **60**:284

Kornilov VG, Lipunov VM (1983b) *Astron. Zh.* **60**:574

Kornilov VG, Lipunov VM (1984) *Astron. Zh.* **61**:686

Krasnobaev KV, Syunyaev RA (1983) *Izv. Akad. Nauk SSSR, Mekh. Zhidk. Gaza*, No. **4**:106

Kruskal M, Schwarzschild M (1954) *Proc. Roy. Soc.* **A223**:348

Kulkarni SR, Clifton TC, Backer DC, Foster RS, Fruchter AS, Taylor JH (1988) *Nature* **331**:50

Kulkarni SR, Hester JJ (1988) *Nature* **335**:801

Kulsrud RM (1971) *Astrophys. J.* **163**:567

Kulsrud RM, Ostriker JP, Gunn JE (1972) *Phys. Rev. Lett.* **28**:636

Kundt W (1976) *Phys. Lett.* **A57**:195

Kundt W (1981) *Vistas in Astron.* **25**:153

Kundt W (1984) In: *Proc. COSPAR Symp.* Graz, Austria, edited by Koch-Miramond L, Lee MA (Pergamon, Oxford)

Kundt W (1985) *Astron. Astrophys.* **150**:216

Kundt W, Robnik M (1980) *Astron. Astrophys.* **91**:305

Kurth WS, Gurnett DA, Scart FL, Poynter RL (1985) *Nature* **312**:27

Lamb DQ, Lamb FK (1976) *Astrophys. J.* **204**:168

Lamb DQ, Lamb FK (1977) *Ann. N.Y. Acad. Sci.* **302**:261

Lamb DQ, Lamb FK (1978) *Astrophys. J.* **220**:291

Lamb DQ, Lamb FK, Pines D, Shaham J (1975) *Astrophys. J. Lett.* **198**:L 21

Lamb FK (1979) In: *Compact Galactic X-ray Sources* edited by Lamb F, Pines D (Univ. of Illinois Press, USA)

Lamb FK, Pethick CJ, Pines D (1973) *Astrophys. J.* **184**:271

Lamb FK, Pines D, Shaham J (1978a) *Astrophys. J.* **224**:969

Lamb FK, Pines D, Shaham J (1978b) *Astrophys. J.* **225**:582

Lamb FK, Aly JJ, Cook MC, Lamb DQ (1983) *Astrophys. J. Lett.* **271**:L 71

Lamb FK, Aly JJ, Cook MC, Lamb DQ (1985) In: *Cataclysmic Variables and Low-mass X-ray Binaries* edited by Lamb DQ, Petterson J (Reidel, Dordrecht)

Lamb G (1947) *Fluid Mechanics* (Russian translation) (Ob'ed Gos. Izd., Moscow)

Landau LD (1932) *Phys. Z. Sowjetunion* **1**:285

Landau LD (1938) *Nature* **141**:333

Landau LD, Lifshitz EM (1953) *Physics of the Continuous Media* (in Russian) (Gostekhizdat, Moscow) 2nd Ed.

Landau LD, Lifshitz EM (1973) *Field Theory* (in Russian) (Nauka, Moscow)

Landau LD, Lifshitz EM (1974) *Quantum Mechanics. Nonrelativistic Theory* (in Russian) (Nauka, Moscow)

Landau LD, Lifshitz EM (1976) *Statistical Physics* (in Russian) (Nauka, Moscow)

Landau LD, Lifshitz EM (1982) *Continuum Electrodynamics* (in Russian) (Nauka, Moscow)

Langer SH, Rappaport S (1982) *Astrophys. J.* **257**:733

Langmeier A, Sztajno M, Trumpler J (1985) *Adv. Space Res.* **5**:121

Lavrent'ev MA, Shabat BV (1973) *Methods in the Functional Theory of Complex Variables* (in Russian) (Nauka, Moscow)

Leahy DA, Darbo W, Elsner RF, Weisskopf MC, Sutherland PG, Kahn S, Grindlay JE (1983) *Astrophys. J.* **266**:160

Lebedinskii AI (1953) In: *Proc. II Conf. on Problems in Cosmogony* (Izd. Akad. Nauk, Moscow, USSR)

Lewin WHG (1979) *Adv. Space Explor.* **3**:133

Lewin WHG, Clark GW (1980) *Ann. N.Y. Acad. Sci.* **336**:451

Lewin WHG, Hoffman JA, Doty J, Clark GW, Swank JH, Becker RH, Pravdo SH, Serlemitsos PJ (1977) *Nature* **267**:28

Lewin WHG, Hoffman JA, Marshall H et al. (1978) IAU Circ. No. **3190**

Lightman AP (1974) *Astrophys. J.* **194**:419

Lightman AP, Eardly DM (1974) *Astrophys. J. Lett.* **187**:L 1

Lightman AP, Hertz P, Grindlay JE (1980) *Astrophys. J.* **241**:367

Lipunov VM (1978a) *Astromet. Astrofiz.* **36**:8

Lipunov VM (1978b) *Astron. Zh.* **55**:1233

Lipunov VM (1978c) *Astron. Tsirkul.* No. **993**:1

Lipunov VM (1980a) *Astron. Zh.* **57**:1253

Lipunov VM (1980b) *Astron. Tsirkul.* No. **1092**:2

Lipunov VM (1981a) *Astron. Zh.* **58**:663

Lipunov VM (1981b) In: *Stars and Stellar Systems* (in Russian) edited by Martynov DY (Nauka, Moscow)

Lipunov VM (1982a) *Astron. Space Sci.* **85**:451

Lipunov VM (1982b) *Astron. Zh.* **59**:87

Lipunov VM (1982c) *Astron. Zh.* **59**:888

Lipunov VM (1982d) *Astron. Space Sci.* **82**:343

Lipunov VM (1982e) *Pis'ma Astron. Zh.* **8**:358
Lipunov VM (1983a) *Astron. Astrophys.* **127**:L1
Lipunov VM (1983b) *Astron. Space Sci.* **97**:121
Lipunov VM (1984) *Adv. Space Res.* **3**:323
Lipunov VM (1987a) *Astron. Space Sci.* **132**:1
Lipunov VM (1987b) *Astron. Zh.* **64**:321
Lipunov VM, Moskalenko EI, Shakura NI (1982) *Astron. Space Sci.* **85**:459
Lipunov VM, Postnov KA (1984) *Astron. Space Sci.* **106**:103
Lipunov VM, Postnov KA (1985) *Astron. Astrophys.* **144**:L13
Lipunov VM, Postnov KA (1987) *Astron. Zh.* **64**:548
Lipunov VM, Postnov KA (1988) *Astrophys. and Space Sci.* **145**:1
Lipunov VM, Postnov KA (1988) *Astron. Astrophys.* **206**:L15
Lipunov VM, Postnov KA, Prokhorov ME (1986) *Astrophys. Lett.* **11**:25
Lipunov VM, Prokhorov ME (1984) *Astron. Space Sci.* **98**:221
Lipunov VM, Prokhorov ME (1987) *Astron. Zh.* **64**:1189
Lipunov VM, Semenov ES, Shakura NI (1981) *Astron. Zh.* **58**:765
Lipunov VM, Shakura NI (1976) *Pis'ma Astron. Zh.* **2**:343
Lipunov VM, Shakura NI (1980) *Pis'ma Astron. Zh.* **6**:28
Lipunov VM, Shakura NI (1982) *Astron. Zh.* **59**:631 (in Russian) Engl. translation *Sov. Astron*
    25(4)
Livio M (1984) *Astron. Astrophys.* **141**:L4
Lloyd-Evans J, Coy RN, Lambert A et al. (1983) *Nature* **305**:784
Lozinskaya TA (1986) *Supernova Stars and Stellar Wind* (in Russian) (Nauka, Moscow)
Lynden-Bell D (1969) *Nature* **233**:690
Lynden-Bell D, Pringle JE (1974) *Mon. Not. Roy. Astron. Soc.* **168**:603
Lyne AG, Anderson B, Salter MJ (1982) *Mon. Nat. Roy Astron. Soc.* **201**:503
Lyubarskii YE (1979) *Pis'ma Astron. Zh.* **5**:601
Lyubarskii YE (1984) *Astron. Zh.* **61**:100
Lyubarskii YE, Syunyaev RA (1982) *Pis'ma Astron Zh.* **8**:612
Lyubarskii YE, Shakura NI (1987) *Pis'ma Astron. Zh.* **13**:917 (in Russian)
Macy WW Jr. (1974) *Astrophys. J.* **190**:153
Maguire JJ, Carovillano RL (1966) *J. Geophys. Res.* **71**:5533
Malov IF (1985) *Astron. Zh.* **62**:252
Manchester RN, Taylor JH (1981) *Astron. J.* **86**:1953
Manchester RN, Taylor JH (1978) *Pulsars* (Freeman, San Francisco)
Maraschi L, Cavaliere A (1977) In: *Highlights of Astronomy* edited by Müller EA (Reidel, Dordrecht) Vol. 4, Part I
Maraschi L, Traversini R, Treves A (1983) *Mon. Not. Roy. Astron. Soc.* **204**:1179
Margon B (1984) *Ann. Rev. Astron. Astrophys.* **22**:507
Margon B, Ford HC, Katz JI, Kwitter KB, Ulrich RK, Stone RPS, Klemola A (1979) *Astrophys. J. Lett.* **230**:L41
Maxwell GD (1873) *A Treatise on Electricity and Magnetism* (Cambridge Univ. Press Cambridge)
Mazets EP, Golenetskii SV, Aptekar' RL, Gur'yan A (1980) *Pis'ma Astron. Zh.* **6**:706
McCrea WH (1953) *Mon. Not. Roy. Astron. Soc.* **113**:162
Mead GD, Beard DD (1964) *J. Geophys. Res.* **69**:1169
Medgley J, Davis LJ (1962) *J. Geophys. Res.* **67**:499
Mestel L (1954) *Mon. Not. Roy. Astron. Soc.* **114**:437
Michalas D (1978) *Stellar Atmospheres* (Freeman, San Francisco)
Michel FC (1971) *Comm. Astrophys. Space Sci.* **3**:80
Michel FC (1977) *Astrophys. J.* **2**:836
Michel FC (1982) *Rev. Mod. Phys.* **54**:1
Middleditch J, Priedhorsky W (1985) IAU Circ. No. 4060
Middleditch J, Mason KO, Nelson JE, White N (1981) *Astrophys. J.* **244**:1001
Middleditch J et al. (1989) IAU Circ. No. 4735
Migdal AB (1959) *Zh. Eksp. Teor. Fiz.* **37**:249

Mikhailovskii AB (1975, 1977) *Theory of Plasma Instabilities* (in Russian) (Atomizdat, Moscow) Vol. 1 and 2

Milgrom M (1978) Astron. Astrophys. **67**:L25

Milgrom M (1979) *Astron. Astrophys.* **76**:L3

Misner CW, Zapolsky HS (1964) *Phys. Rev. Lett.* **12**:635

Molnar LA, Reid MJ, Grindlay JE (1984) *Nature* **310**:662

Mitrofanov IG, Pavlov GG, Gnedin YN (1977) *Pis'ma Astron. Zh.* **3**:341 (in Russian)

Morfill GE, Trümper J, Bodenheimer P, Tenorio-Tagle G (1984) *Astron. Astrophys.* **139**:7

Morozov AI, Solov'ev LS (1963) In: *Problems in Plasma Theory* (in Russian) (Gos. Izd. po Atom Nauke i Tekhnike, Moscow)

Nagase F (1985) *Adv. Space Res.* **5**:95

Nelson J, Hills R, Cudaback D, Wampler J (1970) *Astrophys. J. Lett.* **161**:L235

Nicolson GD (1984) In: *VLBI and Compact Radio Sources* edited by Fanti R (IAU)

Nishida A (1978) *Geomagnetic Diagnosis of Magnetosphere* (Springer, New York)

Nityananda R, Narayan R (1984) *Adv. Space Res.* **3**:29

Nomoto K, Tsuruta S (1981) *Astrophys. J. Lett.* **250**:L19

Northrop TG (1956) *Phys. Rev.* **103**:1150

Novikov ID, Zel'dovich YB (1966) *Nuovo Cim. Suppl. (I)* **4**:810

Novikov ID, Perevodchikova TV (1984) *Astron. Zh.* **61**:935

Novikov ID, Thorne KS (1972) In: *Black Holes edited by DeWitt C, DeWitt B* (Gordon and Breach, New York)

Nulson PEJ, Fabian AC (1984) *Nature* **312**:48

Oda M (1981) *Astron. Space Sci. Libr.* **87**:61

Oppenheimer JR, Volkoff GM (1939) *Phys. Rev.* **55**:374

Ostriker JP, Bodenheimer P (1968) *Astrophys. J.* **151**:1089

Ostriker JP, Gunn JE (1969) *Astrophys. J.* **157**:1395

Oyama Y, Arisaka K, Kajita T (1986) *Phys. Rev. Lett.* **56**:991

Pacini F (1967) *Nature* **216**:567

Paczynski B (1965) *Acta Astron.* **15**:89

Paczynski B (1967) *Acta Astron.* **17**:287

Paczynski B (1971) *Ann. Rev. Astron. Astrophys.* **9**:183

Paczynski B (1976) *Commun. Astron. Space Phys.* **6**:95

Paczynski B (1976) In: *IAU Symp. 73 on Structure and Evolution of Close Binary Systems* edited by Eggleton P, Mitton S, Whelan JA (Reidel, Dordrecht)

Paczynski B (1977) *Astrophys. J.* **216**:822

Paczynski B, Wiita PJ (1980) *Astron. Astrophys.* **88**:23

Papaloizou J, Pringle JE (1977) *Mon. Not. Roy. Astron. Soc.* **181**:441

Parenago PP (1950) *Astron Zh.* **27**:41

Pavlov GG, Gnedin YN (1983) *Itogi Nauki i Tekhniki* (in Russian) *Astronomiya* **22**:172

Petterson JA (1977) *Astrophys. J.* **216**:827

Pikel'ner SB (1956) *Astron. Zh.* **33**:785

Pikel'ner SB (1966) *Fundamentals of Cosmic Electrodynamics* (in Russian) (Nauka, Moscow)

Pines D (1980) *J. Phys. Colloq.* **41**:111

Pines D, Shaham J (1972) *Nature, Phys. Sci.* **235**:43

Pines D, Shaham J, Ruderman M (1972) *Nature, Phys. Sci.* **237**:83

Plesset MS, Hsieh D-Y (1964) *Phys. Fluids* **7**:1099

Pozdnyakov LA, Sobol' IM, Syunyaev RA (1982) *Itogi Nauki i Tekhniki* (in Russian) *Astronomiya* **21**:238

Poston T, Stewart I (1978) *Catastrophe Theory and Applications* (Pitman, London)

Pravdo SH, Boldt EA, Holt SS, Serlemitsos PJ (1977) *Astrophys. J. Lett.* **216**:L23

Prendergast K (1960) *Astrophys. J.* **132**:162

Pringle JE, Rees MJ (1972) *Astron. Astrophys.* **21**:1

Pringle JE, Savonije GJ (1979) *Mon. Not. Roy. Astron. Soc.* **187**:777

Protheroe RJ, Clay RW, Gerhardy PR (1984) *Astrophys. J. Lett.* **280**:L47

Radhakrishnan V, Cooke DJ, Komesaroff MM, Morris D (1969) *Nature* **221**:443

Rappaport SA, Joss PC (1977) *Nature* **266**:683

Rappaport SA, Joss PC (1983) In: *Accretion-driven Stellar X-ray Sources* edited by Lewin WHG, van den Heuvel EPJ (Cambridge Univ. Press, Cambridge)

Rees MJ, Gunn JE (1974) *Mon. Not. Roy. Astron. Soc.* **167**:1

Rosenbluth MN, Ruderman M, Dyson F, Bahcall JN, Shaham J, Ostriker J (1973) *Astrophys. J.* **184**:907

Ruderman M (1969) *Nature* **223**:597

Ruderman MA, Sutherland PG (1975) *Astrophys. J.* **196**:51

Ruffini R, Wilson J (1971) Phys. Rev. Lett. **31**:1362

Sadeh D, Byram ET, Chubb TA, Friedman H, Hedler RL, Meekins JF, Wood KS, Yentis DJ (1982) *Astrophys. J.* **257**:214

Salpeter EE (1964) *Astrophys. J.* **140**:796

Samorski M, Stamm W (1983) *Astrophys. J. Lett.* **268**:L17

Savonije GJ, van den Heuvel EPJ (1977) *Astrophys. J. Lett.* **214**:L19

Schmidt M (1963) *Nature* **197**:1040

Schreier E, Levinson R, Gursky H, Kellog E, Tananbaum H, Giacconi R (1972) *Astrophys. J. Lett.* **172**:L79

Schwarzschild M (1958) *Structure and Evolution of Stars* (Princeton Univ. Press, New Jersey)

Seward FD, Harnden FR (1984) In: *X-ray Astronomy 84* edited by Oda M, Giacconi R (Inst. Space Astron. Sci. Press, Bologna, Italy)

Shabad AE, Usov VV (1982) *Nature* **295**:215

Shakura NI (1972) *Astron. Zh.* **49**:921

Shakura NI (1974) *Astron. Zh.* **51**:441

Shakura NI (1975) *Pis'ma Astron. Zh.* **1**:23

Shakura NI, Syunyaev RA (1973) *Astron Astrophys.* **24**:337

Shakura NI, Syunyaev RA (1976) *Mon. Not. Roy. Astron. Soc.* **175**:613

Shakura NI, Syunyaev RA, Zilitinkevich SS (1978) *Astron. Astrophys.* **62**:179

Shapiro SL, Salpeter EE (1975) *Astrophys. J.* **198**:671

Shapiro SL, Teukolsky SA (1985) *Black Holes, White Dwarfs and Neutron Stars*, (Wiley, New York)

Sharleman ET (1978) *Astrophys. J.* **219**:617

Shvartsman VF (1970a) *Radiofizika* **13**:1852

Shvartsman VF (1970b) *Astron. Zh.* **47**:824

Shvartsman VF (1970c) *Astron. Zh.* **47**:660

Shvartsman VF (1970d) *Preprint of the Institute of Applied Mathematics* (in Russian) (Moscow, No. 43)

Shvartsman VF (1971a) *Astron. Zh.* **48**:438

Shvartsman VF (1971b) *Astron. Zh.* **48**:479

Shklovskii IS (1953) *Astron. Zh.* **30**:15

Shklovskii IS (1967) *Astron. Zh.* **44**:930

Shklovskii IS (1970) *Astrophys. J.* **159**:L77

Shklovskii IS (1971) *Astrophys. Lett.* **8**:101

Shklovskii IS (1976) *Pis'ma Astron. Zh.* **2**:119

Shklovskii IS (1979) *Pis'ma Astron. Zh.* **5**:644 (in Russian); Engl. translation *Sov. Astron. Lett.* **5**

Shklovskii IS (1981) *Astron. Zh.* **58**:554

Sibgatullin NR (1984) *Oscillations and Waves in Strong Gravitational and Electromagnetic Fields* (in Russian) (Nauka, Moscow)

Skinner GK, Bedford DK, Elsner RF, Leahy D, Weisskopf MC, Grindlay J (1982) *Nature* **297**:568

Smak J (1971) *Acta Astron.* **21**:15

Smith FG (1977) *Pulsars* (Cambridge Univ. Press, Cambridge)

Snezhko LI (1967) *Peremen. Zvezdy* **16**:253

Spiegel EA (1970) In: *IAU Symp. 39 on Interstellar Gas Dynamics* edited by Habing HJ (Reidel, Dordrecht, Holland)

Srinivasan G, van den Heuvel EPJ (1982) *Astron. Astrophys.* **108**:143

Stepanyan AA (1984) *Adv. Space Res.* **3**:123

Sturock PA (1971) *Astrophys. J.* **164**:529

Syunyaev RA (1978) *Pis'ma Astron Zh.* **4**:75

Syunyaev RA, Shakura NI (1975) *Pis'ma Astron. Zh.* **1**:6
Syunyaev RA, Shakura NI (1977a) *Pis'ma Astron Zh.* **3**:216
Syunyaev RA, Shakura NI (1977b) *Pis'ma Astron. Zh.* **3**:262
Taam RE (1980) *Astrophys. J.* **241**:358
Taam RE, Piklum RE (1979) *Astrophys. J.* **233**:327
Tademaru E, Harrison ER (1975) *Nature* **254**:39
Tanaka Y (1984) In: *X-ray Astronomy* edited by Oda M, Giacconi R, Inst. Space Astron Sci. Press, Bologna, Italy
Tassoul JL (1978) *Theory of Rotating Stars* (Princeton Univ. Press, Princeton, New Jersey)
Taylor JH, Weisberg JM (1989) *Astrophys. J.* **345**:434
Ten Haar D (1972) *Phys. Reports* **3** No. 2
Ternov IM, Khalilov VR, Rodionov VN (1982) *Interaction of Charged Particles with Strong Electric Fields* (in Russian) (Moscow State Univ. Press)
Terrell J, Priedhorsky WC (1984) *Astrophys. J. Lett.* **285**:L15
Trümper J, Peitsch W, Reppin C, Voges W, Staubert R, Kendziora E (1978) *Astrophys. J. Lett.* **219**:L105
Tsakadze DS, Tsakadze SD (1975) *Usp. Fiz. Nauk.* **115**:503
Tsiolkovskii KE (1964) *Collected Works* (in Russian) (Nauka, Moscow) Vol. IV
Tsuruta S, Cameron AGW (1965) *Nature* **207**:364
Tsuruta S, Cameron AGW (1966) *Nature* **211**:356
Tsygan AI (1974) *Astron. Zh.* **51**:1339
Tsygan AI (1976) Preprint No. 518, A.F. Ioffe Physicotechnical Institute, Leningrad
Tsygan AI (1981) D. Sc. Thesis, A.F. Ioffe Physicotechnical Institute, Leningrad
Tutukov AV (1980) D. Sc. Thesis
Tutukov AV, Yungel'son LR (1973) *Nauch. Inform. Astron. Sovet* **27**:70
Tutukov AV, Chugai NN, Yungel'son LR (1984) *Pis'ma Astron. Zh.* **10**:586
Usov VV (1977) *Itogi Nauki i Tekhniki* **9**:5
Van den Heuvel EPJ (1977) *Ann. N.Y. Acad. Sci* **302**:13
Van den Heuvel EPJ, Heise J (1972) *Nature, Phys. Sci.* **239**:67
Van den Heuvel EPJ (1983) *Accretion-driven Stellar X-ray Sources*, edited by Lewin WHG, Van den Heuvel EPJ (Cambridge Univ. Press) p. 303
Van Horn HM, Hansen CJ (1974) *Astrophys. J.* **191**:479
Van der Klis M, Jansen F, Van Paradijs J., Lewin WHG, Van den Heuvel EPJ, Trümper JE, Sztajno M (1985) *Nature* **316**:225
Van Paradijs J (1978) *Nature* **274**:650
Vashakidze MA (1954) *Astron. Tsirkul.* No. **147**:11
Velusamy T, Pramesh Rao A, Sukumar S (1985) *Mon. Not. Roy. Astron. Soc.* **213**:735
Walter FM, Bowyer S, Mason KO, Clarke JT, Henry JP, Halpern J, Grindlay JE (1982) *Astrophys. J. Lett.* **253**:L67
Wang Y-M (1975) *Nature* **253**:249
Wang Y-M (1981) *Astron. Astrophys.* **102**:36
Wang Y-M, Frank J (1981) *Astron. Astrophys.* **93**:255
Wang Y-M, Robnik M (1982) *Astron. Astrophys.* **107**:222
Wang Y-M, Nepveu M (1983) *Astron. Astrophys.* **118**:267
Wang Y-M, Nepveu M, Robertson JA (1984) *Astron. Astrophys.* **135**:66
Wang Y-M, Robertson JA (1985) *Astron. Astrophys.* **151**:361
Wang Y-M, Welter GL (1982) *Astron. Astrophys.* **113**:113
Wdowczyk J, Wolfendale AW (1983) *Nature* **305**:609
Wheaton WA, Howe SK, Goldman A, Cooke BA, Lewin WHG, Gruber DE, Matteson JL (1978) *Bull. Amer. Astron. Soc.* **10**:506
White NE, Swank JH (1982) *Astrophys. J. Lett.* **253**:L61
White NE, Parmar AN, Mason KO (1983) IAU Circ. No. 3882
Wickramasinghe DT, Whelan JAJ (1975) *Nature* **258**:502
Woltjer L (1964) *Astrophys. J.* **140**:1309
Woosley SE, Taam RE (1976) *Nature* **263**:101
Yahel RZ (1980) *Astron. Astrophys.* **90**:26

Yungel'son LR, Masevich AG (1983) *Astrophys. Space Phys. Rev.* **2**:29

Zel'dovich YB (1956) *Zh. Eksp. Teor. Fiz.* **31**:154

Zel'dovich YB (1964) *Dokl. Akad. Nauk USSR* **155**:67

Zel'dovich YB, Shakura NI (1969) *Astron. Zh.* **46**:225 (in Russian) Engl. translation *Sov. Astron.* **13**(2)

Zel'dovich YB, Ivanova LN, Nadezhin DK (1972) *Astron. Zh.* **49**:253

Zel'dovich YB, Myshkis AD (1972) *Elements of Applied Mathematics* (in Russian) (Nauka, Moscow)

Zel'dovich YB, Myshkis AD (1973) *Elements of Mathematical Physics* (in Russian) (Nauka, Moscow)

Zel'dovich YB, Novikov ID (1967) *Relativistic Astrophysics* (in Russian) (Nauka, Moscow)

Zel'dovich YB, Novikov ID (1971) *Theory of Gravitation and Evolution of Stars* (in Russian) (Nauka, Moscow)

Zel'dovich YB, Raizer YP (1966) *Physics of Shock Waves and High-Temperature Hydrodynamic Effects* (in Russian) (Nauka, Moscow)

Zel'dovich YB, Ruzmaikin AA (1982) *Itogi Nauki i Tekhniki* (in Russian) *Astronomia* **21**:151

Zhigulev VN, Romishevskii AA (1959) *Dokl. Akad. Nauk SSSR* **127**:1001

Subject Index

Accretion 4
- column 154
- cylindrical 52
- disk 57, 61, 66, 124, 147, 220
- of magnetic field 79
- nonstationary 76, 198, 222
- spherically symmetrical 40, 42, 49, 115, 198
- supercritical 65, 90, 218
- two stream 78, 135, 152, 218
- wind-fed 73
Accretor 97, 138, 298
Alfvén surface 95
Alfvén zone 106

Bernoulli equation 42
Black hole 97, 298
Bifurcation 37, 200
Burster
- gamma 20
- radio 251
- X-ray 17, 184

Catestrophic equilibrium 163
Chandrasekhar limit 2
Collapse anisotropy 261, 302
Compton scattering 49, 143, 168
Crab Nebula 6, 232, 244
Cusp 122
Cyclotron frequency 168

Dispersion measure 233
Doppler shift 235

Eddington limit 47
Ejector 97, 227, 299
Equilibrium period 162, 171

Faraday rotation 265
Fokker-Planck equation 178

Georotator 97, 277
Glitches 9, 254
Gravi-magnetic parameter 92
Gravitational wave 200

Gyrofrequency 14 *see* cyclotron frequency

Her X-1 15, 96, 173, 175

Instability
- Kelvin-Helmholtz 145, 308
- Rayleigh-Taylor 144, 199, 304

Jeans wavelength 55

Kerr parameter 203

Lagrangian point 69
Langevin equation 178
Light cylinder 84
Liouville equation 259, 290

Mach number 56
Magnetic dipole 83
- radiation 85, 243
Magnetosphere 105, 111
Magnetor 97, 100, 277
Mass function 36
Maxwell equation 110
Magnetic field
- X-ray pulsars 175
- radiopulsars 252
Monte Carlo method 294

Neutrino pulsar 270

Oppenheimer-Volkoff limit 31

Plasma frequency 87
Poisson equation 41
Polar column 153
Polars 279
Precession 275
Porpeller 97, 208, 298
- non-gravitating 216
- relativistic 214
- subsonic 214
- supersonic 212
- very rapid 215

Pulsars
— binary X-ray   166, 297
— binary radio   235, 262, 300

QPO   204, 275

Rotator
— gravi-magnetic   83
— fast   175
— slow   175
Radius
— Alfvén   89, 114, 115, 209, 269
— Bondi   44
— capture   56, 87, 96
— corotation   94
— Debye   107
— Larmor   109
— Light cylinder   84
— Shwarzschild   5
— Shvartsman   87
— spherisation   65
— stopping   86, 90

Scalar potential   161

Sco X-1   5, 275
Single Wolf-Rayet stars   302
Spin-down   8, 111, 159, 170, 172, 217, 271, 288
Spin-up   11, 12, 156, 170, 172, 271, 288
Spin-down index   229
SS 433   272, 301
Star quakes   9, 254
Superaccretor   97, 268, 298
Superejector   97, 272, 298
Superpropeller   97, 272, 298
Superfluid   4, 254
Supernovae 1987a   38

Tensor viscosity   59
Theorem of equipartition   79
Transients   180, 182
Turbulent envelopes   210
Turbulent parameter   59
Twisted disk   151

Uhuru   11

Viscous sresses   60